Biology:
Exploring Life

SECOND EDITION

Biology: Exploring Life

Gil Brum
California State Polytechnic University, Pomona

Larry McKane
California State Polytechnic University, Pomona

Gerry Karp
Formerly of the University of Florida, Gainesville

JOHN WILEY & SONS, INC.

New York / Chichester / Brisbane / Toronto / Singapore

Acquisitions Editor   *Sally Cheney*
Developmental Editor   *Rachel Nelson*
Marketing Manager   *Catherine Faduska*
Associate Marketing Manager   *Deb Benson*
Senior Production Editor   *Katharine Rubin*
Text Designer/"Steps" Illustration Art Direction   *Karin Gerdes Kincheloe*
Manufacturing Manager   *Andrea Price*
Photo Department Director   *Stella Kupferberg*
Senior Illustration Coordinator   *Edward Starr*
"Steps to Discovery" Art Illustrator   *Carlyn Iverson*
Cover Design   *Meryl Levavi*
Text Illustrations   *Network Graphics/Blaize Zito Associates, Inc.*
Photo Editor/Cover Photography Direction   *Charles Hamilton*
Photo Researchers   *Hilary Newman, Pat Cadley, Lana Berkovitz*
Cover Photo   *James H. Carmichael, Jr./The Image Bank*

This book was set in New Caledonia by Progressive Typographers and printed and bound by Von Hoffmann Press. The cover was printed by Lehigh. Color separations by Progressive Typographers and Color Associates, Inc.

Recognizing the importance of preserving what has been written, it is a policy of John Wiley & Sons, Inc. to have books of enduring value published in the United States printed on acid-free paper, and we exert our best efforts to that end.

The paper in this book was manufactured by a mill whose forest management programs include sustained yield harvesting of its timberlands. Sustained yield harvesting principles ensure that the number of trees cut each year does not exceed the amount of new growth.

Library of Congress Cataloging in Publication Data:
Brum, Gilbert D.
    Biology : exploring life/Gil Brum, Larry McKane, Gerry Karp.—2nd ed.
      p.   cm.
    Includes bibliographical references and index.
    ISBN 0-471-54408-6 (cloth)
    1. Biology.   I. McKane, Larry.   II. Karp, Gerald.   III. Title.
  QH308.2.B78   1993
  574—dc20
                                  93-23383
                                  CIP

Unit I ISBN 0-471-01827-9 (pbk)
Unit II ISBN 0-471-01831-7 (pbk)
Unit III ISBN 0-471-01830-9 (pbk)
Unit IV ISBN 0-471-01829-5 (pbk)
Unit V ISBN 0-471-01828-7 (pbk)
Unit VI ISBN 0-471-01832-5 (pbk)

Printed in the United States of America

10 9 8 7 6 5 4 3 2 1

For the Student, we hope this book helps you discover the thrill of exploring life and helps you recognize the important role biology plays in your everyday life.

To Margaret, Jan, and Patsy, who kept loving us even when we were at our most unlovable.

To our children, Jennifer, Julia, Christopher, and Matthew, whose fascination with exploring life inspires us all. And especially to Jenny—we all wish you were here to share the excitement of this special time in life.

Preface to the Instructor

Biology: Exploring Life, Second Edition is devoted to the process of investigation and discovery. The challenge and thrill of understanding how nature works ignites biologists' quests for knowledge and instills a desire to share their insights and discoveries. The satisfactions of knowing that the principles of nature can be understood and sharing this knowledge are why we teach. These are also the reasons why we created this book.

Capturing and holding student interest challenges even the best of teachers. To help meet this challenge, we have endeavored to create a book that makes biology relevant and appealing, that reveals biology as a dynamic process of exploration and discovery, and that emphasizes the widening influence of biologists in shaping and protecting our world and in helping secure our futures. We direct the reader's attention toward principles and concepts to dispel the misconception of many undergraduates that biology is nothing more than a very long list of facts and jargon. Facts and principles form the core of the course, but we have attempted to show the *significance* of each fact and principle and to reveal the important role biology plays in modern society.

From our own experiences in the introductory biology classroom, we have discovered that

- emphasizing principles, applications, and scientific exploration invigorates the teaching and learning process of biology and helps students make the significant connections needed for full understanding and appreciation of the importance of biology; and

- students learn more if a book is devoted to telling the story of biology rather than a recitation of facts and details.

Guided by these insights, we have tried to create a process-oriented book that still retains the facts, structures, and terminology needed for a fundamental understanding of biology. With these goals in mind, we have interwoven into the text

1. an emphasis on the ways that science works,
2. the underlying adventure of exploration,
3. five fundamental biological themes, and
4. balanced attention to the human perspective.

This book should challenge your students to think critically, to formulate their own hypotheses as possible explanations to unanswered questions, and to apply the approaches learned in the study of biology to understanding (and perhaps helping to solve) the serious problems that affect every person, indeed every organism, on this planet.

THE DEVELOPMENT STORY

The second edition of *Biology: Exploring Life* builds effectively on the strengths of the First Edition by Gil Brum and Larry McKane. For this edition, we added a third author, Gerry Karp, a cell and molecular biologist. Our complementary areas of expertise (genetics, zoology, botany, ecology, microbiology, and cell and molecular biology) as well as awards for teaching and writing have helped us form a balanced team. Together, we exhaustively revised and refined each chapter until all three of us, each with our different likes and dislikes, sincerely believed in the result. What evolved from this process was a satisfying synergism and a close friendship.

THE APPROACH

The elements of this new approach are described in the upcoming section "To the Student: A User's Guide." These pedagogical features are embedded in a book that is written in an informal, accessible style that invites the reader to explore the process of biology. In addition, we have tried to keep the narrative focused on *processes*, rather than on static facts, while creating an underlying foundation that helps students make the connections needed to tie together the information into a greater understanding than that which comes from memorizing facts alone. One way to help students make these connections is to relate the fundamentals of biology to humans, revealing the human perspective in each biological principle, from biochemicals to ecosystems. With each such insight, students take a substantial step toward becoming the informed citizens that make up responsible voting public.

We hope that, through this textbook, we can become partners with the instructor and the student. The biology teacher's greatest asset is the basic desire of students to understand themselves and the world around them. Unfortunately, many students have grown detached from this natural curiosity. Our overriding objective in creating this book was to arouse the students' fascination with exploring life, building knowledge and insight that will enable them to make real-life judgments as modern biology takes on greater significance in everyday life.

THE ART PROGRAM

The diligence and refinement that went into creating the text of *Biology: Exploring Life,* Second Edition characterizes the art program as well. Each photo was picked specifically for its relevance to the topic at hand and for its aesthetic and instructive value in illustrating the narrative concepts. The illustrations were carefully crafted under the guidance of the authors for accuracy and utility as well as aesthetics. The value of illustrations cannot be overlooked in a discipline as filled with images and processes as biology. Through the use of cell icons, labeled illustrations of pathways and processes, and detailed legends, the student is taken through the world of biology, from its microscopic chemical components to the macroscopic organisms and the environments that they inhabit.

SUPPLEMENTARY MATERIALS

In our continuing effort to meet all of your individual needs, Wiley is pleased to offer the various topics covered in this text in customized paperback "splits." For more details, please contact your local Wiley sales representative. We have also developed an integrated supplements package that helps the instructor bring the study of biology to life in the classroom and that will maximize the students' use and understanding of the text.

The *Instructor's Manual,* developed by Michael Leboffe and Gary Wisehart of San Diego City College, contains lecture outlines, transparency references, suggested lecture activities, sample concept maps, section concept map masters (to be used as overhead transparencies), and answers to study guide questions.

Gary Wisehart and Mark Mandell developed the test bank, which consists of four types of questions: fill-in questions, matching questions, multiple-choice questions, and critical thinking questions. A computerized test bank is also available.

A comprehensive visual ancillary package includes four-color transparencies (200 figures from the text), *Process of Science* transparency overlays that break down various biological processes into progressive steps, a video library consisting of tapes from Coronet MTI, and the *Bio Sci* videodisk series from Videodiscovery, covering topics in biochemistry, botany, vertebrate biology, reproduction, ecology, animal behavior, and genetics. Suggestions for integrating the videodisk material in your classroom discussions are available in the instructor's manual.

A comprehensive study guide and lab manual are also available and are described in more detail in the User's Guide section of the preface.

Acknowledgments

*I*t was a delight to work with so many creative individuals whose inspiration, artistry, and vital steam guided this complex project to completion. We wish we were able to acknowledge each of them here, for not only did they meet nearly impossible deadlines, but each willingly poured their heart and soul into this text. The book you now hold in your hands is in large part a tribute to their talent and dedication.

There is one individual whose unique talent, quick intellect, charm, and knowledge not only helped to make this book a reality, but who herself made an enormous contribution to the content and pedagogical strength of this book. We are proud to call Sally Cheney, our biology editor, a colleague. Her powerful belief in this textbook's new approaches to teaching biology helped instill enthusiasm and confidence in everyone who worked on it. Indeed, Sally is truly a force of positive change in college textbook publishing – she has an uncommon ability to think both like a biologist and an editor; she knows what biologists want and need in their classes and is dedicated to delivering it; she recognizes that the future of biology education is more than just publishing another look-alike text; and she is knowledgeable and persuasive enough to convince publishers to stick their necks out a little further for the good of educational advancement. Without Sally, this text would have fallen short of our goal. With Sally, it became even more than we envisioned.

Another individual also helped make this a truly special book, as well as made the many long hours of work so delightful. Stella Kupferberg, we treasure your friendship, applaud your exceptional talent, and salute your high standards. Stella also provided us with two other important assets, Charles Hamilton and Hilary Newman. Stella and Charles tirelessly applied their skill, and artistry to get us images of incomparable effectiveness and beauty, and Hilary's diligent handling helped to insure there were no oversights.

Our thanks to Rachel Nelson for her meticulous editing, for maintaining consistency between sometimes dissimilar writing styles of three authors, and for keeping track of an incalculable number of publishing and biological details; to Katharine Rubin for expertly and gently guiding this project through the myriad levels of production, and for putting up with three such demanding authors; to Karin Kincheloe for a stunningly beautiful design; to Ishaya Monokoff and Ed Starr for orchestrating a brilliant art program; to Network Graphics, especially John Smith and John Hargraves, who executed our illustrations with beauty and style without diluting their conceptual strength or pedagogy, and to Carlyn Iverson, whose artistic talent helped us visually distill our "Steps to Discovery" episodes into images that bring the process of science to life.

We would also like to thank Cathy Faduska and Alida Setford, their creative flair helped us to tell the story behind this book, as well as helped us convey what we tried to accomplish. And to Herb Brown, thank you for your initial confidence and continued support. A very special thank you to Deb Benson, our marketing manager. What a joy to work with you, Deb, your energy, enthusiasm, confidence, and pleasant personality bolstered even our spirits.

We wish to acknowledge Diana Lipscomb of George Washington University for her invaluable contributions to the evolution chapters, and Judy Goodenough of the University of Massachusetts, Amherst, for contributing an outstanding chapter on Animal Behavior.

To the reviewers and instructors who used the First Edition, your insightful feedback helped us forge the foundation for this new edition. To the reviewers, and workshop and conference participants for the Second Edition, thank you for your careful guidance and for caring so much about your students.

Dennis Anderson, *Oklahoma City Community College*
Sarah Barlow, *Middle Tennessee State University*
Robert Beckman, *North Carolina State University*
Timothy Bell, *Chicago State University*
David F. Blaydes, *West Virginia University*
Richard Bliss, *Yuba College*
Richard Boohar, *University of Nebraska, Lincoln*
Clyde Bottrell, *Tarrant County Junior College*
J. D. Brammer, *North Dakota State University*
Peggy Branstrator, *Indiana University, East*
Allyn Bregman, *SUNY, New Paltz*
Daniel Brooks, *University of Toronto*

Gary Brusca, *Humboldt State University*
Jack Bruk, *California State University, Fullerton*
Marvin Cantor, *California State University, Northridge*
Richard Cheney, *Christopher Newport College*
Larry Cohen, *California State University, San Marcos*
David Cotter, *Georgia College*
Robert Creek, *Eastern Kentucky University*
Ken Curry, *University of Southern Mississippi*
Judy Davis, *Eastern Michigan University*
Loren Denny, *southwest Missouri State University*
Captain Donald Diesel, *U. S. Air Force Academy*
Tom Dickinson, *University College of the Cariboo*

Mike Donovan, *Southern Utah State College*
Robert Ebert, *Palomar College*
Thomas Emmel, *University of Florida*
Joseph Faryniarz, *Mattatuck Community College*
Alan Feduccia, *University of North Carolina, Chapel Hill*
Eugene Ferri, *Bucks County Community College*
Victor Fet, *Loyola University, New Orleans*
David Fox, *Loyola University, New Orleans*
Mary Forrest, *Okanagan University College*
Michael Gains, *University of Kansas*
S. K. Gangwere, *Wayne State University*
Dennis George, *Johnson County Community College*
Bill Glider, *University of Nebraska*
Paul Goldstein, *University of North Carolina, Charlotte*
Judy Goodenough, *University of Massachusetts, Amherst*
Nathaniel Grant, *Southern Carolina State College*
Mel Green, *University of California, San Diego*
Dana Griffin, *Florida State University*
Barbara L. Haas, *Loyola University of Chicago*
Richard Haas, *California State University, Fresno*
Fredrick Hagerman, *Ohio State University*
Tom Haresign, *Long Island University, Southampton*
Jane Noble-Harvey, *University of Delaware*
W. R. Hawkins, *Mt. San Antonio College*
Vernon Hedricks, *Brevard Community College*
Paul Hertz, *Barnard College*
Howard Hetzle, *Illinois State University*
Ronald K. Hodgson, *Central Michigan University*
W. G. Hopkins, *University of Western Ontario*
Thomas Hutto, *West Virginia State College*
Duane Jeffrey, *Brigham Young University*
John Jenkins, *Swarthmore College*
Claudia Jones, *University of Pittsburgh*
R. David Jones, *Adelphi University*
J. Michael Jones, *Culver Stockton College*
Gene Kalland, *California State University, Dominiquez Hills*
Arnold Karpoff, *University of Louisville*
Judith Kelly, *Henry Ford Community College*
Richard Kelly, *SUNY, Albany*
Richard Kelly, *University of Western Florida*
Dale Kennedy, *Kansas State University*
Mirium Kittrell, *Kingsboro Community College*
John Kmeltz, *Kean College New Jersey*
Robert Krasner, *Providence College*
Susan Landesman, *Evergreen State College*
Anton Lawson, *Arizona State University*
Lawrence Levine, *Wayne State University*
Jerri Lindsey, *Tarrant County Junior College*
Diana Lipscomb, *George Washington University*
James Luken, *Northern Kentucky University*

Ted Maguder, *University of Hartford*
Jon Maki, *Eastern Kentucky University*
Charles Mallery, *University of Miami*
William McEowen, *Mesa Community College*
Roger Milkman, *University of Iowa*
Helen Miller, *Oklahoma State University*
Elizabeth Moore, *Glassboro State College*
Janice Moore, *Colorado State University*
Eston Morrison, *Tarleton State University*
John Mutchmor, *Iowa State University*
Douglas W. Ogle, *Virginia Highlands Community College*
Joel Ostroff, *Brevard Community College*
James Lewis Payne, *Virginia Commonwealth University*
Gary Peterson, *South Dakota State University*
MaryAnn Phillippi, *Southern Illinois University, Carbondale*
R. Douglas Powers, *Boston College*
Robert Raikow, *University of Pittsburgh*
Charles Ralph, *Colorado State University*
Aryan Roest, *California State Polytechnic Univ., San Luis Obispo*
Robert Romans, *Bowling Green State University*
Raymond Rose, *Beaver College*
Richard G. Rose, *West Valley College*
Donald G. Ruch, *Transylvania University*
A. G. Scarbrough, *Towsow State University*
Gail Schiffer, *Kennesaw State University*
John Schmidt, *Ohio State University*
John R. Schrock, *Emporia State University*
Marilyn Shopper, *Johnson County Community College*
John Smarrelli, *Loyola University of Chicago*
Deborah Smith, *Meredith College*
Guy Steucek, *Millersville University*
Ralph Sulerud, *Augsburg College*
Tom Terry, *University of Connecticut*
James Thorp, *Cornell University*
W. M. Thwaites, *San Diego State University*
Michael Torelli, *University of California, Davis*
Michael Treshow, *University of Utah*
Terry Trobec, *Oakton Community College*
Len Troncale, *California State Polytechnic University, Pomona*
Richard Van Norman, *University of Utah*
David Vanicek, *California State University, Sacramento*
Terry F. Werner, *Harris-Stowe State College*
David Whitenberg, *Southwest Texas State University*
P. Kelly Williams, *University of Dayton*
Robert Winget, *Brigham Young University*
Steven Wolf, *University of Missouri, Kansas City*
Harry Womack, *Salisbury State University*
William Yurkiewicz, *Millersville University*

Gil Brum
Larry McKane
Gerry Karp

Brief Table of Contents

P A R T 1 / *Biology: The Study of Life*   *1*

Chapter 1   Biology: Exploring Life   3
Chapter 2   The Process of Science   29

P A R T 2 / *Chemical and Cellular Foundations of Life*   *45*

Chapter 3   The Atomic Basis of Life   47
Chapter 4   Biochemicals: The Molecules of Life   65
Chapter 5   Cell Structure and Function   87
Chapter 6   Energy, Enzymes, and Metabolic Pathways   115
Chapter 7   Movement of Materials Across Membranes   135
Chapter 8   Processing Energy: Photosynthesis and Chemosynthesis   151
Chapter 9   Processing Energy: Fermentation and Respiration   171
Chapter 10   Cell Division: Mitosis   193
Chapter 11   Cell Division: Meiosis   211

P A R T 3 / *The Genetic Basis of Life*   *227*

Chapter 12   On the Trail of Heredity   229
Chapter 13   Genes and Chromosomes   245
Chapter 14   The Molecular Basis of Genetics   265
Chapter 15   Orchestrating Gene Expression   291
Chapter 16   DNA Technology: Developments and Applications   311
Chapter 17   Human Genetics: Past, Present, and Future   333

P A R T 4 / *Form and Function of Plant Life*   *355*

Chapter 18   Plant Tissues and Organs   357
Chapter 19   The Living Plant: Circulation and Transport   383

Chapter 20   Sexual Reproduction of Flowering Plants   399
Chapter 21   How Plants Grow and Develop   423

P A R T 5 / *Form and Function of Animal Life*   *445*

Chapter 22   An Introduction to Animal Form and Function   447
Chapter 23   Coordinating the Organism: The Role of the Nervous System   465
Chapter 24   Sensory Perception: Gathering Information about the Environment   497
Chapter 25   Coordinating the Organism: The Role of the Endocrine System   517
Chapter 26   Protection, Support, and Movement: The Integumentary, Skeletal, and Muscular Systems   539
Chapter 27   Processing Food and Providing Nutrition: The Digestive System   565
Chapter 28   Maintaining the Constancy of the Internal Environment: The Circulatory and Excretory Systems   585
Chapter 29   Gas Exchange: The Respiratory System   619
Chapter 30   Internal Defense: The Immune System   641
Chapter 31   Generating Offspring: The Reproductive System   661
Chapter 32   Animal Growth and Development: Acquiring Form and Function   685

P A R T 6 / *Evolution*   *711*

Chapter 33   Mechanisms of Evolution   713
Chapter 34   Evidence for Evolution   741
Chapter 35   The Origin and History of Life   759

P A R T 7 / *The Kingdoms of Life: Diversity and Classification* 781

Chapter 36   The Monera Kingdom and Viruses   783
Chapter 37   The Protist and Fungus Kingdoms   807
Chapter 38   The Plant Kingdom   829
Chapter 39   The Animal Kingdom   857

P A R T 8 / *Ecology and Animal Behavior* 899

Chapter 40   The Biosphere   901
Chapter 41   Ecosystems and Communities   931

Chapter 42   Community Ecology: Interactions between Organisms   959
Chapter 43   Population Ecology   983
Chapter 44   Animal Behavior   1005

Appendix A   Metric and Temperature Conversion Charts   A-1
Appendix B   Microscopes: Exploring the Details of Life   B-1
Appendix C   The Hardy-Weinberg Principle   C-1
Appendix D   Careers in Biology   D-1

Contents

PART 1
Biology: The Study of Life   1

Chapter 1 / Biology: Exploring Life   3

Steps to Discovery:
Exploring Life—The First Step   4

Distinguishing the Living from the Inanimate   6
Levels of Biological Organization   11
What's in a Name   11
Underlying Themes of Biology   15
Biology and Modern Ethics   25

Reexamining the Themes   25
Synopsis   26

Chapter 2 / The Process of Science   29

Steps to Discovery:
What Is Science?   30

The Scientific Approach   32
Applications of the Scientific Process   35
Caveats Regarding "The" Scientific Method   40

Synopsis   41

BIOETHICS

Science, Truth, and Certainty:
Is a Theory "Just a Theory"?   40

PART 2
Chemical and Cellular Foundations of Life   45

Chapter 3 / The Atomic Basis of Life   47

Steps to Discovery:
The Atom Reveals Some of Its Secrets   48

The Nature of Matter   50
The Structure of the Atom   51
Types of Chemical Bonds   55
The Life-Supporting Properties of Water   59
Acids, Bases, and Buffers   61

Reexamining the Themes   62
Synopsis   63

BIOLINE

Putting Radioisotopes to Work   54

THE HUMAN PERSPECTIVE

Aging and Free Radicals   56

Chapter 4 / Biochemicals: The Molecules of Life   65

Steps to Discovery:
Determining the Structure of Proteins   66

The Importance of Carbon in Biological Molecules   68
Giant Molecules Built From Smaller Subunits   69
Four Biochemical Families   69
Macromolecular Evolution and Adaptation   83

Reexamining the Themes   84
Synopsis   85

THE HUMAN PERSPECTIVE

Obesity and the Hungry Fat Cell   76

Chapter 5 / Cell Structure and Function   87

Steps to Discovery:
The Nature of the Plasma Membrane   88

Discovering the Cell   90
Basic Characteristics of Cells   91
Two Fundamentally Different Types of Cells   91
The Plasma Membrane: Specialized for
Interaction with the Environment   93
The Eukaryotic Cell: Organelle
Structure and Function   95
Just Outside the Cell   108
Evolution of Eukaryotic Cells: Gradual
Change or an Evolutionary Leap?   110

Reexamining the Themes   111
Synopsis   112

THE HUMAN PERSPECTIVE

Lysosome-Related Diseases   101

Chapter 6 / Energy, Enzymes, and Metabolic Pathways   115

Steps to Discovery:
The Chemical Nature of Enzymes   116

Acquiring and Using Energy   118
Enzymes   123
Metabolic Pathways   127
Macromolecular Evolution and Adaptation
of Enzymes   131

Reexamining the Themes   132
Synopsis   132

BIOLINE

Saving Lives by Inhibiting Enzymes   126

THE HUMAN PERSPECTIVE

The Manipulation of Human Enzymes   131

Chapter 7 / Movement of Materials Across Membranes   135

Steps to Discovery:
Getting Large Molecules Across Membrane Barriers   136

Membrane Permeability   138
Diffusion: Depending on the Random
Movement of Molecules   138
Active Transport   142
Endocytosis   144

Reexamining the Themes   148
Synopsis   148

THE HUMAN PERSPECTIVE

LDL Cholesterol, Endocytosis, and Heart Disease   146

Chapter 8 / Processing Energy: Photosynthesis and Chemosynthesis   151

Steps to Discovery:
Turning Inorganic Molecules into Complex Sugars   152

Autotrophs and Heterotrophs:
Producers and Consumers   154
An Overview of Photosynthesis   155
The Light-Dependent Reactions: Converting Light
Energy into Chemical Energy   156
The Light-Independent Reactions   163
Chemosynthesis: An Alternate Form of Autotrophy   167

Reexamining the Themes   168
Synopsis   169

BIOLINE

Living on the Fringe of the Biosphere   167

THE HUMAN PERSPECTIVE

Producing Crop Plants Better Suited
to Environmental Conditions   165

Chapter 9 / Processing Energy: Fermentation and Respiration   171

Steps to Discovery:
The Machinery Responsible for ATP Synthesis   172

Fermentation and Aerobic Respiration:
A Preview of the Strategies   174
Glycolysis   175
Fermentation   178
Aerobic Respiration   180

Balancing the Metabolic Books: Generating and
Yielding Energy   185
Coupling Glucose Oxidation to Other Pathways   188

Reexamining the Themes   190
Synopsis   190

BIOLINE

The Fruits of Fermentation   179

THE HUMAN PERSPECTIVE

The Role of Anaerobic and Aerobic Metabolism
in Exercise   186

Chapter 10 / Cell Division: Mitosis   193

Steps to Discovery:
Controlling Cell Division   194

Types of Cell Division   196
The Cell Cycle   199
The Phases of Mitosis   202
Cytokinesis: Dividing the Cell's Cytoplasm
and Organelles   206

Reexamining the Themes   207
Synopsis   208

THE HUMAN PERSPECTIVE

Cancer: The Cell Cycle Out of Control   202

Chapter 11 / Cell Division: Meiosis   211

Steps to Discovery:
Counting Human Chromosomes   212

Meiosis and Sexual Reproduction: An Overview   214
The Stages of Meiosis   216
Mitosis Versus Meiosis Revisited   222

Reexamining the Themes   223
Synopsis   223

THE HUMAN PERSPECTIVE

Dangers That Lurk in Meiosis   221

PART 3
The Genetic Basis of Life   227

Chapter 12 / On the Trail of Heredity   229

Steps to Discovery:
The Genetic Basis of Schizophrenia   230

Gregor Mendel: The Father of Modern Genetics   232
Mendel Amended   239
Mutation: A Change in the Genetic Status Quo   241

Reexamining the Themes   242
Synopsis   242

Chapter 13 / Genes and Chromosomes   245

Steps to Discovery:
The Relationship Between Genes and Chromosomes   246

The Concept of Linkage Groups   248
Lessons from Fruit Flies   249
Sex and Inheritance   253
Aberrant Chromosomes   257
Polyploidy   260

Reexamining the Themes   261
Synopsis   262

BIOLINE

The Fish That Changes Sex   254

THE HUMAN PERSPECTIVE

Chromosome Aberrations and Cancer   260

**Chapter 14 / The Molecular Basis
of Genetics   265**

Steps to Discovery:
The Chemical Nature of the Gene   266

Confirmation of DNA as the Genetic Material   268
The Structure of DNA   268
DNA: Life's Molecular Supervisor   272
The Molecular Basis of Gene Mutations   286

Reexamining the Themes   287
Synopsis   288

THE HUMAN PERSPECTIVE

The Dark Side of the Sun   280

Chapter 15 / Orchestrating Gene Expression   291

Steps to Discovery:
Jumping Genes: Leaping into the Spotlight   292

Why Regulate Gene Expression?   294
Gene Regulation in Prokaryotes   298
Gene Regulation in Eukaryotes   302
Levels of Control of Eukaryotic Gene Expression   303

Reexamining the Themes   308
Synopsis   309

BIOLINE

RNA as an Evolutionary Relic   306

THE HUMAN PERSPECTIVE

Clones: Is There Cause for Fear?   296

Chapter 16 / DNA Technology: Developments and Applications   311

Steps to Discovery:
DNA Technology and Turtle Migration   312

Genetic Engineering   314
DNA Technology I: The Formation and Use of Recombinant DNA Molecules   321
DNA Technology II: Techniques That Do Not Require Recombinant DNA Molecules   325
Use of DNA Technology in Determining Evolutionary Relationships   329

Reexamining the Themes   329
Synopsis   330

BIOLINE

DNA Fingerprints and Criminal Law   328

THE HUMAN PERSPECTIVE

Animals That Develop Human Diseases   320

BIOETHICS

Patenting a Genetic Sequence   315

Chapter 17 / Human Genetics: Past, Present, and Future   333

Steps to Discovery:
Developing a Treatment for an Inherited Disorder   334

The Consequences of an Abnormal Number of Chromosomes   336
Disorders that Result from a Defect in a Single Gene   338
Patterns of Transmission of Genetic Disorders   343
Screening Humans for Genetic Defects   346

Reexamining the Themes   350
Synopsis   351

BIOLINE

Mapping the Human Genome   340

THE HUMAN PERSPECTIVE

Correcting Genetic Disorders by Gene Therapy   348

PART 4
Form and Function of Plant Life   355

Chapter 18 / Plant Tissues and Organs   357

Steps to Discovery:
Fruit of the Vine and the French Economy   358

The Basic Plant Design   360
Plant Tissues   362
Plant Tissue Systems   365
Plant Organs   372

Reexamining the Themes   379
Synopsis   379

BIOLINE

Nature's Oldest Living Organisms   366

THE HUMAN PERSPECTIVE

Agriculture, Genetic Engineering, and Plant Fracture Properties   375

Chapter 19 / The Living Plant: Circulation and Transport   383

Steps to Discovery:
Exploring the Plant's Circulatory System   384

Xylem: Water and Mineral Transport   386
Phloem: Food Transport   393

Reexamining the Themes   396
Synopsis   397

BIOLINE

Mycorrhizae and Our Fragile Deserts   388

THE HUMAN PERSPECTIVE

Leaf Nodules and World Hunger: An Unforeseen Connection   396

Chapter 20 / Sexual Reproduction of Flowering Plants   399

Steps to Discovery:
What Triggers Flowering?   400

Flower Structure and Pollination   402
Formation of Gametes   407
Fertilization and Development   408
Fruit and Seed Dispersal   412
Germination and Seedling Development   416
Asexual Reproduction   417
Agricultural Applications   418

Reexamining the Themes   419
Synopsis   420

BIOLINE
The Odd Couples: Bizarre Flowers and Their Pollinators   404

THE HUMAN PERSPECTIVE
The Fruits of Civilization   414

Chapter 21 / How Plants Grow and Develop   423

Steps to Discovery:
Plants Have Hormones   424

Levels of Control of Growth and Development   426
Plant Hormones: Coordinating Growth and Development   427
Timing Growth and Development   434
Plant Tropisms   438

Reexamining the Themes   440
Synopsis   440

THE HUMAN PERSPECTIVE
Saving Tropical Rain Forests   428

PART 5
Form and Function of Animal Life   445

Chapter 22 / An Introduction to Animal Form and Function   447

Steps to Discovery:
The Concept of Homeostasis   448

Homeostatic Mechanisms   450
Unity and Diversity Among Animal Tissues   451
Unity and Diversity Among Organ Systems   455
The Evolution of Organ Systems   457
Body Size, Surface Area, and Volume   460

Reexamining the Themes   462
Synopsis   463

Chapter 23 / Coordinating the Organism: The Role of the Nervous System   465

Steps to Discovery:
A Factor Promoting the Growth of Nerves   466

Neurons and Their Targets   468
Generating and Conducting Neural Impulses   470
Neurotransmission: Jumping the Synaptic Cleft   474
The Nervous System   479
Architecture of the Human Central Nervous System   481
Architecture of the Peripheral Nervous System   489
The Evolution of Complex Nervous Systems from Single-Celled Roots   491

Reexamining the Themes   493
Synopsis   494

BIOLINE
Deadly Meddling at the Synapse   478

THE HUMAN PERSPECTIVE
Alzheimer's Disease: A Status Report   484

BIOETHICS
Blurring the Line Between Life and Death   489

Chapter 24 / Sensory Perception: Gathering Information About the Environment   497

Steps to Discovery:
Echolocation: Seeing in the Dark   498

The Response of a Sensory Receptor to a Stimulus   500
Somatic Sensation   500
Vision   502
Hearing and Balance   506

Chemoreception: Taste and Smell 509

The Brain: Interpreting Impulses into Sensations 510

Evolution and Adaptation of Sense Organs 511

Reexamining the Themes 514

Synopsis 514

BIOLINE

Sensory Adaptations to an Aquatic Environment 512

THE HUMAN PERSPECTIVE

Two Brains in One 504

Chapter 25 / Coordinating the Organism: The Role of the Endocrine System 517

Steps to Discovery: The Discovery of Insulin 518

The Role of the Endocrine System 520

The Nature of the Endocrine System 521

A Closer Look at the Endocrine System 523

Action at the Target 533

Evolution of the Endocrine System 534

Reexamining the Themes 536

Synopsis 537

BIOLINE

Chemically Enhanced Athletes 528

THE HUMAN PERSPECTIVE

The Mysterious Pineal Gland 533

Chapter 26 / Protection, Support, and Movement: The Integumentary, Skeletal, and Muscular Systems 539

Steps to Discovery: Vitamin C's Role in Holding the Body Together 540

The Integument: Covering and Protecting the Outer Body Surface 542

The Skeletal System: Providing Support and Movement 544

The Muscles: Powering the Motion of Animals 551

Body Mechanics and Locomotion 558

Evolution of the Skeletomuscular System 560

Reexamining the Themes 561

Synopsis 562

THE HUMAN PERSPECTIVE

Building Better Bones 550

Chapter 27 / Processing Food and Providing Nutrition: The Digestive System 565

Steps to Discovery: The Battle Against Beri-Beri 566

The Digestive System: Converting Food into a Form Available to the Body 568

The Human Digestive System: A Model Food-Processing Plant 568

Evolution of Digestive Systems 576

Reexamining the Themes 582

Synopsis 583

BIOLINE

Teeth: A Matter of Life and Death 570

THE HUMAN PERSPECTIVE

Human Nutrition 578

Chapter 28 / Maintaining the Constancy of the Internal Environment: the Circulatory and Excretory Systems 585

Steps to Discovery: Tracing the Flow of Blood 586

The Human Circulatory System: Form and Function 588

Evolution of Circulatory Systems 603

The Lymphatic System: Supplementing the Functions of the Circulatory System 606

Excretion and Osmoregulation: Removing Wastes and Maintaining the Composition of the Body's Fluids 606

Thermoregulation: Maintaining a Constant Body Temperature 614

Reexamining the Themes 615

Synopsis 615

BIOLINE

The Artificial Kidney 611

THE HUMAN PERSPECTIVE

The Causes of High Blood Pressure 591

xviii • *Contents*

Chapter 29 / Gas Exchange: The Respiratory System   619

Steps to Discovery:
Physiological Adaptations in Diving Mammals   620

Properties of Gas Exchange Surfaces   622
The Human Respiratory System: Form and Function   623
The Exchange of Respiratory Gases and the
   Role of Hemoglobin   627
Regulating the Rate and Depth of Breathing
   to Meet the Body's Needs   629
Adaptations for Extracting Oxygen from
   Water Versus Air   632
Evolution of the Vertebrate Respiratory System   636

Reexamining the Themes   637
Synopsis   637

THE HUMAN PERSPECTIVE
Dying for a Cigarette   630

Chapter 30 / Internal Defense: The Immune System   641

Steps to Discovery:
On the Trail of a Killer: Tracking the AIDS Virus   642

Nonspecific Mechanisms: A First Line of Defense   644
The Immune System: Mediator of Specific
   Mechanisms of Defense   646
Antibody Molecules: Structure and Formation   650
Immunization   654
Evolution of the Immune System   655

Reexamining the Themes   658
Synopsis   658

BIOLINE
Treatment of Cancer with Immunotherapy   648

THE HUMAN PERSPECTIVE
Disorders of the Human Immune System   656

Chapter 31 / Generating Offspring: The Reproductive System   661

Steps to Discovery:
The Basis of Human Sexuality   662

Reproduction: Asexual Versus Sexual   664
Human Reproductive Systems   667
Controlling Pregnancy: Contraceptive Methods   677
Sexually Transmitted Diseases   681

Reexamining the Themes   682
Synopsis   682

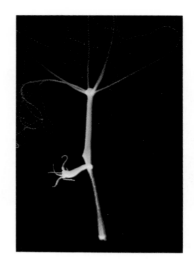

BIOLINE
Sexual Rituals   668

THE HUMAN PERSPECTIVE
Overcoming Infertility   679

BIOETHICS
Frozen Embryos and Compulsive Parenthood   680

Chapter 32 / Animal Growth and Development: Acquiring Form and Function   685

Steps to Discovery:
Genes That Control Development   686

The Course of Embryonic Development   688
Beyond Embryonic Development: Continuing Changes in
   Body Form   699
Human Development   700
Embryonic Development and Evolution   705

Reexamining the Themes   706
Synopsis   707

THE HUMAN PERSPECTIVE
The Dangerous World of a Fetus   704

PART 6
Evolution   711

Chapter 33 / Mechanisms of Evolution   713

Steps to Discovery:
Silent Spring Revisited   714

The Basis of Evolutionary Change   717
Speciation: The Origin of Species   731
Patterns of Evolution   734
Extinction: The Loss of Species   736
The Pace of Evolution   737

Reexamining the Themes 738

Synopsis 738

BIOLINE

A Gallery of Remarkable Adaptations 728

Chapter 34 / Evidence for Evolution 741

Steps to Discovery:
An Early Portrait of the Human Family 742

Determining Evolutionary Relationships 744
Evidence for Evolution 746
The Evidence of Human Evolution:
The Story Continues 752

Reexamining the Themes 756

Synopsis 757

Chapter 35 / The Origin and History of Life 759

Steps to Discovery:
Evolution of the Cell 760

Formation of the Earth: The First Step 762
The Origin of Life and Its Formative Stages 763
The Geologic Time Scale: Spanning Earth's History 764

Reexamining the Themes 775

Synopsis 778

BIOLINE

The Rise and Fall of the Dinosaurs 772

PART 7
The Kingdoms of Life: Diversity and
Classification 781

Chapter 36 / The Monera Kingdom
and Viruses 783

Steps to Discovery:
The Burden of Proof: Germs and Infectious Diseases 784

Kingdom Monera: The Prokaryotes 787
Viruses 800

Reexamining the Themes 804

Synopsis 805

BIOLINE

Living with Bacteria 799

THE HUMAN PERSPECTIVE

Sexually Transmitted Diseases 794

Chapter 37 / The Protist and Fungus
Kingdoms 807

Steps to Discovery:
Creating Order from Chaos 808

The Protist Kingdom 810
The Fungus Kingdom 820

Reexamining the Themes 826

Synopsis 827

Chapter 38 / The Plant Kingdom 829

Steps to Discovery:
Distinguishing Plant Species: Where Should the Dividing
Lines Be Drawn? 830

Major Evolutionary Trends 833
Overview of the Plant Kingdom 834
Algae: The First Plants 835
Bryophytes: Plants Without Vessels 842
Tracheophytes: Plants with Fluid-Conducting
Vessels 843

Reexamining the Themes 854

Synopsis 854

BIOLINE

The Exceptions: Plants That Don't Photosynthesize 836

THE HUMAN PERSPECTIVE

Brown Algae: Natural Underwater Factories 840

Chapter 39 / The Animal Kingdom 857

Steps to Discovery:
The World's Most Famous Fish 858

Evolutionary Origins and Body Plans 861
Invertebrates 863
Vertebrates 886

Reexamining the Themes 895

Synopsis 895

BIOLINE

Adaptations to Parasitism among Flatworms 872

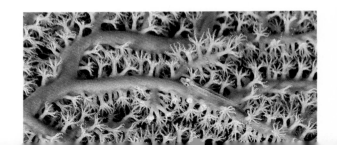

PART 8
Ecology and Animal Behavior   899

Chapter 40 / The Biosphere   901

Steps to Discovery:
The Antarctic Ozone Hole   902

Boundaries of the Biosphere   904
Ecology Defined   905
The Earth's Climates   905
The Arenas of Life   908

Reexamining the Themes   927
Synopsis   928

THE HUMAN PERSPECTIVE

Acid Rain and Acid Snow: Global Consequences
of Industrial Pollution   918

BIOETHICS

Values in Ecology and Environmental Science: Neutrality
or Advocacy?   909

Chapter 41 / Ecosystems and Communities   931

Steps to Discovery:
The Nature of Communities   932

From Biomes to Microcosms   934
The Structure of Ecosystems   934
Ecological Niches and Guilds   939
Energy Flow Through Ecosystems   942
Biogeochemical Cycles: Recycling Nutrients
in Ecosystems   947
Succession: Ecosystem Change and Stability   952

Reexamining the Themes   955
Synopsis   956

BIOLINE

Reverberations Felt Throughout an Ecosystem   936

THE HUMAN PERSPECTIVE

The Greenhouse Effect: Global Warming   950

Chapter 42 / Community Ecology: Interactions Between Organisms   959

Steps to Discovery:
Species Coexistence: The Unpeaceable Kingdom   960

Symbiosis   962
Competition: Interactions That Harm
Both Organisms   963

Interactions that Harm One Organism and
Benefit the Other   965
Commensalism: Interactions that Benefit One Organism
and Have No Affect on the Other   976
Interactions That Benefit Both Organisms   977

Reexamining the Themes   979
Synopsis   979

Chapter 43 / Population Ecology   983

Steps to Discovery:
Threatening a Giant   984

Population Structure   986
Factors Affecting Population Growth   988
Factors Controlling Population Growth   996
Human Population Growth   997

Reexamining the Themes   1002
Synopsis   1002

BIOLINE

Accelerating Species Extinction   997

THE HUMAN PERSPECTIVE

Impacts of Poisoned Air, Land, and Water   993

Chapter 44 / Animal Behavior   1005

Steps to Discovery:
Mechanisms and Functions of Territorial Behavior   1006

Mechanisms of Behavior   1008
Learning   1011
Development of Behavior   1014
Evolution and Function of Behavior   1015
Social Behavior   1018
Altruism   1022

Reexamining the Themes   1026
Synopsis   1027

BIOLINE

Animal Cognition   1024

**Appendix A   Metric and Temperature
Conversion Charts   A-1**
**Appendix B   Microscopes: Exploring the Details of
Life   B-1**
Appendix C   The Hardy-Weinberg Principle   C-1
Appendix D   Careers in Biology   D-1
Glossary   G-1
Photo Credits   P-1
Index   I-1

To The Student:
A User's Guide

*B*iology is a journey of exploration and discovery, of struggle and breakthrough. It is enlivened by the thrill of understanding not only what living things do but also how they work. We have tried to create such an experience for you.

Excellence in writing, visual images, and broad biological coverage form the core of a modern biology textbook. But as important as these three factors are in making difficult concepts and facts clear and meaningful, none of them reveals the excitement of biology—the adventure that unearths what we know about life. To help relate the true nature of this adventure, we have developed several distinctive features for this book, features that strengthen its biological core, that will engage and hold your attention, that reveal the human side of biology, that enable every reader to understand how science works, that stimulate critical thinking, and that will create the informed citizenship we all hope will make a positive difference in the future of our planet.

Steps to Discovery

*T*he process of science enriches all parts of this book. We believe that students, like biologists, themselves, are intrigued by scientific puzzles. Every chapter is introduced by a "Steps to Discovery" narrative, the story of an investigation that led to a scientific breakthrough in an area of biology which relates to that chapter's topic. The "Steps to Discovery" narratives portray biologists as they really are: human beings, with motivations, misfortunes, and mishaps, much like everyone experiences. We hope these narratives help you better appreciate biological investigation, realizing that it is understandable and within your grasp.

Throughout the narrative of these pieces, the writing is enlivened with scientific work that has provided knowledge and understanding of life. This approach is meant not just to pay tribute to scientific giants and Nobel prize winners, but once again to help you realize that science does not grow by itself. Facts do not magically materialize. They are the products of rational ideas, insight, determination, and, sometimes, a little luck. Each of the "Steps to Discovery" narratives includes a painting that is meant primarily as an aesthetic accompaniment to the adventure described in the essay and to help you form a mental picture of the subject.

S TEPS TO DISCOVER Y
A Factor Promoting the Growth of Nerves

*R*ita Levi-Montalcini received her medical degree from the University of Turin in Italy in 1936, the same year that Benito Mussolini began his anti-Semitic campaign. By 1939, as a Jew, Levi-Montalcini had been barred from carrying out research and practicing medicine, yet she continued to do both secretly. As a student, Levi-Montalcini had been fascinated with the structure and function of the nervous system. Unable to return to the university, she set up a simple laboratory in her small bedroom in her family's home. As World War II raged throughout Europe, and the Allies systematically bombed Italy, Levi-Montalcini studied chick embryos in her bedroom, discovering new information about the growth of nerve cells from the spinal cord into the nearby limbs. In her autobiography *In Praise of Imperfection*, she writes: "Every time the alarm sounded, I would carry down to the precarious safety of the cellars the Zeiss binocular microscope and my most precious silver-stained embryonic sections." In September 1943, German troops arrived in Turin to support the Italian Fascists. Levi-Montalcini and her family fled southward to Florence, where they remained in hiding for the remainder of the war.

After the war ended, Levi-Montalcini continued research at the University of Turin. In 1946, she accepted an invitation from Viktor Hamburger, a leading expert on the development of the chick nervous system, to come to Washington University in St. Louis to work with him for a semester; she remained at Washington University for 30 years.

A chick embryo and one of its nerve cells helped scientists discover nerve growth factor (NGF).

One of Levi-Montalcini's first projects was the reexamination of a previous experiment of Elmer Bueker, a former student of Hamburger's. Bueker had removed a limb from a chick embryo, replaced it with a fragment of a mouse connective tissue tumor, and found that nerve fibers grew into this mass of implanted tumor cells. When Levi-Montalcini repeated the experiment she made an unexpected discovery: One part of the nervous system of these experimental chick embryos—the sympathetic nervous system—had grown five to six times larger than had its counterpart in a normal chick embryo. (The sympathetic nervous system helps control the activity of internal organs, such as the heart and digestive tract.) Close examination revealed that the small piece of tumor tissue that had been grafted onto the embryo had caused sympathetic nerve fibers to grow "wildly" into all of the chick's internal organs, even causing some of the blood vessels to become obstructed by the invasive fibers. Levi-Montalcini hypothesized that the tumor was releasing some soluble substance that induced the remarkable growth of this part of the nervous system. Her hypothesis was soon confirmed by further experiments. She called the active substance **nerve growth factor (NGF)**.

The next step was to determine the chemical nature of NGF, a task that was more readily performed by growing the tumor cells in a culture dish rather than an embryo. But Hamburger's laboratory at Washington University did not have the facilities for such work. To continue the project, Levi-Montalcini boarded a plane, with a pair of tumor-bearing mice in the pocket of her overcoat, and flew to Brazil, where she had a friend who operated a tissue culture laboratory. When she placed sympathetic nervous tissue in the proximity of the tumor cells in a culture dish, the nervous tissue sprouted a halo of nerve fibers that grew toward the tumor cells. When the tissue was cultured in the absence of NGF, no such growth occurred.

For the next 2 years, Levi-Montalcini's lab was devoted to characterizing the substance in the tumor cells that possessed the ability to cause nerve outgrowth. The work was carried out primarily by a young biochemist, Stanley Cohen, who had joined the lab. One of the favored approaches to studying the nature of a biological molecule is to determine its sensitivity to enzymes. In order to determine if nerve growth factor was a protein or a nucleic acid, Cohen treated the active material with a small amount of snake venom, which contains a highly active enzyme that degrades nucleic acid. It was then that chance stepped in.

Cohen expected that treatment with the venom would either destroy the activity of the tumor cell fraction (if NGF was a nucleic acid) or leave it unaffected (if NGF was a protein). To Cohen's surprise, treatment with the venom *increased* the nerve-growth promoting activity of the material. In fact, treatment of sympathetic nerve tissue with the venom alone (in the absence of the tumor extract) induced the growth of a halo of nerve fibers! Cohen soon discovered why: The snake venom possessed the same nerve growth factor as did the tumor cells, but at much higher concentration. Cohen soon demonstrated that NGF was a protein.

Levi-Montalcini and Cohen reasoned that since snake venom was derived from a *modified* salivary gland, the other salivary glands might prove to be even better sources of the protein. This hypothesis proved to be correct. When Levi-Montalcini and Cohen tested the salivary glands from male mice, they discovered the richest source of NGF yet, a source 10,000 times more active than the tumor cells and ten times more active than snake venom.

A crucial question remained: Did NGF play a role in the normal development of the embryo, or was its ability to stimulate nerve growth just an accidental property of the molecule? To answer this question, Levi-Montalcini and Cohen injected embryos with an antibody against NGF, which they hoped would inactivate NGF molecules wherever they were present in the embryonic tissues. The embryos developed normally, with one major exception: They virtually lacked a sympathetic nervous system. The researchers concluded that NGF must be important during normal development of the nervous system; otherwise, inactivation of NGF could not have had such a dramatic effect.

By the early 1970s, the amino acid sequence of NGF had been determined, and the protein is now being synthesized by recombinant DNA technology. During the past decade, Fred Gage, of the University of California, has found that NGF is able to revitalize aged or damaged nerve cells in rats. Based on these studies, NGF is currently being tested as a possible treatment of Alzheimer's disease. For their pioneering work, Rita Levi-Montalcini and Stanley Cohen shared the 1987 Nobel Prize in Physiology and Medicine.

Many students are overwhelmed by the diversity of living organisms and the multitude of seemingly unrelated facts that they are forced to learn in an introductory biology course. Most aspects of biology, however, can be thought of as examples of a small number of recurrent themes. Using the thematic approach, the details and principles of biology can be assembled into a body of knowledge that makes sense, and is not just a collection of disconnected facts. Facts become ideas, and details become parts of concepts as you make connections between seemingly unrelated areas of biology, forging a deeper understanding.

All areas of biology are bound together by evolution, the central theme in the study of life. Every organism is the product of evolution, which has generated the diversity of biological features that distinguish organisms from one another and the similarities that all organisms share. From this basic evolutionary theme emerge several other themes that recur throughout the book:

- **Relationship between Form and Function**
- **Biological Order, Regulation, and Homeostatis**
- **Acquiring and Using Energy**
- **Unity Within Diversity**
- **Evolution and Adaptation**

We have highlighted the prevalent recurrence of each theme throughout the text with an icon, shown above. The icons can be used to activate higher thought processes by inviting you to explore how the fact or concept being discussed fits the indicated theme.

Reexamining the Themes

*E*ach chapter concludes with a "Reexamining the Themes" section, which revisits the themes and how they emerge within the context of the chapter's concepts and principles. This section will help you realize that the same themes are evident at all levels of biological organization, whether you are studying the molecular and cellular aspects of biology or the global characteristics of biology.

When two organisms have the same protein, the difference in amino acid sequence of that protein can be correlated with the evolutionary relatedness of the organisms. The amino acid sequence of hemoglobin, for example, is much more similar between humans and monkeys—organisms that are closely related—than between humans and turtles, who are only distantly related. In fact, the evolutionary tree that emerges when comparing the structure of specific proteins from various animals very closely matches that previously constructed from fossil evidence.

The fact that the amino acid sequences of proteins change as organisms diverge from one another reflects an underlying change in their genetic information. Even though a DNA molecule from a mushroom, a redwood tree, and a cow may appear superficially identical, the sequences of nucleotides that make up the various DNA molecules are very different. These differences reflect evolutionary changes resulting from natural selection (Chapter 34).

Virtually all differences among living organisms can be traced to evolutionary changes in the structure of their various macromolecules, originating from changes in the nucleotide sequences of their DNA. (See CTQ #7.)

REEXAMINING THE THEMES

Relationship between Form and Function

The structure of a macromolecule correlates with a particular function. The unbranched, extended nature of the cellulose molecule endows it with resistance to pulling forces, an important property of plant cell walls. The hydrophobic character of lipids underlies many of their biological roles, explaining, for example, how waxes are able to provide plants with a waterproof covering. Protein function is correlated with protein shape. Just as a key is shaped to open a specific lock, a protein is shaped for a particular molecular interaction. For example, the shape of each polypeptide chain of hemoglobin enables a molecule of oxygen to fit perfectly into its binding site. A single alteration in the amino acid sequence of a hemoglobin chain can drastically reduce the molecule's oxygen-carrying capacity.

Biological Order, Regulation, and Homeostasis

Both blood sugar levels and body weight in humans are controlled by complex homeostatic mechanisms. The level of glucose in your blood is regulated by factors acting on the liver, which stimulate either glycogen breakdown (which increases blood sugar) or glycogen formation (which decreases blood sugar). Your body weight is, at least partly, determined by factors emanating from fat cells which either increase metabolic rate (which tends to decrease body weight) or slow down metabolic rate (which tends to increase body weight).

Acquiring and Utilizing Energy

The chemical energy that fuels biological activities is stored primarily in two types of macromolecules: polysaccharides and fats. Polysaccharides, including starch in plants and glycogen in animals, function primarily in the short-term storage of chemical energy. These polysaccharides can be rapidly broken down to sugars, such as glucose, which are readily metabolized to release energy. Gram-for-gram, fats contain even more energy than polysaccharides and function primarily as a long-term storage of chemical energy.

Unity within Diversity

All organisms, from bacteria to humans, are composed of the same four families of macromolecules, illustrating the unity of life—even at the biochemical level. The precise nature of these macromolecules and the ways they are organized into higher structures differ from organism to organism, thereby building diversity. Plants, for example, polymerize glucose into starch and cellulose, while animals polymerize glucose into glycogen. Similarly, many proteins (such as hemoglobin) are present in a variety of organisms, but the precise amino acid sequence of the protein varies from one species to the next.

Evolution and Adaptation

Evolution becomes very apparent at the molecular level when we compare the structure of macromolecules among diverse organisms. Analysis of the amino acid sequences of proteins and the nucleotide sequences of nucleic acids reveals a gradual change over time in the structure of macromolecules. Organisms that are closely related have proteins and nucleic acids whose sequences are similar than are those of distantly related organisms. To a large degree, the differences observed among diverse organisms derives from the evolutionary differences in nucleic acid and protein sequences.

The segregation of alleles and their independent assortment during meiosis increase genotype diversity by promoting new combinations of genes. But the shuffling of existing genes alone does not explain the presence of such a vast diversity of life. If all organisms descended from a common ancestor, with its relatively small complement of genes, where did all the genes present in today's millions of species come from? The answer is mutation.

Most mutant alleles are detrimental; that is, they are more likely to disrupt a well-ordered, smoothly functioning organism than to increase the organism's fitness. For example, a mutation might change a gene so that it produces an inactive enzyme needed for a critical life function. Occasionally, however, one of these stable genetic changes creates an advantageous characteristic that fitness of the offspring. In this way, mutation raw material for evolution and the diversification earth.

One of the requirements for genes is stability remain basically the same from generation to ge the fitness of organisms would rapidly deterio same time, there must be some capacity fo change; otherwise, there would be no potentia tion. Alterations in genes do occur, albeit rarely changes (mutations) represent the raw materia tion. (See CTQ #7.)

REEXAMINING THE THEMES

Biological Order, Regulation, and Homeostasis

Mendel discovered that the transmission of genetic factors followed a predictable pattern, indicating that the processes responsible for the formation of gametes, including the segregation of alleles, must occur in a highly ordered manner. This orderly pattern can be traced to the process of meiosis and the precision with which homologous chromosomes are separated during the first meiotic division. Mendel's discovery of independent assortment can also be connected with the first meiotic division, when each pair of homologous chromosomes becomes aligned at the metaphase plate in a manner that is independent of other pairs of homologues.

Unity within Diversity

All eukaryotic, sexually reproducing organisms follow the same "rules" for transmitting inherited traits. Although Mendel chose to work with peas, he could have come to the same conclusions had he studied fruit flies or mice or had he scrutinized a family's medical records on the transmission

of certain genetic diseases, such as cystic fibrosis. Al the mechanism by which genes are transmitted from the genes themselves are highly diverse from one org to the next. It is this genetic difference among specie forms the very basis of biological diversity.

Evolution and Adaptation

Mendel's findings provided a critical link in our kn edge of the mechanism of evolution. A key tenet in theory of evolution is that favorable genetic variations crease the likelihood that an individual will survive to productive age and that its offspring will exhibit these sa favorable characteristics. Mendel's demonstration th units of inheritance pass from parents to offspring witho being blended revealed the means by which advantageo traits could be preserved in a species over many genera tions. The subsequent discovery of genetic change by mu tation revealed how new genes appeared in a population thus providing the raw material for evolution.

SYNOPSIS

Gregor Mendel discovered the pattern by which inherited traits are transmitted from parents to offspring. Mendel discovered that inherited traits were controlled by pairs of factors (genes). The two factors for a given trait in an individual could be identical (homozygous) or different (heterozygous). In heterozygotes, one of the gene variants (alleles) may be dominant over the other, recessive allele. Because of dominance, the appearance (phenotype) of the heterozygote (genotype of *Aa*) is identical to that of the homozygote with two dominant alleles

Students will naturally find many ways in which the material presented in any biology course relates to them. But it is not always obvious how you can use biological information for better living or how it might influence your life. Your ability to see yourself in the course boosts interest and heightens the usefulness of the information. This translates into greater retention and understanding.

To accomplish this desirable outcome, the entire book has been constructed with you—the student—in mind. Perhaps the most notable feature of this approach is a series of boxed essays called "The Human Perspective" that directly reveals the human relevance of the biological topic being discussed at that point in the text. You will soon realize that human life, including your own, is an integral part of biology.

PART 2 / *Chemical and Cellular Foundations of Life*

◁ THE HUMAN PERSPECTIVE ▷
Obesity and the Hungry Fat Cell

FIGURE 1
Actor Robert DeNiro in (*left*) a scene from the movie *Raging Bull* and (*right*) a recent photograph.

It has become increasingly clear in recent years that people who are exceedingly overweight—that is, obese—are at increased risk of serious health problems, including heart disease and cancer. By most definitions, a person is obese if he or she is about 20 percent above "normal" or desirable body weight. Approximately 35 percent of adults in the United States are considered obese by this definition, twice as many as at the turn of the century. Among young adults, high blood pressure is five times more prevalent and diabetes three times more prevalent in a group of obese people than in a group of people who are at normal weight. Given these statistics, together with the social stigma facing the obese, there would seem to be strong motivation for maintaining a "normal" body weight. Why, then, are so many of us so overweight? And, why is it so hard to lose unwanted pounds and yet so easy to gain them back? The answers go beyond our fondness for high-calorie foods.

Excess body fat is stored in fat cells (*adipocytes*) located largely beneath the skin. These cells can change their volume more than a hundredfold, depending on the amount of fat they contain. As a person gains body fat, his or her fat cells become larger and larger, accounting for the bulging, sagging body shape. If the person becomes sufficiently overweight, and their fat cells approach their maximum fat-carrying capacity, chemical messages are sent through the blood, causing formation of new fat cells that are "hungry" to begin accumulating their own fat. Once a fat cell is formed, it may expand or contract in volume, but it appears to remain in the body for the rest of the person's life.

◔ Although the subject remains controversial, current research findings suggest that body weight is one of the properties subject to physiologic regulation in humans. Apparently, each person has a particular weight that his or her body's regulatory machinery acts to maintain. This particular value—whether 40 kilograms (80 pounds) or 200 kilograms (400 pounds)—is referred to as the person's **set-point.**

People maintain their body weight at a relatively constant value by balancing energy intake (in the form of food calories) with energy expenditure (in the form of calories burned by metabolic activities or excreted). Obese individuals are thought to have a higher set-point than do persons of normal weight. In many cases, the set-point value appears to have a strong genetic component. For instance, studies reveal there is no correlation between the body mass of adoptees and their adoptive parents, but there is a clear relationship between adoptees and their biological parents, with whom they have not lived.

The existence of a body-weight set-point is most evident when the body weight of a person is "forced" to deviate from the regulated value. Individuals of normal body weight who are fed large amounts of high-calorie foods under experimental conditions tend to gain increasing amounts of weight. If these people cease their energy-rich diets, however, they return quite rapidly to their previous levels, at which point further weight loss stops. This is illustrated by actor Robert DeNiro, who reportedly gained about 50 pounds for the filming of the movie "Raging Bull" (Figure 1), and then lost the weight prior to his next acting role. Conversely, a person who is put on a strict, low-calorie diet will begin to lose weight. The drop in body weight soon triggers a decrease in the person's resting metabolic rate, that is, the amount of calories burned when the person is not engaged in physical activity. The drop in metabolic rate is the body's compensatory measure for the decreased food intake. In other words, it is the body's attempt to halt further weight loss. This effect is particularly pronounced among obese people who diet and lose large amounts of weight. Their pulse rate and blood pressure drop markedly, their fat cells shrink to "ghosts" of their former selves, and they tend to be continually hungry. If these obese individuals go back to eating a *normal* diet, they tend to regain the lost weight rapidly. The drive of these formerly obese persons to increase their food intake is probably a response to chemical signals emanating from the fat cells as they shrink below their previous size.

---

630 • PART 5 / *Form and Function of Animal Life*

◁ THE HUMAN PERSPECTIVE ▷
Dying for a Cigarette?

average, smoking cigarettes will cut imately 6 to 8 years off your life. an 5 minutes for every cigarette Cigarette smoking is the greatest preventable death in the United according to a 1991 report by the for Disease Control (CDC), ,000 Americans die each year ng-related causes. Smoking ac-87 percent of all lung-cancer smokers are more susceptible e esophagus, larynx, mouth, bladder than are nonsmok- eased incidence of lung mong smokers compared to shown in Figure 1a, and d by quitting is shown in effects of smoking on lung Figure 2. Atherosclero- and peptic ulcers also greater frequency than rs. For example, long- 5 times more likely to terial disease than are sema (a condition ction of lung tissue, culty in breathing) mmation of the air- prevalent among

ger other people. sponsible for the nnocent bystand- re the same air passive (invol-own: second-seriously ill rs have dou-ry infections osed to to-ng married us: 20 per-mong non-ibutable to inhaling other

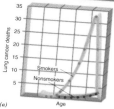

(a)

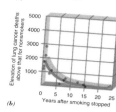

(b)

people's tobacco smoke. Another "innocent bystander" is a fetus developing in the uterus of a woman who smokes. Smoking increases the incidence of miscarriage and stillbirth and decreases the birthweight of the infant. Once born, these babies suffer twice as many respiratory infections as do babies of nonsmoking mothers.

Why is smoking so bad for your health? The smoke emitted from a burning cigarette contains more than 2,000 identifiable substances, many of which are either irritants or carcinogens. These compounds include carbon monoxide, sulfur dioxide, formaldehyde, nitrosamines, toluene, ammonia, and radioactive isotopes. Autopsies of respiratory tissues from smokers (and from nonsmokers who have lived for long periods with smokers) show widespread cellular changes, including the presence of precancerous cells (cells that may become malignant, given time) and a marked reduction in the number of cilia that play a vital role in the removal of bacteria and debris from the airways.

Of all the compounds found in tobacco (including smokeless varieties), the most important is nicotine, not because it is carcinogenic, but because it is so addictive. Nicotine is addictive because it acts like a neurotransmitter by binding to certain acetylcholine receptors (page 477), stimulating postsynaptic neurons. The physiological effects of this stimulation include the release of epinephrine, an increase in blood sugar, an elevated heart rate, and the constriction of blood vessels, causing elevated blood pressure. A smoker's nervous system becomes "accustomed" to the presence of nicotine and decreases the output of the natural neurotransmitter. As a result, when a person tries to stop smoking, the sudden absence of nicotine, together with the decreased level of the natural transmitter, decreases stimulation of postsynaptic neurons, which creates a craving for a cigarette—a "nicotine fit." Ex-smokers may be so conditioned to the act of smoking that the craving for cigarettes can continue long after the physiological addiction disappears.

*T*he "Biolines" are boxed essays that highlight fascinating facts, applications, and real-life lessons, enlivening the mainstream of biological information. Many are remarkable stories that reveal nature to be as surprising and interesting as any novelist could imagine.

◁ B I O L I N E ▷
DNA Fingerprints and Criminal Law

On February 5, 1987, a woman and her 2-year-old daughter were found stabbed to death in their apartment in the New York City borough of the Bronx. Following a tip, the police questioned a resident of a neighboring building. A small bloodstain was found on the suspect's watch, which was sent to a laboratory for DNA fingerprint analysis. The DNA from the white blood cells in the stain was amplified using the PCR technique and was digested with a restriction enzyme. The restriction fragments were then separated by electrophoresis, and a pattern of labeled fragments was identified with a radioactive probe. The banding pattern produced by the DNA from the suspect's watch was found to be a perfect match to the pattern produced by DNA taken from one of the victims. The results were provided to the opposing attorneys, and a pretrial hearing was called in 1989 to discuss the validity of the DNA evidence.

During the hearing, a number of expert witnesses for the prosecution explained the basis of the DNA analysis. According to these experts, no two individuals, with the exception of identical twins, have the same nucleotide sequence in their DNA. Moreover, differences in DNA sequence can be detected by comparing the lengths of the fragments produced by restriction-enzyme digestion of different DNA samples. The patterns produce a DNA fingerprint" (Figure 1) that is as unique to an individual as is a set of conventional fingerprints lifted from a glass. In 1989 DNA fingerprints had already been used in more than 200 criminal cases in the United States and had been hailed as the most important development in forensic science (the application of medical facts

FIGURE 1
Alec Jeffreys of the University of Leicester, England, examining a DNA fingerprint. Jeffreys was primarily responsible for developing the DNA fingerprint technique and was the scientist who confirmed the death of Josef Mengele.

to legal problems) in decades. The widespread use of DNA fingerprinting evidence in court had been based on its general acceptability in the scientific community. According to a report from the company performing the DNA analysis, the likelihood that the same banding patterns could be obtained by chance from two *different* individuals in the community was only one in 100 million.

What made this case (known as the Castro case, after the defendant) memorable and distinct from its predecessors was that the defense also called on expert witnesses to scrutinize the data and to present

their opinions. While these experts confirmed the capability of DNA fingerprinting to identify an individual out of a huge population, they found serious technical flaws in the analysis of the DNA samples used by the prosecution. In an unprecedented occurrence, the experts who had earlier testified *for the prosecution* agreed that the DNA analysis in this case was unreliable and should not be used as evidence! The problem was not with the technique itself but in the way it had been carried out in this particular case. Consequently, the judge threw out the evidence.

In the wake of the Castro case, the use of DNA fingerprinting to decide guilt or innocence has been seriously questioned. Several panels and agencies are working to formulate guidelines for the licensing of forensic DNA laboratories and the certification of their employees. In 1992, a panel of the National Academy of Sciences released a report endorsing the general reliability of the technique but called for the institution of strict standards *to be set by scientists.*

Meanwhile, another issue regarding DNA fingerprinting has been raised and hotly debated. Two geneticists, Richard Lewontin of Harvard University and Daniel Hartl of Washington University, coauthored a paper published in December 1991, suggesting that scientists do not have enough data on genetic variation within different racial or ethnic groups to calculate the odds that two individuals—a suspect and a perpetrator of the crime—are one and the same on the basis of an identical DNA fingerprint. The matter remains an issue of great concern in both the scientific and legal communities and has yet to be resolved.

◁ B I O L I N E ▷
The Fish That Changes Sex

In vertebrates, gender is generally a biologically inflexible commitment: An individual develops into either a male or a female as dictated by the sex chromosomes acquired from one's parents. Yet, even among vertebrates, there are organisms that can reverse their sexual commitment. The Australian cleaner fish (Figure 1), a small animal that sets up "cleaning stations" to which larger fishes come for parasite removal, can change its gender in response to environmental demands. Most male cleaner fish travel alone rather than with a school. Except for a single male, schools of cleaner fish are comprised entirely of females. Although it might seem logical to conclude that maleness engenders solo travel, it is actually the other way around: Being alone fosters maleness. A cleaner fish that develops away from a school *becomes* a male, whereas the same fish developing in a school would have become a female.

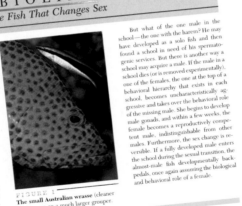

FIGURE 1
The small Australian wrasse (cleaner fish) is seen on a much larger grouper.

But what of the one male in the school—the one with the harem? He may have developed as a solo fish and then found a school in need of his spermatogenic services. But there is another way a school may acquire a male. If the male in a school dies (or is removed experimentally), one of the females, the one at the top of a behavioral hierarchy that exists in each school, becomes uncharacteristically aggressive and takes over the behavioral role of the missing male. She begins to develop male gonads, and within a few weeks, the female becomes a reproductively competent male, indistinguishable from other males. Furthermore, the sex change is reversible. If a fully developed male enters the school during the sexual transition, the almost-male fish developmentally backpedals, once again assuming the biological and behavioral role of a female.

⚠ Not all organisms follow the mammalian pattern of sex determination. In some animals, most notably birds, the opposite pattern is found: The female's cells have an X and a Y chromosome, while the male's cells have two Xs. An exception to this rule of a strict relation between sex and chromosomes is discussed in the Bioline: The Fish That Changes Sex. Although some plants possess sex chromosomes and gender distinctions between individuals, most have only autosomes; consequently, each individual produces both male and female parts.

SEX LINKAGE

For fruit flies and humans alike, there are hundreds of genes on the X chromosome that have no counterpart on the smaller Y chromosome. Most of these genes have nothing to do with determining gender, but their effect on phenotype usually *depends* on gender. For example, in females, a recessive allele on one X chromosome will be masked (and not expressed) if a dominant counterpart resides on the other X chromosome. In males, it only takes one recessive allele on the single X chromosome to determine the individual's phenotype since there is no corresponding allele on the Y chromosome. Inherited characteristics determined by genes that reside on the X chromosome are called **X-linked characteristics.**

So far, some 200 human X-linked characteristics have been described, many of which produce disorders that are found almost exclusively in men. These include a type of heart-valve defect (*mitral stenosis*), a particular form of mental retardation, several optical and hearing impairments, muscular dystrophy, and red-green colorblindness (Figure 13-8).

One X-linked recessive disorder has altered the course of history. The disease is **hemophilia,** or "bleeder's disease," a genetic disorder characterized by the inability to produce a clotting factor needed to halt blood flow quickly following an injury. Nearly all hemophiliacs are males. Although females can inherit two recessive alleles for hemophilia, this occurrence is extremely rare. In general, women who have acquired the rare defective allele are heterozygous **carriers** for the disease. The phenotype of a carrier

Several ethical issues are discussed in the Bioethics essays which add provocative pauses throughout the text. Biological Science does not operate in a vacuum but has profound consequences on the general community. Because biologists study life, the science is peppered with ethical considerations. The moral issues discussed in these essays are neither simple nor easy to resolve, and we do not claim to have any certain answers. Our goal is to encourage you to consider the bioethical issues that you will face now and in the future.

◁ B I O E T H I C S ▷
Blurring the Line between Life and Death
By ARTHUR CAPLAN
Division of the Center for Biomedical Ethics at the University of Minnesota

Theresa Ann Campo Pearson didn't have a very long life. When she died in 1992, she was only 10 days old. Despite her short life, she became the center of a very strange, sad, and wrenching ethical controversy. Theresa died because her brain had failed to form. She had anencephaly, a condition in which only the brainstem, located at the top of the spinal cord, is present. Her parents wanted to donate Theresa's organs; the courts said no. Some people found it strange that Theresa's parents, Laura Campo and Justin Pearson, did not get their way. Why not allow donation, when every day in North America a baby dies because there is no heart, lung, or liver available for transplantation?

Anencephaly is best described as completely "unabling," not disabling. Children born with anencephaly cannot think, feel, sense, or be aware of the world. Many are stillborn; the majority of the rest die within days of birth. A mere handful live for a few weeks. Theresa's parents

knew all this. But rather than abort the pregnancy, they chose to have their baby. In fact, the baby was born by Caesarean section, at least partly in the hope that it would be born alive, thereby making organ donation possible. When Theresa died at Broward General Medical Center in Fort Lauderdale, Florida, however, no organs were taken. Two Florida courts ruled that the baby could not be used as a source of organs unless she was brain-dead, and Theresa Ann Campo was never pronounced brain-dead.

Brain death refers to a situation in which the brain has irreversibly lost all function and activity. Babies born with anencephaly have some brain function in their brainstem so, while they cannot think or feel, they are alive. According to Florida law—and the law in more than 40 other states—only those individuals declared brain-dead can donate organs. The courts of Florida had no other option but to deny the request for organ donation.

One obvious solution is to change the law so that states could decide that organs can be removed upon parental consent from either those who are born brain-dead or from babies who are born with anencephaly. Another solution is to rewrite the definition of death to say that death occurs either when the brain has totally ceased to function or if a baby is born anencephalic. Do you feel that either of these changes should be made? Some may argue that medicine will fudge the line between life and death in order to get organs for transplant. Do you agree with this concern? How do you think redefining death will affect a person's decision to check off the donation box on the back of a driver's license? Do you think people may worry that if they are known to be potential donors they won't be aggressively treated at the hospital? In your opinion, would changing the definition of death to include anencephaly be beneficial or deleterious?

Like the brain, the spinal cord is composed of white matter (myelinated axons) and gray matter (dendrites and cell bodies). However, the arrangement of these types of matter is reversed in the spinal cord, compared to their arrangement in the brain: The spinal cord's white matter surrounds the gray matter (Figure 23-16).

The human central nervous system is the most complex and highly evolved assembly of matter. Among its functions are the processing of sensory information collected from both the external and internal environment; the regulation of internal physiological activities; the coordination of complex motor activities; and the endowment of such intangible "mental" qualities as emotions, creativity, language, and the ability to think, learn, and remember. (See CTQ #6.)

ARCHITECTURE OF THE PERIPHERAL NERVOUS SYSTEM

The peripheral nervous system provides the neurological bridge between the central nervous system and the various parts of the body. The peripheral nervous system is made up of paired nerves that extend into the periphery from the CNS at various levels along the body. Each nerve is composed of a large bundle of myelinated axons surrounded by a connective tissue sheath. Twelve pairs of **cranial nerves** emerge from the central stalk of the human brain, and 31 pairs of **spinal nerves** extend from the spinal cord out between the vertebrae of humans (Figure 23-16). For the most part, the cranial nerves *innervate* (supply nerves to) tissues and organs of the head and neck, whereas the spinal nerves innervate the chest, abdomen, and limbs.

Additional Pedagogical Features

We have worked to assure that each chapter in this book is an effective teaching and learning instrument. In addition to the pedagogical features discussed above, we have included some additional tried-and-proven-effective tools.

KEY POINTS

Key points follow each major section and offer a condensation of the relevant facts and details as well as the concepts discussed. You can use these key points to reaffirm your understanding of the previous reading or to alert you to misunderstood material before moving on to the next topic. Each key point is tied to a Critical Thinking Question found at the end of the chapter; together, they encourage you to analyze the information, taking it beyond mere memorization.

---

*Plant Tissues and Organs / * C H A P T E R   18 • 361

➠ Many plants replenish old and dying cells with vigorous new cells. But since each plant cell has a surrounding cell wall (Chapter 7) old plant cells do not just wither and disappear when they die. Instead, dead plant cells leave cellular "skeletons" where they once lived. As a result, the longer a plant lives, the more complex its anatomy becomes. **Annuals** are plants that live for 1 year or less, such as corn and marigolds. Because they live for such a brief period, these plants do not completely replace old cells. As a result, annuals are anatomically less complex than are **biennials**—plants that live for 2 years—and **perennials**—herbs, shrubs, and trees that live longer than 2 years. Biennials (carrots, Queen Anne's lace) and perennials (rosebushes, apple trees) are able to live longer than annuals because they produce new cells to replace those that cease functioning or die, providing a continual supply of young, vigorous cells.

▣ In this chapter, we will focus on the body construction of flowering plants, the most familiar, most evolutionarily advanced, and structurally complex of any group in the plant kingdom. All flowering plants are **vascular plants;** that is, they contain specialized cells that circulate water, minerals, and food (organic molecules) throughout the plant. Botanists divide flowering plants into two main groups: **dicotyledons**, or dicots (*di* = two, *cotyledon* = embryonic seed leaf), and **monocotyledons**, or monocots (*mono* = one). Table 18-1 illustrates the many differences that distinguish dicots from monocots and will be used as a reference throughout the chapter.

SHOOTS AND ROOTS

The flowering plant body is a study in contradictions. A typical plant grows through the soil and the air simultaneously, two very different habitats with very different conditions. As a result, the two main parts of the plant differ dramatically in form (anatomy) and function (physiology): The underground **root system** anchors the plant in the soil and absorbs water and nutrients, while the aerial **shoot system** absorbs sunlight and gathers carbon dioxide for photosynthesis (Figure 18-2). The shoot system also produces stems, leaves, flowers, and fruits. Interconnected vascular tissues transport materials between the aerial shoot system and the underground root system. These connections allow water and minerals absorbed by the root to be conducted to shoot tissues, and for food produced by the shoot to be transported to root tissues. We will discuss the various components of these two systems in more detail later in the chapter.

Over 90 percent of all plant species are flowering plants. Flowering plants are the most recently evolved plant group, having undergone rapid evolution during the past 1 million to 2 million years as environmental conditions on land became more variable. (See CTQ #2.)

TABLE 18-1

Dicot and Monocot Comparison		
	DICOT	MONOCOT
Embryo	2 Cotyledons	1 Cotyledon
Flowers	Parts in 4s or 5s	Parts in 3s
Leaves	Net veined	Veins parallel
Leaf anatomy	Two types of photosynthetic cells	One type of photosynthetic
Roots	One main root (tap root system)	Many main roots (fibrous root system)
Stem anatomy	Vascular bundles in rings	Scattered vascular bundles
Root anatomy	Xylem in center	Pith in center
Secondary growth	Yes	No

---

the corresponding polypeptide. The cumulative effect of gradual changes in polypeptides over evolutionary time has been the generation of life's diversity.

Evolution and Adaptation

➠ Evolutionary change from generation to generation depends on genetic variability. Much of this variability arises from reshuffling maternal and paternal genes during meiosis, but somewhere along the way *new* genetic information must be introduced into the population. Ne[...] netic information arises from mutations in existing g[...] Some of these mutations arise during replication; o[...] occur as the result of unrepaired damage as the DNA i[...] "sitting" in a cell. Mutations that occur in an individ[...] germ cells can be considered the raw material on w[...] natural selection operates; whereas harmful mutations [...] duce offspring with a reduced fitness, beneficial mutat[...] produce offspring with an increased fitness.

S Y N O P S I S

Experiments in the 1940s and 1950s established conclusively that DNA is the genetic material. These experiments included the demonstration that DNA was capable of transforming bacteria from one genetic strain to another; that bacteriophages injected their DNA into a host cell during infection; and that the injected DNA was transmitted to the bacteriophage progeny.

DNA is a double helix. DNA is a helical molecule consisting of two chains of nucleotides running in opposite directions, with their backbones on the outside, and the nitrogenous bases facing inward like rungs on a ladder. Adenine-containing nucleotides on one strand always pair with thymine-containing nucleotides on the other strand, likewise for guanine- and cytosine-containing nucleotides. As a result, the two strands of a DNA molecule are complementary to one another. Genetic information is encoded in the specific linear sequence of nucleotides that make up the strands.

DNA replication is semiconservative. During replication, the double helix separates, and each strand serves as a template for the formation of a new, complementary strand. Nucleotide assembly is carried out by the enzyme DNA polymerase, which moves along the two strands in opposite directions. As a result, one of the strands is synthesized continuously, while the other is synthesized in segments that are covalently joined. Accuracy is maintained by a proofreading mechanism present within the polymerase.

Information flows in a cell from DNA to RNA to protein. Each gene consists of a linear sequence of nucleotides that determines the linear sequence of amino acids in a polypeptide. This is accomplished in two maj[...] steps: transcription and translation.

During transcription, the information spelled out b[...] the gene's nucleotide sequence is encoded in a mole[...] cule of messenger RNA (mRNA). The mRNA contain[...] a series of codons. Each codon consists of three nucleotides. Of the 64 possible codons, 61 specify an amino acid, and the other 3 stop the process of protein synthesis.

During translation, the sequence of codons in the mRNA is used as the basis for the assembly of a chain of specific amino acids. Translating mRNA messages occurs on ribosomes and requires tRNAs, which serve as decoders. Each tRNA is folded into a cloverleaf structure with an anticodon at one end—which binds to a complementary codon in the mRNA—and a specific amino acid at the other end—which becomes incorporated into the growing polypeptide chain. Amino acids are added to their appropriate tRNAs by a set of enzymes. The sequential interaction of charged tRNAs with the mRNA results in the assembly of a chain of amino acids in the precise order dictated by the DNA.

Mutation is a change in the genetic message. Gene mutations may occur as a single nucleotide substitution, which leads to the insertion of an amino acid different from that originally encoded. In contrast, the addition of one or two nucleotides throws off the reading frame of the ribosome as it moves along the mRNA, leading to the incorporation of incorrect amino acids "downstream" from the point of mutation. Exposure to mutagens increases the rate of mutation.

SYNOPSIS

The synopsis section offers a convenient summary of the chapter material in a readable narrative form. The material is summarized in concise paragraphs that detail the main points of the material, offering a useful review tool to help reinforce recall and understanding of the chapter's information.

REVIEW QUESTIONS

Along with the synopsis, the Review Questions provide a convenient study tool for testing your knowledge of the facts and processes presented in the chapter.

---

224 • PART 2 / *Chemical and Cellular Foundations of Life*

Key Terms

zygote (p. 214)
meiosis (p. 214)
life cycle (p. 214)
germ cell (p. 214)
somatic cell (p. 214)
meiosis I (p. 216)

reduction division (p. 216)
synapsis (p. 216)
tetrad (p. 216)
crossing over (p. 216)
genetic recombination (p. 216)
synaptonemal complex (p. 218)

maternal chromosome (p. 219)
paternal chromosome (p. 219)
independent assortment (p. 219)
meiosis II (p. 219)

Review Questions

1. Match the activity with the phase of meiosis in which it occurs.

   a. synapsis
   b. crossing over
   c. kinetochores split
   d. independent assortment
   e. homologous chromosomes separate
   f. cytokinesis

   1. prophase I
   2. metaphase I
   3. anaphase I
   4. telophase I
   5. prophase II
   6. anaphase II
   7. telophase II

2. How do crossing over and independent assortment increase the genetic variability of a species?

3. Why is meiosis I (and not meiosis II) referred to as the reduction division?

4. Suppose that one human sperm contains x amount of DNA. How much DNA would a cell just entering meiosis contain? A cell entering meiosis II? A cell just completing meiosis II? Which of these three cells would have a haploid number of chromosomes? A diploid number of chromosomes?

Critical Thinking Questions

1. Why are disorders, such as Down syndrome, that arise from abnormal chromosome numbers, characterized by a number of seemingly unrelated abnormalities?

2. A gardener's favorite plant had white flowers and long seed pods. To add some variety to her garden, she transplants some plants of the same type, but with pink flowers and short seed pods from her neighbor's garden. To her surprise, in a few generations, she grows plants with white flowers and short seed pods and plants with pink flowers and long seed pods, as well as the original combinations. What are two ways in which these new combinations could have arisen?

3. Set up the meiosis template in the diagram below on a large sheet of paper. Then use pieces of colored yarn or pipe cleaners to simulate chromosomes and make a model of the phases of meiosis. (*See template on opposite page*)

4. Would you expect two genes on the same chromosome, such as yellow flowers and short stems, always to be exchanged during crossing over? How might they remain together *in spite of* crossing over?

5. Suppose paternal chromosomes always lined up on the same side of the metaphase plate of cells in meiosis I. How would this affect genetic variability of offspring? Would they all be identical? Why or why not?

Additional Readings

Chandley, A. C. 1988. Meiosis in man. *Trends in Gen.* 4:79–83. (Intermediate)

Hsu, T. C. 1979. *Human and mammalian cytogenetics.* New York: Springer-Verlag. (Intermediate)

John, B. 1990. *Meiosis.* New York: Cambridge University Press. (Advanced)

Moens, P. B. 1987. *Meiosis.* Orlando: Academic. (Advanced)

Patterson, D. 1987. The causes of Down syndrome. *Sci. Amer.* Feb:52–60. (Intermediate-Advanced)

White, M. J. D. 1973. *The chromosomes.* Halsted. (Advanced)

STIMULATING CRITICAL THINKING

Each chapter contains as part of its end material a diverse mix of Critical Thinking Questions. These questions ask you to apply your knowledge and understanding of the facts and concepts to hypothetical situations in order to solve problems, form hypotheses, and hammer out alternative points of view. Such exercises provide you with more effective thinking skills for competing and living in today's complex world.

ADDITIONAL READINGS

Supplementary readings relevant to the Chapter's topics are provided at the end of every chapter. These readings are ranked by level of difficulty (introductory, intermediate, or advanced) so that you can tailor your supplemental readings to your level of interest and experience.

Careers in Biology

*T*he appendices of this edition include "Careers in Biology," a frequently overlooked aspect of our discipline. Although many of you may be taking biology as a requirement for another major (or may have yet to declare a major), some of you are already biology majors and may become interested enough to investigate the career opportunities in life sciences. This appendix helps students discover how an interest in biology can grow into a livelihood. It also helps the instructor advise students who are considering biology as a life endeavor.

APPENDIX
◄ D ►

Careers in Biology

Although many of you are enrolled in biology as a requirement for another major, some of you will become interested enough to investigate the career opportunities in life sciences. This interest in biology can grow into a satisfying livelihood. Here are some facts to consider:

- Biology is a field that offers a very wide range of possible science careers

- Biology offers high job security since many aspects of it deal with the most vital human needs: health and food

- Each year in the United States, nearly 40,000 people obtain bachelor's degrees in biology. But the number of newly created and vacated positions for biologists is increasing at a rate that exceeds the number of new graduates. Many of these jobs will be in the newer areas of biotechnology and bioservices.

Biologists not only enjoy job satisfaction, their work often changes the future for the better. Careers in medical biology help combat diseases and promote health. Biologists have been instrumental in preserving the earth's life-supporting capacity. Biotechnologists are engineering organisms that promise dramatic breakthroughs in medicine, food production, pest management, and environmental protection. Even the economic vitality of modern society will be increasingly linked to biology.

Biology also combines well with other fields of expertise. There is an increasing demand for people with backgrounds or majors in biology complexed with such areas as business, art, law, or engineering. Such a distinct blend of expertise gives a person a special advantage.

The average starting salary for all biologists with a Bachelor's degree is $22,000. A recent survey of California State University graduates in biology revealed that most were earning salaries between $20,000 and $50,000. But as important as salary is, most biologists stress job satisfaction, job security, work with sophisticated tools and scientific equipment, travel opportunities (either to the field or to scientific conferences), and opportunities to be creative in their job as the reasons they are happy in their career.

Here is a list of just a few of the careers for people with degrees in biology. For more resources, such as lists of current openings, career guides, and job banks, write to Biology Career Information, John Wiley and Sons, 605 Third Avenue, New York, NY 10158.

A SAMPLER OF JOBS THAT GRADUATES HAVE SECURED IN THE FIELD OF BIOLOGY*

Agricultural Biologist	Bioanalytical Chemist	Brain Function	Environmental Center
Agricultural Economist	Biochemical/Endocrine	Researcher	Director
Agricultural Extension	Toxicologist	Cancer Biologist	Environmental Engineer
Officer	Biochemical Engineer	Cardiovascular Biologist	Environmental Geographer
Agronomist	Pharmacology Distributor	Cardiovascular/Computer	Environmental Law Specialist
Amino-acid Analyst	Pharmacology Technician	Specialist	Farmer
Analytical Biochemist	Biochemist	Chemical Ecologist	Fetal Physiologist
Anatomist	Biogeochemist	Chromatographer	Flavorist
Animal Behavior	Biogeographer	Clinical Pharmacologist	Food Processing Technologist
Specialist	Biological Engineer	Coagulation Biochemist	Food Production Manager
Anticancer Drug Research	Biologist	Cognitive Neuroscientist	Food Quality Control
Technician	Biomedical	Computer Scientist	Inspector
Antiviral Therapist	Communication Biologist	Dental Assistant	Flower Grower
Arid Soils Technician	Biometerologist	Ecological Biochemist	Forest Ecologist
Audio-neurobiologist	Biophysicist	Electrophysiology/	Forest Economist
Author, Magazines & Books	Biotechnologist	Cardiovascular Technician	Forest Engineer
Behavioral Biologist	Blood Analyst	Energy Regulation Officer	Forest Geneticist
Bioanalyst	Botanist	Environmental Biochemist	Forest Manager

Study Guide

Written by Gary Wisehart and Michael Leboffe of San Diego City College, the *Study Guide* has been designed with innovative pedagogical features to maximize your understanding and retention of the facts and concepts presented in the text. Each chapter in the *Study Guide* contains the following elements.

Concepts Maps

In Chapter 1 of the *Study Guide*, the beginning of a concept map stating the five themes is introduced. In each subsequent chapter, the concept map is expanded to incorporate topics covered in each chapter as well as the interconnections between chapters and the five themes. "Connector" phrases are used to link the concepts and themes, and the text icons representing the themes are incorporated into the concept maps.

Go Figure!

In each chapter, questions are posed regarding the figures in the text. Students can explore their understanding of the figures and are asked to think critically about the figures based on their understanding of the surrounding text and their own experiences.

Self-Tests

Each chapter includes a set of matching and multiple-choice questions. Answers to the Study Guide questions are provided.

Concept Map Construction

The student is asked to create concept maps for a group of terms, using appropriate connector phrases and adding terms as necessary.

Laboratory Manual

Biology: Exploring Life, Second Edition is supplemented by a comprehensive *Laboratory Manual* containing approximately 60 lab exercises chosen by the text authors from the National Association of Biology Teachers. These labs have been thoroughly class-tested and have been assembled from various scientific publications. They include such topics as

- Chaparral and Fire Ecology: Role of Fire in Seed Germination (*The American Biology Teacher*)
- A Model for Teaching Mitosis and Meiosis (*American Biology Teacher*)

- Laboratory Study of Climbing Behavior in the Salt Marsh Snail (*Oceanography for Landlocked Classrooms*)
- Down and Dirty DNA Extraction (*A Sourcebook of Biotechnology Activities*)
- Bioethics: The Ice-Minus Case (*A Sourcebook of Biotechnology Activities*)
- Using Dandelion Flower Stalks for Gravitropic Studies (*The American Biology Teacher*)
- pH and Rate of Enzymatic Reactions (*The American Biology Teacher*)

CHAPTER
◄ 18 ►

Plant Tissues and Organs

STEPS
TO
DISCOVERY
Fruit of the Vine and the French Economy

THE BASIC PLANT DESIGN

Shoots and Roots

PLANT TISSUES

Plants Grow from Meristems
Parenchyma Cells Provide Versatility
Collenchyma Cells Support Stems and Leaves
Sclerenchyma Cells Support and Protect

PLANT TISSUE SYSTEMS

The Dermal Tissue System

The Vascular Tissue System
The Ground Tissue System

PLANT ORGANS

The Stem: Growth, Support, and Conduction
The Root: Growth, Absorption, and Conduction
The Leaf: The Plant's Primary Photosynthetic Organ
The Flower: The Site of Sexual Reproduction

BIOLINE
Nature's Oldest Living Organisms

THE HUMAN PERSPECTIVE
Agriculture, Genetic Engineering, and Plant Fracture Properties

Fruit of the Vine and the French Economy

*T*he devastation was swift. Within just 10 years, nearly two-thirds of all the vineyards in Europe had been ravaged. The widespread ruin not only destroyed the livelihood of thousands of families but threatened the economies of entire nations. Each year, people stood helpless and watched the circle of death widen, not knowing the cause or the solution.

It would take European scientists more than 40 years to find a solution to the problem. Because of the huge impact on the economy of France, this monumental research task was funded and guided by the French government itself. In 1868, the cause of the devastation was finally identified by

Professor J. E. Planchon of the University of Montpellier as the prolific work of an insect he named *Phylloxera vastatrix.* Every detail of the insect's life cycle was investigated while the French government funded selective breeding and grafting experiments with the hope of producing resistant vines. The costs of the research were staggering.

Humans have been cultivating grapes for wine and food since before 4000 B.C. After nearly 6,000 years of cultivation, it was reasonable to ask why there hadn't been an earlier outbreak of infestation and how so much destruction by one insect could have occurred within such a short period of time. Part of the answer to these questions is that

Native North American grape vines are resistant to attacks by the wingless larva of **Phylloxera. The French wine industry**

only one species of grape, *Vitis vinifera*, had been cultivated for commercial purposes, so all European vineyards were planted with the same species of vine; any insect that attacked this species had the power to wipe out the entire wine industry. In addition, until the 1860s, the insect had been restricted to the United States. (American vines apparently were naturally resistant to the insect's attacks.) Between 1860 and 1870, however, native U.S. vines were being transplanted to France; along with the vines came the insect and, consequently, the rapid destruction. Planchon's research revealed that the insect took on four different forms in its life cycle and that it was the wingless, root-lice form that attacked small vine roots, causing them to swell and eventually rot.

Desperate to stop the expanding destruction, the French government sent specialists to the United States to investigate why the American vines were resistant to *Phylloxera* and to find a resistant vine to replace the ones being damaged in Europe. The French specialists eventually identified three species of grape—*V. riparia*, *V. rupestris*, and *V. berlandieri*—as being the most resistant to *Phylloxera* because of their thicker root bark and more vigorous health. Because *V. berlandieri* proved to be better suited to European soils than were the other two species, the specialists recommended that it be crossbred with *V. vinifera*, in the hope of producing a resistant hybrid vine. Unfortunately, although the hybrid root stocks were indeed resistant to *Phylloxera*, the fruits were of such low quality that they could not be used for quality wine. In addition, crossbreeding for resistance requires a great deal of time. The plants need time to grow, produce offspring, and develop mature flowers and fruits. They must then be crossbred again and again until a resistant strain is found. The French government wanted a much quicker solution.

Ironically, the solution to the *Phylloxera* problem did not emerge from one of the many government-supported scientists. Instead, a practical remedy came from Pierre Marie Alexis Millardet, a physician who had given up a successful medical practice to study botany and was teaching at the University of Bordeaux. In 1872, Millardet discovered that not all grape roots were destroyed by *Phyllox-*

era. He suggested that instead of undergoing the long and expensive process of trying to breed resistant hybrids, *V. vinifera stems* could be grafted directly onto the roots of resistant vines from the United States. In this way, quality grapes would be produced from the stems of *V. vinifera* on rootstocks that are resistant to *Phylloxera*.

The French government immediately began importing American rootstocks to France in 1878. Millardet, along with Professors L. Ravaz and P. Viala, also of the University of Bordeaux, initiated studies to evaluate the resistance of the imported rootstocks. They discovered that some rootstocks were more resistant to *Phylloxera* than were others and that resistance varied with soil type. Viala and Ravaz later devised a resistance scale based on soil type and rated American rootstocks from 0 to 20. According to their scale, "The numbers 16–20 correspond to a sufficient resistance for all soils; the numbers 15 and 14 express resistance sufficient for sandy and deep soils where the *Phylloxera* does little harm; the numbers 13 and under should be totally discarded for vineyards."

In order to rebuild the European vineyards, all the old vines had to be dug up and replanted with the resistant American rootstocks above value 13. The stems of *V. vinifera* then had to be grafted onto the resistant rootstock, a massive and very expensive undertaking. Many rootstocks perished before grafting; many others died a short time after grafting. Some of the rootstocks flourished, however, yielding even heavier crops than those previously harvested from *V. vinifera* grapes alone.

After such an immense effort and expense, it would be logical to assume that the entire wine industry would now be well protected from *Phylloxera* infestations. Unfortunately, that isn't the case. *Phylloxera* now threatens many vineyards in California and South Africa which, like the European vineyards in the late 1800s, are planted with only *V. vinifera* vines. Unless scientists can develop a resistant hybrid or come up with a way of destroying the pest without damaging the plants, many Californian and South African vineyards will have to be entirely replanted.

was saved by grafting these resistant rootstocks to the vines in their vineyards.

At 8:30 A.M. on May 18, 1980, in southwestern Washington, Mount St. Helens erupted with a force estimated to equal that of 750 atomic bombs detonated simultaneously. The enormous blast shot 4 km³ (1 mi.³) of earth 20,000 kilometers (about 65,000 feet) into the air, lowering the elevation of Mount St. Helens by more than 300 km (1,000 feet). (The colossal volume of earth fired up into the air is approximately equal to the size of Mt. Rushmore, including the busts of all four presidents.) Where a dense forest once stood, a barren "moonscape" remained. Extending 4 kilometers (14 miles) out from this lifeless center was a progression of "damage" zones—areas populated only by the stumps of incinerated trees; heaps of toppled trees piled one on top of the other like toothpicks dropped out of a box; and zones of scorched, standing trees. As destructive as the May eruption was, within just 1 year, biologists began documenting the return of life to the slopes of the volcano (Figure 18-1). The first organisms to return were more than 150 species of plants. Plants precede animals in all "new" environments; plants create the habitats and provide the food on which all animal life depends. Eventually, plants once again will become the most abundant form of life on the slopes of Mount St. Helens.

▼ ▼ ▼

THE BASIC PLANT DESIGN

The earth's 400,000 known plant species display an enormous range of external body forms and internal cell structures. Some plants, such as mosses, are simply collections of nearly identical cells, whereas others, such as the giant redwood, are composed of billions of specialized cells. In other words, there is no "typical" plant design.

(a) *(b)*

FIGURE 18-1

Mount St. Helens. *(a)* Within minutes, thousands of acres of forest were incinerated following the May eruption. *(b)* Within only 1 year of the eruption, plants began to return to the slopes of Mount St. Helens; soon after, animals returned.

▐▶ Many plants replenish old and dying cells with vigorous new cells. But since each plant cell has a surrounding cell wall (Chapter 7) old plant cells do not just wither and disappear when they die. Instead, dead plant cells leave cellular "skeletons" where they once lived. As a result, the longer a plant lives, the more complex its anatomy becomes. **Annuals** are plants that live for 1 year or less, such as corn and marigolds. Because they live for such a brief period, these plants do not completely replace old cells. As a result, annuals are anatomically less complex than are **biennials**—plants that live for 2 years—and **perennials**—herbs, shrubs, and trees that live longer than 2 years. Biennials (carrots, Queen Anne's lace) and perennials (rosebushes, apple trees) are able to live longer than annuals because they produce new cells to replace those that cease functioning or die, providing a continual supply of young, vigorous cells.

In this chapter, we will focus on the body construction of flowering plants, the most familiar, most evolutionarily advanced, and structurally complex of any group in the plant kingdom. All flowering plants are **vascular plants;** that is, they contain specialized cells that circulate water, minerals, and food (organic molecules) throughout the plant. Botanists divide flowering plants into two main groups: **dicotyledons,** or dicots (*di* = two, *cotyledon* = embryonic seed leaf), and **monocotyledons,** or monocots (*mono* = one). Table 18-1 illustrates the many differences that distinguish dicots from monocots and will be used as a reference throughout the chapter.

SHOOTS AND ROOTS

The flowering plant body is a study in contradictions. A typical plant grows through the soil and the air simultaneously, two very different habitats with very different conditions. As a result, the two main parts of the plant differ dramatically in form (anatomy) and function (physiology): The underground **root system** anchors the plant in the soil and absorbs water and nutrients, while the aerial **shoot system** absorbs sunlight and gathers carbon dioxide for photosynthesis (Figure 18-2). The shoot system also produces stems, leaves, flowers, and fruits. Interconnected vascular tissues transport materials between the aerial shoot system and the underground root system. These connections allow water and minerals absorbed by the root to be conducted to shoot tissues, and for food produced by the shoot to be transported to root tissues. We will discuss the various components of these two systems in more detail later in the chapter.

Over 90 percent of all plant species are flowering plants. Flowering plants are the most recently evolved plant group, having undergone rapid evolution during the past 1 million to 2 million years as environmental conditions on land became more variable. (See CTQ #2.)

TABLE 18-1

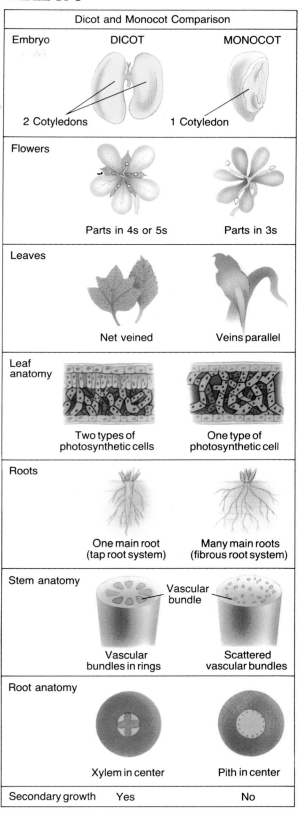

Dicot and Monocot Comparison

	DICOT	MONOCOT
Embryo	2 Cotyledons	1 Cotyledon
Flowers	Parts in 4s or 5s	Parts in 3s
Leaves	Net veined	Veins parallel
Leaf anatomy	Two types of photosynthetic cells	One type of photosynthetic cell
Roots	One main root (tap root system)	Many main roots (fibrous root system)
Stem anatomy	Vascular bundles in rings	Scattered vascular bundles (Vascular bundle)
Root anatomy	Xylem in center	Pith in center
Secondary growth	Yes	No

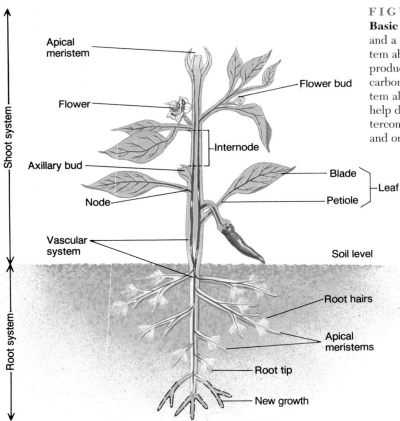

Shoot system

Apical meristem

Flower

Axillary bud

Node

Vascular system

Root system

Flower bud

Internode

Blade

Petiole

Leaf

Soil level

Root hairs

Apical meristems

Root tip

New growth

FIGURE 18-2

Basic plant design. A plant body has an aerial shoot system and a subterranean root system. The highly branched root system absorbs water and minerals from the soil, while the shoot produces stems and leaves for gathering sunlight and extracting carbon dioxide from the air for photosynthesis. The shoot system also produces flowers for sexual reproduction and fruits to help disperse seeds. All flowering plants have a network of interconnecting vascular tissues for transporting water, minerals, and organic molecules.

PLANT TISSUES

As in all multicellular organisms, groups of plant cells work together to form tissues. **Simple tissues,** such as those that store starch in potatoes, contain the same type of cell, whereas **complex tissues,** such as a plant's vascular tissues, are composed of different kinds of cells working together to carry out a particular function. Plants are composed of four simple tissues: *meristems, parenchyma, collenchyma,* and *sclerenchyma.*

PLANTS GROW FROM MERISTEMS

Meristems are clusters of cells devoted solely to cell division. Meristematic cells divide by mitosis to produce new cells for increased and continual growth. All shoots and roots have **apical meristems**—groups of meristematic cells located at their apexes, or tips (Figure 18-2). Divisions of apical meristem cells increase the *length* of these structures. All growth that originates from apical meristems is termed **primary growth,** and the cells and tissues produced by apical meristems form the plant's **primary tissues.** Annual plants experience only primary growth and thus are composed entirely of primary tissues. Such plants

are called *herbaceous* plants. In contrast, biennials and perennials also develop lateral meristems called **cambia** (*cambium* = singular) in addition to primary tissues. Cambia are rings or clusters of meristematic cells that increase the *width* of a plant's stems and roots as they divide. Cambia produce **secondary growth,** and the cells produced by these lateral meristems form the plant's **secondary tissues.** Thus, a pine, oak, or any other perennial undergoes both primary and secondary growth at the same time (Figure 18-3). Such plants are called *woody* plants.

Perennials develop two types of cambia: The **vascular cambium** produces new vascular tissues that increase transport capacity, keeping pace with the plant's increase in size and replacing aging or damaged conducting cells. The **cork cambium** produces cork cells that make up the outer covering of the bark and repair damaged tissues, such as outer tissues that have been damaged as the plant's stems and roots expand in width as a result of secondary growth.

Together, apical and lateral meristems impart potential immortality to perennial plants because aging and nonfunctioning cells are continuously replaced with new ones. This potential for continuous growth is called *open growth,* and it enables some plants to live for hundreds—and even thousands—of years (see Bioline: Nature's Oldest Living Organisms). Oddly enough, some of the oldest plants on

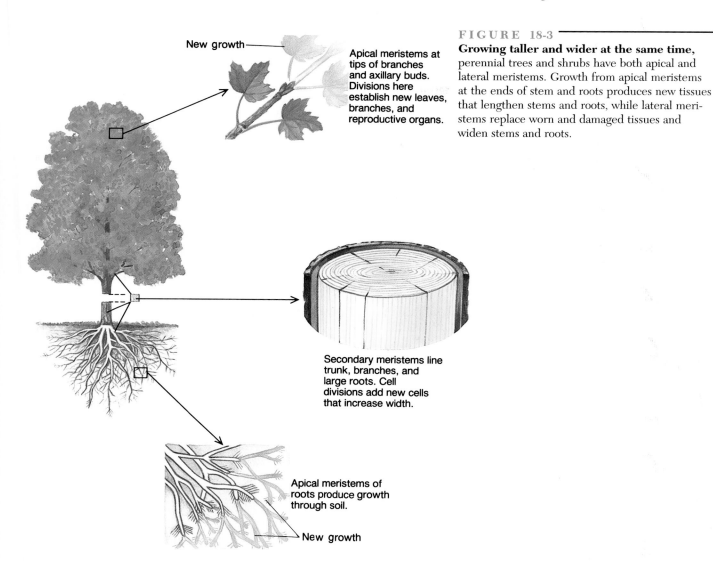

New growth

Apical meristems at tips of branches and axillary buds. Divisions here establish new leaves, branches, and reproductive organs.

Secondary meristems line trunk, branches, and large roots. Cell divisions add new cells that increase width.

Apical meristems of roots produce growth through soil.

New growth

FIGURE 18-3

Growing taller and wider at the same time, perennial trees and shrubs have both apical and lateral meristems. Growth from apical meristems at the ends of stem and roots produces new tissues that lengthen stems and roots, while lateral meristems replace worn and damaged tissues and widen stems and roots.

earth live in some of the harshest habitats, such as some of the perennial shrubs that grow in the earth's deserts. In the desert, the combination of favorable conditions for reproduction (adequate water supplies, moderate temperatures, availability of insects for pollen transfer and seed dispersal) are rarely met from one year to the next. It often takes several years for all the conditions to be favorable for plant reproduction. As a result, the longer a plant lives, the greater are its chances of encountering the proper conditions for successful reproduction.

PARENCHYMA CELLS PROVIDE VERSATILITY

The most prevalent type of cell in an herbaceous plant is **parenchyma** (Figure 18-4). Parenchyma cells provide both structural and functional flexibility, enabling the plant to adjust and respond to its immediate surroundings as well as to changing conditions.

Structurally, parenchyma cells form expanses of simple tissues and join with other types of cells to form complex tissues. Functionally, parenchyma cells have many roles. Among them, parenchyma are responsible for

- photosynthesis
- storage of food, water, and pigments
- transport
- healing of wounds
- production of new roots, stems, and other plant parts.

With such a variety of functions, it's not surprising that parenchyma cells vary a great deal in size, shape, and cell wall characteristics, although most are generally isodiametric (roundish) and have thin walls (Figure 18-4). Parenchyma cells are unique in their ability to differentiate into any other kind of plant cell, making them one of the most versatile of all plant cells.

COLLENCHYMA CELLS SUPPORT STEMS AND LEAVES

The shape of a plant cell and the nature of its cell wall contribute to its function. For example, support cells are usually elongated and have thick cell walls. One particular type of support cell—**collenchyma**—are cigar-shaped cells with unevenly thickened cell walls that help strengthen growing plant parts (Figure 18-4). The strands of a celery stalk are a prime example of collenchyma cells. In stems and leaves, collenchyma tissues lie just beneath the epidermis (the outer cell covering). Collenchyma cells keep the plant and its leaves upright and have chloroplasts to carry out photosynthesis.

SCLERENCHYMA CELLS SUPPORT AND PROTECT

The thick, secondary walls of **sclerenchyma** cells make the tissues they form even stronger and more supportive than collenchyma tissues. Some sclerenchyma also physically protect a plant from hungry animals. The critical component of a sclerenchyma cell is its wall, not the living cytoplasm inside. The secondary wall contains lignin, the hardening substance that gives the cell its toughness and strength. Because the secondary wall is left behind after the cell dies, sclerenchyma tissues continue functioning even after all of the cells die.

Sclerenchyma cells are classified as either sclerenchyma fibers or sclereids, based on their shape. **Sclerenchyma fibers** are extremely long, narrow cells with uniformly thickened walls (Figure 18-4). The elastic cell walls of a sclerenchyma fiber allow the cell to bend and then spring back to its original position. Elasticity is particularly important in leaves, stems, and flowers for proper orientation. For example, after strong gusts of wind die down, sclerenchyma fibers enable a plant's leaves to reorient quickly, exposing the maximum surface to sunlight for photosynthesis. The sclerenchyma fibers that make up hemp and flax are used to make rope and linen.

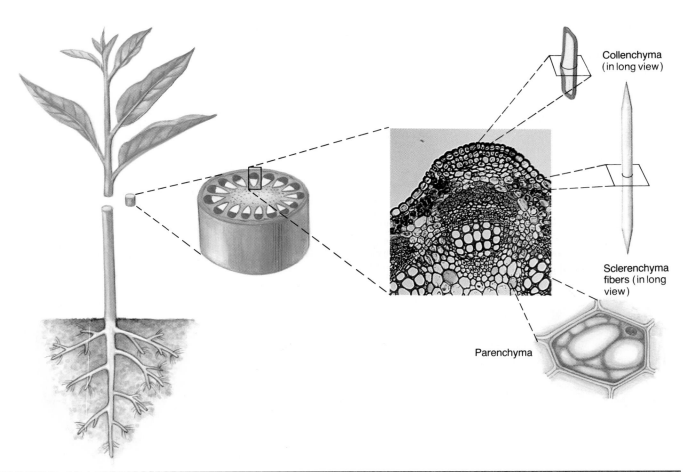

FIGURE 18-4

Simple tissues. Roundish parenchyma cells fill most of the stem of an annual plant. Elongated collenchyma cells with thickened walls help support the stem, and are found immediately beneath the epidermis. Dense collections of sclerenchyma fibers form a cap of supportive cells over each vascular bundle.

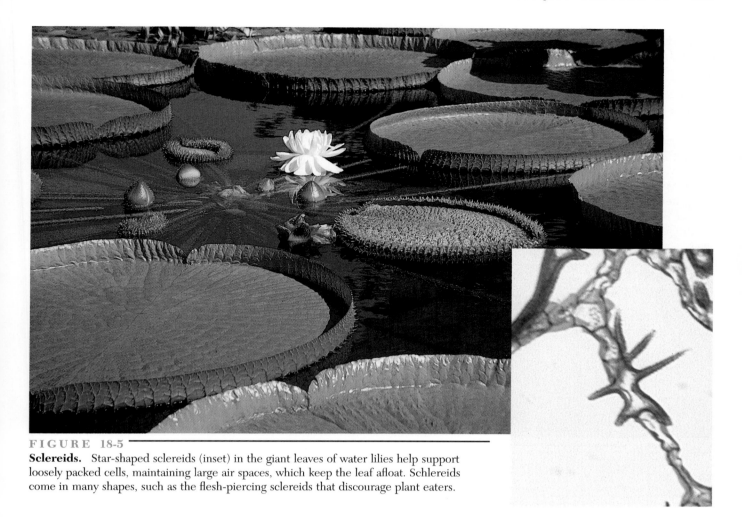

FIGURE 18-5

Sclereids. Star-shaped sclereids (inset) in the giant leaves of water lilies help support loosely packed cells, maintaining large air spaces, which keep the leaf afloat. Schlereids come in many shapes, such as the flesh-piercing sclereids that discourage plant eaters.

The second type of sclerenchyma cells—**sclereids**—have hard, thick secondary walls. Sclereids come in various shapes; the shape of a sclereid determines its function. For example, sharp, star-shaped sclereids not only help support loosely arranged leaf cells, but they also discourage even the hungriest browsing animal by puncturing the gums and tongue of the predator that nibbles on the leaf (Figure 18-5). Other sclereids include rectangular-shaped cells. Densely packed, rectangular sclereids form the "shell" of a walnut, which protects the plant's embryo from injury as it falls from a tree or tumbles along the ground. The gritty texture of a pear is the result of clumps of stone-shaped sclereids that help support huge parenchyma cells filled with sweet pear juices.

Organization of plant cells into tissues localizes functions and improves efficiency. For example, groups of dividing cells maintain coordinated plant growth, and clusters of support cells bolster strength sufficiently to hold stems, leaves, and flowers upright. (See CTQ #3.)

PLANT TISSUE SYSTEMS

Continuous groups of cells, including those that we just identified, connect the plant's root, stem, leaf, and flowers. Without such linkages, the plant could not grow or develop normally and would be unable to supply necessary provisions to each organ or to share in their products. Interconnecting tissues form one of three plant **tissue systems** (Figure 18-6):

• The protective **dermal tissue system,** or **epidermis,** forms the outer cell covering of the plant.

• The **vascular tissue system** contains conducting cells that transport food, water, and minerals throughout the plant.

• The **ground tissue system** comprises all remaining cells and tissues.

Each tissue system has a primary function: The dermal tissue system forms a seamless protective covering; the vascular tissue system transports essential materials between

◁ B I O L I N E ▷
Nature's Oldest Living Organisms

FIGURE 1

FIGURE 2

Imagine living for 5,000 years. Imagine having been born around 3000 B.C., at the beginning of human civilization. You are just starting to practice agriculture, build cities, develop art, and scratch out the first forms of writing. During your 5,000-year lifetime, you will live through the grandeur of the Roman Empire and see it fall into ruin; you will survive the constraints of the

Middle Ages to experience the glory of the Renaissance; you will survive tragic world wars, hear the first report of people walking on the moon, and experience the beginning of the age of computers and biotechnology.

The scenario is intriguing but impossible. Humans cannot live that long; vital cells simply wear out and are not replaced. But perennial plants are continually able to replace worn-out cells with new ones, maintaining a constant supply of young, efficient cells. Unlike humans, perennials have the potential to live forever.

For many years, the bristlecone pine (*Pinus longaeva*) has been the reigning "oldest living organism," with a documented age of 4,900 years (see Figure 1). In fact, a bristlecone pine actually lived through all of the milestones in the development of human civilization discussed above. Strangely, these ancient plants do not live in a rich environmental setting. Instead, they grow on the wind-beaten, desertlike slopes of the White Mountains in eastern California.

The reign of the bristlecone is currently being threatened by an unimposing, common neighbor, the creosote bush (*Larrea divaricata*) (Figure 2). Some creosote bushes have been growing identical clones of themselves for an estimated 11,000 years; the bristlecones are middle

aged in comparison. The clones form a ring of new individuals that surround the parent plant, each clone having developed asexually from a branch that contacted the soil surface. A number of botanists argue that the creosote's clone production is analogous to the way the bristlecone produces rings of new secondary tissues year after year. Either way—producing rings of new secondary tissues or rings of new individuals—the bristlecone pine and the creosote bush perpetuate cells over thousands of years, cells that originated from a single individual plant. If more botanists begin to agree that perpetuation by cloning is the same as perpetuation by secondary growth, the common creosote bush will likely assume the throne as the world's "oldest living organism."

---

all plant organs; and the ground tissue system provides support and storage continuity.

THE DERMAL TISSUE SYSTEM

The dermal tissue system (*derma* = skin) has two critical, yet opposing, functions: The epidermis must form a barrier to protect the plant's delicate internal tissues, yet it must allow for exchanges of essential materials between the plant and its surrounding environment. In herbaceous plants,

such as peas, the dermal tissue system is a single sheet of cells; it is the only thing that stands between the fragile interior cells and the often hostile external environment. For protection, the outer cell walls of the shoot epidermis are covered by a **cuticle,** a waxy layer that retards water loss and helps prevent dehydration. The cuticle is such an effective water barrier that it has been used to protect some of our valued possessions. For example, the cuticle of the Brazilian wax palm is harvested to make carnuba wax for safeguarding furniture and cars (Figure 18-7).

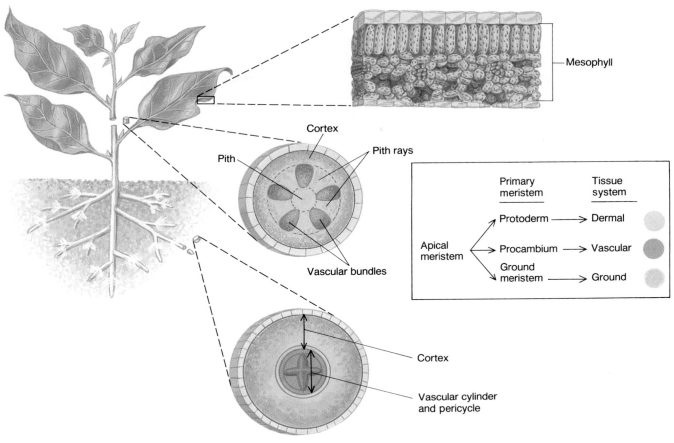

FIGURE 18-6

Plant tissue systems. Three tissue systems run continuously throughout the body of a plant: the dermal tissue system, the vascular tissue system, and the ground tissue system. The ground tissue system found in the stem consists of the cortex (the area between the epidermis and vascular tissues), pith (the area from inside the vascular tissues to the stem or root center), and pith rays between the vascular tissues; the mesophyll found in leaves; and the cortex found in roots.

FIGURE 18-7

The cuticle that protects Brazilian wax palm leaves also protects automobile finishes from damage.

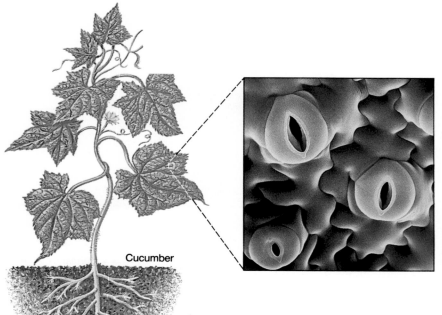

Cucumber

FIGURE 18-8
Stomatal pores between adjacent guard cells enable gases to diffuse into and out of air spaces between internal cells. Three stomates are shown in the photograph. Each stomatal pore is flanked by two guard cells. Carbon dioxide mainly diffuses inward through the stomatal pore for photosynthesis, while water vapor diffuses out. As internal turgor pressure increases, guard cells bend and enlarge the size of the stomatal pore; decreasing turgor pressure narrows or closes the stomatal pore, blocking further gas exchange.

The epidermis of the shoot is riddled with microscopic pores called **stomates** (Figure 18-8), which allow the exchange of gases between the plant and the outside environment. The stomate is flanked by two specialized **guard cells** which regulate the rate of gas diffusion—carbon dioxide for photosynthesis, oxygen for aerobic respiration, and water vapor for cooling. By regulating water vapor, guard cells not only moderate temperature, they also help prevent dehydration in plants that grow in hot, dry environments.

FIGURE 18-9
Creating the effect. A sculptured epidermis (inset) gives marigold petals their velvety texture. Bright pigments stored in the vacuoles of these epidermal cells give the petals their brilliant orange and yellow color.

Cork

Cork
cambium

Parenchyma

(b)

FIGURE 18-10

The periderm not only protects internal tissues during secondary growth but allows for gas exchange through lenticels. *(a)* Birch periderm with thousands of elongate lenticels on the surface. *(b)* Section through a single lenticel.

(a)

In general, the cells in the dermal tissue system are referred to as *epidermal cells.* Some epidermal cells, such as those found in begonias, become meristems, asexually producing buds that grow into complete new plants. In the petals of many colorful flowers, the vacuoles of epidermal cells are engorged with water-soluble pigments that help attract animals to the flower for sexual reproduction (see Figure 18-9 and Chapter 20).

In woody plants, the **periderm** takes over the protective and regulating functions of the epidermis when the epidermis has become damaged as the stem and root increase in width. The periderm is commonly referred to as "bark," but it is actually only the outer part of bark. **Bark** includes all those plant tissues outside the wood (secondary xylem) and is made up of many distinct tissues, including the vascular cambium, phloem, cells in the cortex, cork cambium, and cork. The periderm is composed only of cork cells, the cork cambium, and a layer of parenchyma cells immediately inside the cork cambium (Figure 18-10).

Like sclerenchyma, cork cells continue to function even when the cell is dead. Each cell is composed of layers of secondary walls that are impregnated with a waterproof wax called suberin. These waxy cell layers effectively seal off internal tissues, protecting them against excess water-vapor loss, disease, extreme weather, and foraging insects. To allow oxygen from the atmosphere to reach living cells beneath the periderm, random eruptions in the periderm, called **lenticels,** form air channels that allow the transfer of carbon dioxide and oxygen (Figure 18-10).

THE VASCULAR TISSUE SYSTEM

The vascular tissue system forms an internal circulatory network—the "veins and arteries" of a vascular plant. This tissue system contains two types of complex tissues: **xylem** and **phloem.** Water and dissolved minerals are carried primarily through the xylem; food (carbohydrates produced during photosynthesis) and other organic chemicals syn-

Pits

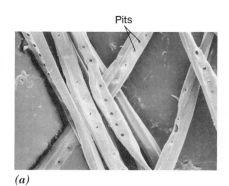

(a)

Perforation

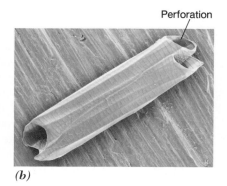

(b)

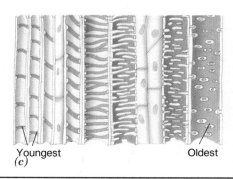

Youngest

Oldest

(c)

FIGURE 18-11

Conducting cells in the xylem transport water and dissolved minerals. They include *(a)* long, thin tracheids that are aligned precisely so that the openings (pits) of one match up with those of other tracheids, and *(b)* vessel members that have pits and perforated ends. Some vessel members *(c)* have characteristic cell wall patterns, the result of being stretched during stem and root growth.

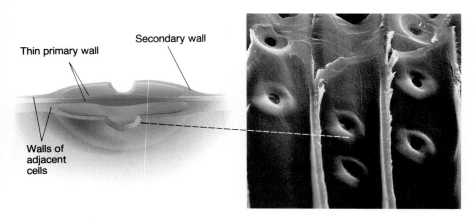

FIGURE 18-12
Pits. Water and dissolved minerals move through connecting pits of adjacent xylem cells.

Thin primary wall

Secondary wall

Walls of adjacent cells

thesized by the plant (such as hormones, amino acids, and proteins) move through the phloem.

Xylem: Water and Mineral Transport and Support

The main conducting cells of the xylem are **tracheids** and **vessel members** (Figure 18-11), both of which are dead cells; water and minerals flow through their hollow interiors. The walls of each cell are riddled with pits—small holes in the secondary wall that are adjacent to thin depressions in the primary wall (Figure 18-12). Xylem cells line up so that the pits of one cell match up exactly with those of a neighbor, forming a pit pair. Water flows from one cell to the next through these pit pairs. Opposite ends of vessel members also have a **perforation plate**—an area where all or large portions of the end wall are absent (Figure 18-11). Vessel members stack one on top of another to form open **vessels** through which water readily travels (Figure 18-13).

Secondary xylem is called **wood.** New secondary xylem cells are produced by the vascular cambium. Toward the end of one growth period (usually at the end of summer), the inside of xylem tracheids and vessel members gets progressively smaller. When water becomes plentiful again (usually not until the following spring), new tracheids and vessel members with large diameters are formed. The abrupt transition from xylem cells with narrow diameters to those with large diameters forms a distinct line, the border of a **growth ring** (Figure 18-14). Because one growth ring is usually produced each year, the number of growth rings is often used to estimate the age of a tree. Nevertheless, many botanists have abandoned using the term "annual ring" because plants sometimes have more than one growth spurt and, therefore, form more than one growth ring in a single year.

Phloem: Food Transport

Phloem runs parallel with the xylem, transporting food and other organic chemicals throughout the plant. Unlike xylem tracheids and vessel members, however, phloem conducting cells function as living cells. In flowering plants, phloem conducting cells are called **sieve-tube members** because the clusters of pores in their cell walls resemble a strainer or sieve (Figure 18-15). These porous areas are located at opposite ends of the sieve-tube member and are called **sieve plates.** A row of sieve-tube members, stacked sieve plate to sieve plate, forms a sieve tube.

Although sieve-tube members are *technically* alive, they lack a nucleus and most of the other cellular organelles of a typical living cell. As these conducting cells mature, their organelles degenerate, leaving only a veneer of cytoplasm against the inside walls and a delicate mesh of protein that extends from sieve-tube member to sieve-tube member. To remain alive and active, sieve-tube members are continually nourished by an adjacent parenchyma cell

Pits

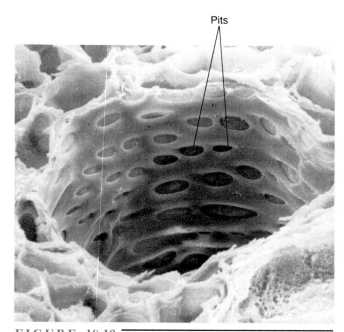

FIGURE 18-13
A vessel pipeline is formed by stacks of vessel members. Pits allow lateral transport.

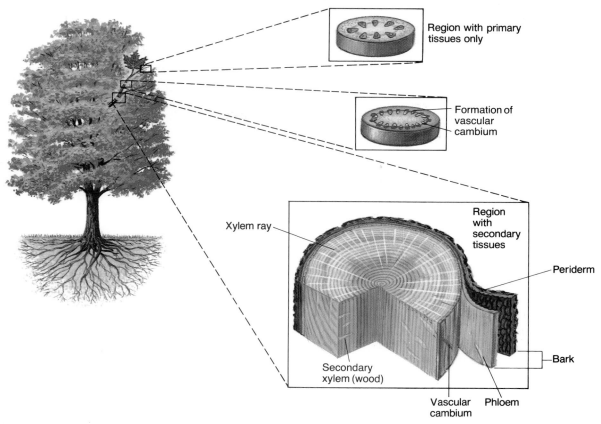

FIGURE 18-14

Sections of a stem and trunk of a tree. Although primary growth continues from apical meristems at the tips of stems, most of the tissue in a tree is secondary xylem, or wood. The vascular cambium divides to produce secondary xylem toward the inside and secondary phloem toward the outside.

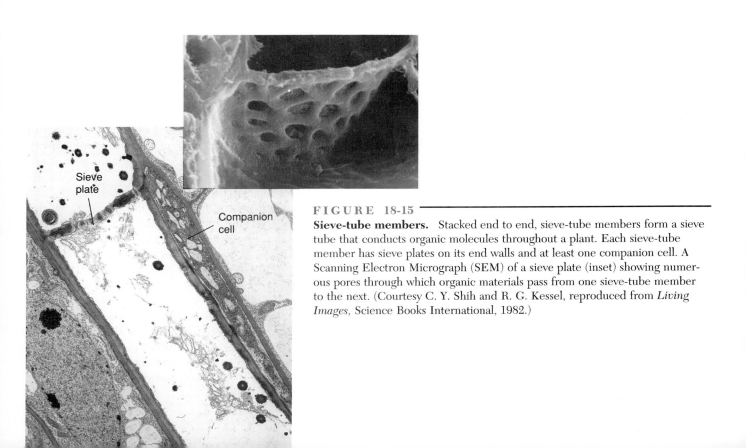

FIGURE 18-15

Sieve-tube members. Stacked end to end, sieve-tube members form a sieve tube that conducts organic molecules throughout a plant. Each sieve-tube member has sieve plates on its end walls and at least one companion cell. A Scanning Electron Micrograph (SEM) of a sieve plate (inset) showing numerous pores through which organic materials pass from one sieve-tube member to the next. (Courtesy C. Y. Shih and R. G. Kessel, reproduced from *Living Images*, Science Books International, 1982.)

called a **companion cell.** Sieve-tube members and companion cells are interdependent; if one dies, its partner dies soon after.

THE GROUND TISSUE SYSTEM

With protection and exchange functions performed by the dermal tissue system, and long-distance transport accomplished by the vascular tissue system, all remaining plant functions are carried out in the ground tissue system, which is composed of a mixture of cells with different functions (see Figure 18-6). The ground tissue system is made up of cells in the **cortex** (the region between the epidermis and the vascular tissues), the **pith** (the area from inside the vascular tissues to the center of the stem or root), and **pith rays** (the areas between bundles of vascular tissues). The photosynthetic cells between the upper and lower epidermis of leaves are also part of the ground tissue system.
◐ The ground tissues are the principal sites of photosynthesis and food and water storage in a plant. Ground tissues also contain collenchyma and sclerenchyma, which provide support for shoot structures.

Plant cells and tissues are organized into three tissue systems, each of which is continuous throughout the entire plant body. Each tissue system has the same unique set of functions in all plant organs. (See CTQ #4.)

PLANT ORGANS

All the plant cells, tissues, and tissue systems we have discussed thus far are organized into four plant organs: the *stem, root, leaf,* and *flower.* By necessity, photosynthesis is restricted to aerial organs that are exposed to light, whereas water and minerals are absorbed by root tissues that grow in the water and mineral reservoirs found in soil.

THE STEM: GROWTH, SUPPORT, AND CONDUCTION

The **stem** is an exquisite example of nature's bioengineering. Not only do stems physically support all of a plant's leaves, flowers, fruits, and even its other stems, they produce the structures they support. The stems of herbaceous plants consist of only primary stem tissues, whereas the stems of woody plants consist of both primary and secondary stem tissues.

Primary Stem Tissues

At the tip of every stem is a dome-shaped apical meristem (Figure 18-16). The end of each stem is divided into three regions:

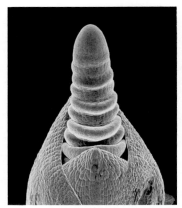

FIGURE 18-16
Apical meristem. A scanning electron micrograph of a stem tip shows the dome-shaped apical meristem and upwardly arched juvenile leaves. Above each young leaf is a developing axillary bud.

- The **meristematic region** contains cells that divide for primary growth. Young leaves and juvenile axillary buds (a bud that is directly above each leaf on a stem) begin developing in this region.

- The **region of elongation** lies just below the meristematic region. This is where cells enlarge and lengthen as they differentiate.

- The **region of maturation,** which lies immediately below the region of elongation, is where nearly all cells have completed enlarging and differentiating.

Monocot and dicot flowering plants differ in the way their stems' vascular tissues are arranged (refer back to Table 18-1). Dicot vascular bundles are organized in distinct rings (Figure 18-17), whereas the more numerous monocot vascular bundles are scattered throughout the ground tissues. Monocots and dicots differ in another important way. Perennial and biennial dicots form cambia for secondary growth, whereas monocots do not.

Secondary Stem Tissues

The tallest organisms on earth are those plants that experience secondary growth. In crowded habitats, taller plants have access to greater amounts of sunlight for photosynthesis than do shorter, shaded plants. A tall plant is able to grow so large because its diameter (and thus the plant's supportive strength) increases as the plant ages. Thus, as apical meristems produce cells that elongate stems, making the plant taller, the vascular and cork cambia increase the stem's support and transport capacities by adding new cells that increase the diameter of the stem. All new secondary vascular cells match up with those in the primary xylem and phloem to maintain a continuous connection of vascular tissues throughout even the tallest plant.

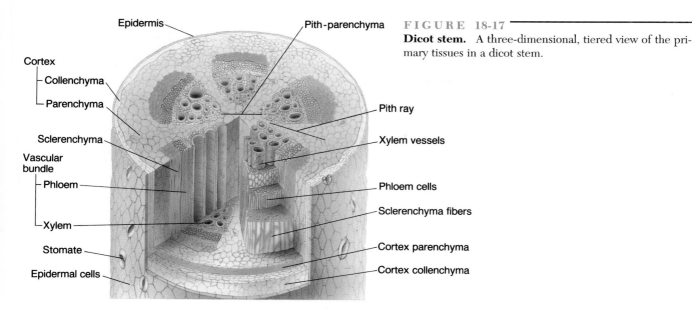

Dicot stem. A three-dimensional, tiered view of the primary tissues in a dicot stem.

The vascular cambium produces two arrangements of conducting cells: One allows substances to move up and down the stem, while the other conducts materials laterally through secondary xylem and phloem rays (Figure 18-14). In plants, the term "ray" is used to describe any structure that runs parallel to a radius line from the stem center to the epidermis.

THE ROOT: GROWTH, ABSORPTION, AND CONDUCTION

Much of a plant lies hidden beneath the ground. The extensive nature of a plant's root system is illustrated by a single, tiny rye plant. A rye plant only 25 centimeters tall (10 inches) has more than 14 million roots, which, if placed end to end, would stretch over 609 kilometers (380 miles). Knowing this, you might assume that a massive plant like a giant sequoia would have even more roots than would the rye; surprisingly, this is not the case. The sequoia has fewer roots than does the rye plant, but the entire root system of a giant sequoia is much larger than is that of the rye. This unexpected disparity exists because the rye and giant sequoia have different types of root systems. The giant sequoia has a **tap root system** (Table 18-1), which is similar to a carrot, with its one main root and many smaller branches, called **lateral roots** (Figure 18-18*a*). Like the giant sequoia, most dicots have a tap root system. In contrast, the rye is a monocot and has a **fibrous root system,** which is composed of many main roots — more than 140 for rye (Figure 18-18*b*). In both tap and fibrous root systems, main roots produce lateral roots; lateral roots form more lateral roots; and so on, until their origin can't be traced any further. This repeated branching of roots generates many root tips — locations of **root hairs,** extended surface cells through which water and minerals are absorbed (Figure 18-19).

Sometimes roots arise in unexpected places — on stems or even leaves. Roots that develop directly from shoot tissues form an **adventitious root system.** For example, corn plants develop prop roots from the base of their stems (Figure 18-18*c*), which support the shoot when it is burdened with heavy clusters of fruits (what we call a "corn cob").

Primary Root Tissues

Unlike stem apical meristems, root apical meristems are not located at the very tip of each root (Figure 18-20). Instead, the tip of a root has a **root cap** — a protective cellular "helmet" that surrounds delicate meristematic cells and shields them from abrasion as the root grows through the soil. The root's apical meristem continuously regenerates its root cap as cells are worn away or pierced by sharp soil particles. Root cap cells also help the root penetrate the soil by manufacturing and secreting a gelatinlike substance that helps lubricate the root. This secretion also creates a favorable habitat for certain soil bacteria that supply the plant with essential elements, especially nitrogen.

In addition to forming its own protective cap, the root apical meristem produces all the cells that will differentiate into the root's primary tissue systems. As in stems, roots have a *meristematic region,* a *region of elongation,* and a *region of maturation* (Figure 18-20). Root hairs form in the region of maturation and are the principle site of water and mineral absorption.

Roots contain two important tissues that are not found in stems — an endodermis and a pericycle — both of which

Tap root system   Fibrous root system   Adventitious root system

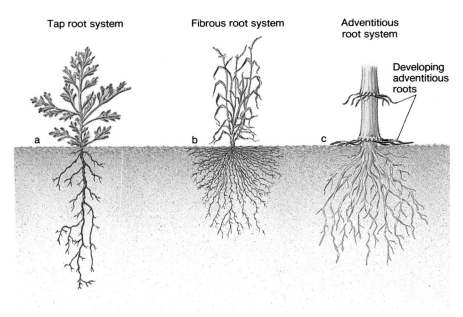

Developing adventitious roots

FIGURE 18-18

Three types of root systems in plants. *(a)* A tap root system consists of one main root with many branching lateral roots. *(b)* A fibrous root system has more than one main root, each with many branching lateral roots. Fibrous root systems are generally shallow and spread out from the stem base. *(c)* In an adventitious root system, roots arise from organs other than the root itself, mainly from stems or leaves. Adventitious roots develop when plant cuttings are made.

are cylinders of cells that encircle the vascular tissues (Figure 18-20 and 18-21). The **endodermis** is the innermost layer of the cortex. Early in its development, each endodermal cell is encircled by a band of waxy suberin, forming the **Casparian strip.** Water-repelling Casparian strips of adjacent endodermal cells are aligned, creating a "gasket" that seals the space between adjacent cells so that water and minerals can only pass through living endodermal cells. By changing their solute concentration, endodermal cells control the uptake of water and minerals and prevent water from leaking out of the root and back into the sometimes dry or salty soil.

FIGURE 18-19

Root hairs give this radish seedling its fuzzy appearance. The enlargement (inset) shows the root cells beginning to expand to form root hairs for absorbing water and minerals from the soil.

◁ THE HUMAN PERSPECTIVE ▷
Agriculture, Genetic Engineering, and Plant Fracture Properties

Attempts to breed or genetically engineer plants with larger fruits, stems, or leaves for improved productivity could fail if the fracture properties of plant structures is not considered. As a plant's structures increase in size, they are more likely to fracture. In order to produce large fruits, stems, or leaves that will not fracture, they would have to be so "tough" (filled with supportive cells and tissues) that they may not be palatable.

Some botanists study the strength of plant structures as they are subjected to varying stresses in an attempt to reach some general conclusions about the dura-

bility of plant parts; such conclusions may have agricultural and horticultural applications. For example, the shoot system of a plant must sometimes withstand slashing winds and pelting rains. The way a plant resists fracturing under such conditions is the result of its mechanical design, from the cellulose in its cell walls, to specialized support and strengthing cells, to complexes of support tissues. Under normal conditions, the parallel arrangement of sclerenchyma fibers in a leaf petiole (such as a celery stalk) resists damage. If winds are strong enough to inflict damage, however, the cellular structure and arrange-

ment control the fracture in a way that reduces injury, an advantage to an agricultural field of celery on a windy day.

Fracture studies point out a basic principle: The smaller the plant or plant part, the greater its inherent toughness and structural integrity. This principle can be of great importance as biologists try to engineer crops that can survive in regions that are exposed to severe weather, greatly expanding the world's agricultural area. Increasing the food supply is the first step in solving world hunger.

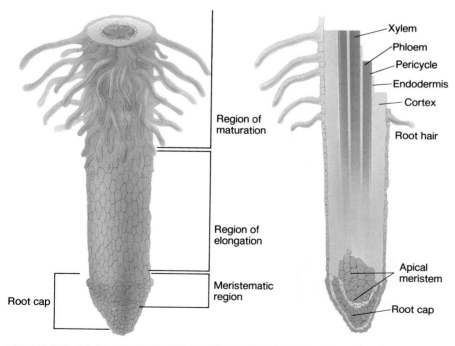

FIGURE 18-20

The three regions of the root apex. The meristematic region of active cell division. The region of elongation, where cells enlarge in length as well as width and begin differentiating. The region of maturation, where most cells, including root hairs, have completed differentiation. Water and minerals are absorbed only in the region of maturation.

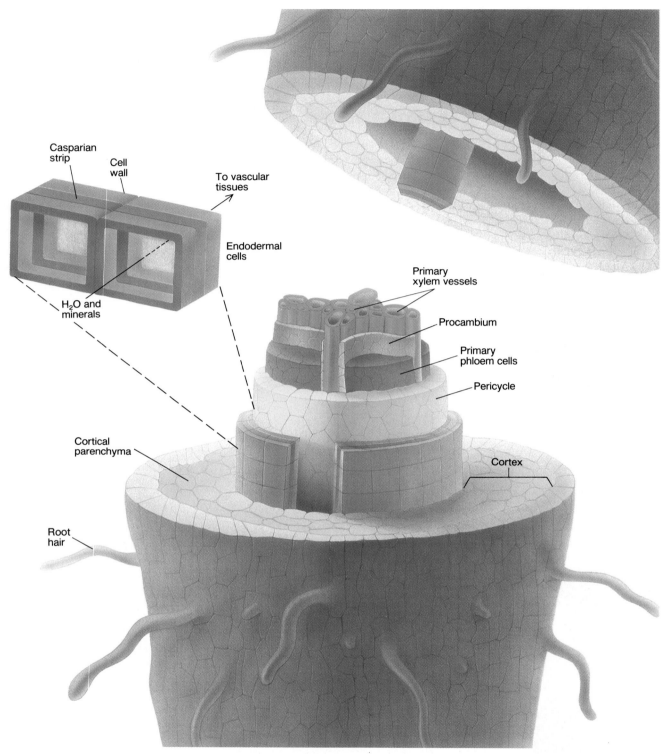

FIGURE 18-21

A three-dimensional, tiered view of the primary tissues found in a dicot root.   The innermost layer of cortex cells is the endodermis. A Casparian strip encircles each endodermal cell, enabling endodermal cells to control water movement. The pericycle is a cylinder of cells that lies immediately inside the endodermis. The pericycle produces lateral roots or forms part of the vascular cambium during secondary growth.

Like the endodermis, the **pericycle** is one cell layer thick. The pericycle lies immediately inside the endodermis (Figure 18-21) and divides to form lateral roots. In a perennial or a biennial plant, the pericycle also contributes to the formation of a vascular cambium in the root for secondary growth.

Secondary Root Tissues

If there is secondary growth in the stem, there will also be secondary growth in the root. As in the stem, secondary tissues arise from the vascular and cork cambia. Except for the absence of a pith in the root, secondary growth in the root and stem appear very similar.

THE LEAF: THE PLANT'S PRIMARY PHOTOSYNTHETIC ORGAN

☀ Leaves are the plant's "solar-collectors," "energy generators," and "energy transmission lines," all rolled into one. Leaves capture solar energy, house chloroplasts for converting the energy in sunlight into chemical energy during photosynthesis, and transport materials through an extensive network of interconnecting vascular tissues.

The leaves of flowering plants typically are made up of a flattened **blade** that collects sunlight for photosynthesis, and a **petiole**—a stalk that connects the blade to the stem. The part of a plant stem where one or more leaves are attached is called the **node;** regions between nodes are **internodes.** Directly above the spot where a leaf joins the stem is an **axillary bud**—an underdeveloped cluster of cells that will either grow into a new stem with additional leaves and more axillary buds or will develop into flowers.

Leaves are classified as one of two types, depending on whether or not the blade is divided (Figure 18-22). A **simple leaf** has a single, undivided blade, although, in some cases the blade is deeply indented (lobed) almost to the central, main vein. In contrast, the blade of a **compound leaf** is divided into a number of clearly separated *leaflets.*

(a) *(b)*

(c) *(d)*

FIGURE 18-22

Simple versus compound leaves. *(a)* Apple (*Malus*) and *(b)* sycamore (*Platanus*) trees both have simple leaves because each blade is undivided, even though the leaves of sycamore are clearly lobed. *(c)* Roses (*Rosa*) and *(d)* poison ivy (*Toxicodendron*) have compound leaves with leaflets.

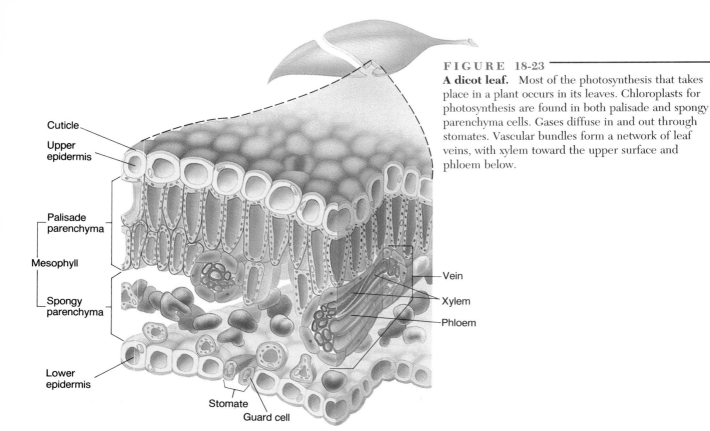

FIGURE 18-23
A dicot leaf. Most of the photosynthesis that takes place in a plant occurs in its leaves. Chloroplasts for photosynthesis are found in both palisade and spongy parenchyma cells. Gases diffuse in and out through stomates. Vascular bundles form a network of leaf veins, with xylem toward the upper surface and phloem below.

Cuticle

Upper epidermis

Palisade parenchyma

Mesophyll

Spongy parenchyma

Lower epidermis

Stomate

Guard cell

Vein

Xylem

Phloem

Leaves have an upper and lower epidermis that sandwich the leaf's **mesophyll**—the layers of cells found between the two epidermises (Figure 18-23). In dicot leaves, the mesophyll contain two types of cells: column-shaped **palisade parenchyma** toward the top, and loosely packed **spongy parenchyma** toward the bottom. Air spaces between these mesophyll cells form corridors for diffusion of carbon dioxide for photosynthesis.

The extensive vascular leaf tissues form veins (Figure 18-24), which create a characteristic **venation** pattern for different plants. Monocots have leaves with larger veins that run parallel to each other, whereas dicots have highly branched venation (Table 18-1).

THE FLOWER: THE SITE OF SEXUAL REPRODUCTION

The balance between form and function is most elegantly displayed in flowers, which are the site of sexual reproduction. Each part of a flower participates in the process of reproduction by either producing gametes or by influencing the transfer of gametes so that fertilization can take place. Following fertilization, all or part of the flower develops (ripens) into a fruit that contains seeds with newly formed plant embryos. The fruit protects these embryos and helps disperse them to new habitats. Because of their complexity and central role in reproduction, we devote an entire chapter (Chapter 21) to the structure and functions of flowers and fruits.

Dividing essential functions among different organs greatly improves the effectiveness of each. Leaf construction facilitates photosynthesis, flower color and form improves sexual reproduction, stems produce new growth and support aerial structures, and roots absorb water and minerals. (See CTQ #6.)

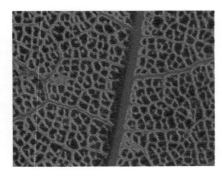

FIGURE 18-24
An extensive network of vascular tissue is revealed when all tissues except the xylem are cleared from a leaf.

R E E X A M I N I N G   T H E   T H E M E S

Relationship between Form and Function

The relationship between form and function is found not only at the more familiar organ level but at the cellular level as well. Specialized cells that help support aerial stems, leaves, flowers, and fruits all have reinforced cell walls. The shape and characteristics of guard cells enable these cells to change shape, thereby regulating the flow of carbon dioxide and water vapor. The hollow interiors of xylem conducting cells form after the cell dies, allowing water and minerals to flow from one cell to the next. The various shapes of thick-walled sclereids enable these cells to carry out specific functions.

Acquiring and Using Energy

The basic plant structure is a consequence of the way plants acquire and use energy. The relatively few structures needed to gather light, water, and carbon dioxide for photosynthesis include leaves, which provide an enlarged surface area for absorbing radiant energy from the sun; stems, which contain vascular tissues that connect with vascular tissues in the root to supply necessary water and minerals and to distribute the products of photosynthesis; stomate pores, which allow gaseous carbon dioxide and oxygen to diffuse to photosynthetic cells; and cells with chloroplasts, the cellular site of photosynthesis.

Evolution and Adaptation

Flowering plants are the most evolutionarily advanced plant group. Judgments about whether organisms are "primitive" or "advanced" are based on many criteria, including, but not limited to, fossil record data, similarities and differences in structures, and the means by which efficient structures carry out specific functions. Flowering plants have a number of structures that are more efficient at completing their tasks than are those found in other plant groups. Examples include the flower, for sexual reproduction; fruits, for protecting and disseminating offspring; sieve-tube members, for translocating organic molecules; and vessel members, for transporting water and dissolved minerals.

S Y N O P S I S

Flowering plants are the most evolutionarily advanced and structurally complex group in the plant kingdom. Whether they are dicots or monocots, flowering plants form a branched root system that absorbs and conducts water and dissolved minerals, and helps support the aerial shoot system. The plant's shoot produces stems with leaves and axillary buds that develop into more stems or into flowers for sexual reproduction. Leaves are adapted for photosynthesis.

Plants can be classified by their growth pattern. Short-lived, annual plants experience only primary growth, whereas longer-lived biennials and perennials experience both primary and secondary growth simultaneously. Primary growth from meristems increases the length of stems and roots, whereas secondary growth from cambia increases stem and root width as older or dead cells are replaced with new, active cells.

Long-lived perennials replace old or damaged cells with new cells from meristems and cambia. As a result, many plants live for hundreds, or even thousands, of years. Growth in width from cambia produces wood (secondary xylem) for increased support and water and mineral transport, as well as an outer protective periderm.

All flowering plants contain specialized vascular tissues. The xylem conducts water and minerals, and the phloem transports food and other organic molecules.

Plant tissues form groups of cells that perform the same function. Meristems (apical and cambia) divide by mitosis to produce complex vascular tissues (xylem and phloem) and simple tissues (parenchyma, collenchyma, sclerenchyma). The cells that make up a plant's tissues vary in shape, size, and cell wall characteristics, enabling the tissues to carry out specific functions.

Some plant cells — mainly xylem and sclerenchyma — function as dead cells. The unique characteristics of the surrounding cell wall enables some plant cells to continue functioning even after the cell dies.

Parenchyma cells are the most prevalent type of cell in many plants. Versatile parenchyma cells carry out many functions, including photosynthesis, storage, and the ability to change into any other type of plant cell.

The organs of a flowering plant are interconnected by three tissue systems that extend throughout the plant's body. The outer epidermis forms the protective dermal tissue system. The internal network of conducting cells forms the vascular tissue system of xylem and phloem, which is surrounded by cells that make up the ground tissue system.

Each of the four plant organs—root, stem, leaf, flower—is adapted for a different function. Roots absorb water and minerals, store food reserves, and help support the shoot system. Stems produce leaves and axillary buds, which develop into new stems with more leaves and more axillary buds or into flowers for sexual reproduction. Most photosynthesis takes place in the plant's leaves and stems.

Key Terms

annual (p. 361)
biennial (p. 361)
perennial (p. 361)
vascular plant (p. 361)
dicotyledon (p. 361)
monocotyledon (p. 361)
root system (p. 361)
shoot system (p. 361)
simple tissue (p. 362)
complex tissue (p. 362)
meristem (p. 362)
apical meristem (p. 362)
primary growth (p. 362)
primary tissue (p. 362)
cambia (p. 362)
secondary growth (p. 362)
secondary tissue (p. 362)
vascular cambium (p. 362)
cork cambium (p. 362)
parenchyma (p. 363)
collenchyma (p. 364)
sclerenchyma (p. 364)
sclerenchyma fiber (p. 364)
sclereid (p. 365)

tissue system (p. 365)
dermal tissue system (p. 365)
epidermis (p. 365)
vascular tissue system (p. 365)
ground tissue system (p. 365)
cuticle (p. 366)
stomate (p. 368)
guard cell (p. 368)
periderm (p. 369)
bark (p. 369)
lenticel (p. 369)
xylem (p. 369)
phloem (p. 369)
tracheid (p. 370)
vessel member (p. 370)
perforation plate (p. 370)
vessel (p. 370)
wood (p. 370)
growth ring (p. 370)
sieve-tube member (p. 370)
sieve plate (p. 370)
companion cell (p. 372)
pith (p. 372)
pith ray (p. 372)

stem (p. 372)
meristematic region (p. 372)
region of elongation (p. 372)
region of maturation (p. 372)
tap root system (p. 373)
lateral root (p. 373)
fibrous root system (p. 373)
root hair (p. 373)
adventitious root system (p. 373)
root cap (p. 373)
endodermis (p. 374)
Casparian strip (p. 374)
pericycle (p. 377)
blade (p. 377)
petiole (p. 377)
node (p. 377)
internode (p. 377)
axillary bud (p. 377)
simple leaf (p. 377)
compound leaf (p. 377)
mesophyll (p. 378)
palisade parenchyma (p. 378)
spongy parenchyma (p. 378)
venation (p. 378)

Review Questions

1. Name an important adaptation that all shoot dermal tissue system cells have in common but is entirely absent in the root's epidermis. Why is this adaptation restricted to the shoot? Why is it so critical to the survival of a land plant?

2. Plants contain cells (such as xylem vessels and sclerenchyma fibers) that continue to function even after the living part of the cell is dead. List the types of functions these dead cells perform. What characteristics do dead, functioning cells have in common that enable them to continue to function even after the cell dies?

3. How are the terms "periderm" and "bark" related? Why do most botanists avoid using the term "bark"?

4. Why are parenchyma cells sometimes called "the most important type of cell in a plant"?

5. Using the following list of terms, identify with an "S" those that are found only in the shoot system, with an "R" those that are found only in the root system, and with a "B" those found in both the shoot and root systems:

apical meristem	collenchyma	sclerenchyma
parenchyma	node	vessel member
bundle sheath	Casparian strip	axillary bud
companion cell	lenticel	vascular cambium

6. Why do biennials and perennials experience both primary and secondary growth simultaneously, while annual plants only experience primary growth?

7. List five differences and five similarities between the tissues in the stem and the tissues in a root.

8. What is the function of the Casparian strip on endodermal cells? Why don't stem cells have such a strip?

9. How are flowers and stems similar? From an evolutionary point of view, what is the significance of the similarity?

10. List the structural adaptations land plants have evolved to help reduce the amount of valuable water they lose to the surrounding air.

Critical Thinking Questions

1. Imagine that you have been asked to help identify the source of destruction of the grape vines in the French vineyards. When you started working on the project, it was not known that the damage was being caused by an insect. How would you go about determining the source of destruction? How would you differentiate damage caused by insects (or some other animal) from that caused by other environmental factors, such as shortage of rainfall, nutrient deficiency, lack of pollinators, and so on? Choose five environmental factors that could account for the devastation. For each, design a test to verify whether or not that factor was responsible for the damage. In nature, often more than one factor may cause damage. How would you test for a combination of factors?

2. The longer a group of organisms exists, the greater the time for new species to evolve from that group. It seems logical to assume, then, that older groups of organisms would have the greatest species diversity. But the most *recently* evolved plant group (the flowering plants) contains the greatest number of species—over 90 percent of all plant species. How do you explain this apparent contradiction? List some factors (both environmental and biological) that may have triggered such rapid evolution in flowering plants.

3. Businesses use organizational charts to explain the working relationships among personnel. Prepare an organizational chart for a flowering plant, showing how it is organized to carry out the life functions.

4. Using the characteristics described for the various types of cells found in flowering plants, construct a dichotomous key that could be used to identify cells microscopically.

5. Sieve-tube members in the phloem and vessel members in the xylem are both transport cells. What features do these cells have in common that aid in transport, and how are they different? Why wouldn't all transport cells have the same structure?

6. Suppose you could design a flowering plant. What modifications would you make to the roots, stems, leaves, or flowers to create a plant that would be adapted to each of the following habitats: a desert, a rocky intertidal zone, the surface of a freshwater pond?

7. On a nature walk, you find a herbaceous plant that has leaves with parallel veins and flowers with 12 petals. There are no fruits or seeds, so it is impossible to determine whether the embryo has one or two cotyledons. Why would you suspect that the plant is a monocot? What other characteristics would you look for to verify your hypothesis?

8. Plants are annuals, biennials, or perennials. What is the adaptive value for each of these life span strategies? Compare a desert, seashore, and mountain environment. Would one strategy have increased survivability and reproduction over the others in each of these different environments?

Additional Readings

Esau, K. 1977. *Anatomy of seed plants.* New York: Wiley. (Advanced)

Evans, M., R. Moore, and K. Hasenstein. 1986. How roots respond to gravity. *Sci. Amer.* Dec:112–119. (Introductory)

Hohn, R. 1980. *Curiosities of the plant kingdom.* New York: Universe Book. (Introductory)

Raven, P., R. Evert, and S. Eichhorn. 1992. *Biology of plants.* New York: Worth. (Intermediate)

Simpson, B. and M. Conner-Ogorzaly. 1986. *Economic botany: Plants in our world.* New York: McGraw-Hill.

The Living Plant:
Circulation and Transport

STEPS
TO
DISCOVERY
Exploring the Plant's Circulatory System

XYLEM: WATER AND MINERAL TRANSPORT

Absorption of Water and Minerals

Transpiration

PHLOEM: FOOD TRANSPORT

Transport of Organic Substances

Translocation Through the Phloem

Pressure-Flow Mechanism For Phloem Transport

BIOLINE

Mycorrhizae and Our Fragile Deserts

THE HUMAN PERSPECTIVE

Leaf Nodules and World Hunger: An Unforeseen Connection

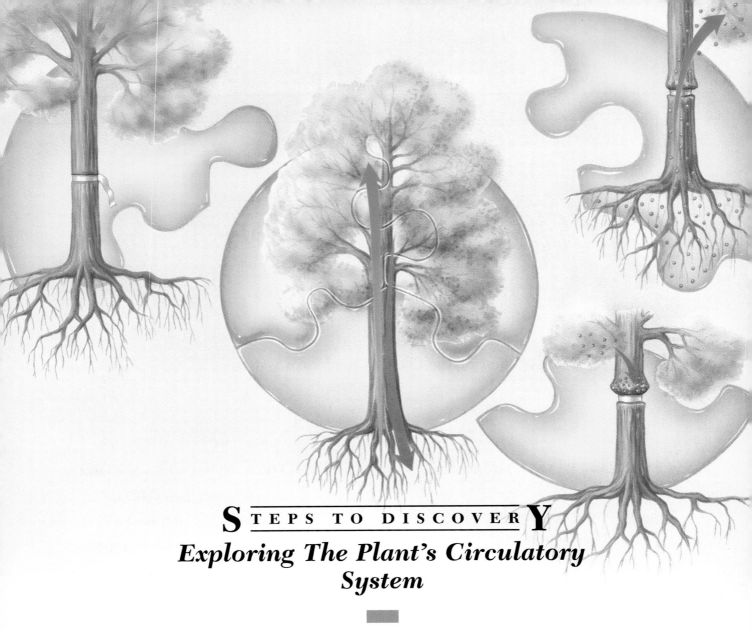

Exploring The Plant's Circulatory System

*F*or 3 centuries, biologists have known that blood surges through the human body through vessels. In 1661, Marcello Malpighi, a professor of medicine at the University of Bologna, described the network of the smallest of these vessels, the capillaries. But Malpighi's interest extended beyond the circulatory system of humans; he was curious about how materials moved through the bodies of *all* organisms, both animals and plants.

Malpighi suspected that, like humans, larger plants must have some type of a circulatory system for carrying materials from one part to another; otherwise, how could materials move across long distances, such as between deep roots and the leaves on the tips of branches? Unlike animal research, little microscopic work had been done on plant tissues, so very little was known about how materials could be transported through a plant. Malpighi decided to test whether the bark of a tree contained the plant's circulatory system. He removed a complete strip of bark from around a tree's trunk, a technique called *girdling*. After a period of time, he observed that the bark *above* the girdle swelled; eventually, the tree died. Because the bark swelled only above the girdle, Malpighi concluded that girdling had

By stripping off a ring of bark, scientists solved the puzzle of how materials flow through plant tissues—essential nutrients

blocked the downward movement of materials. The blockage caused these materials to accumulate above the girdle, which then caused the bark to swell; if the materials had been moving upward, the bark *below* the girdle would have swelled. The fact that the tree eventually died led Malpighi to conclude that the materials that move downward through the bark are essential to the plant's life. But just what were these materials?

In 1928, two British plant physiologists, T. Mason and E. Maskell, repeated Malpighi's girdling experiment. After girdling the tree's trunk, the researchers measured the rate of water-vapor loss from the tree's leaves. Mason and Maskell knew from earlier studies that most of the water that moves through a plant is lost by evaporation from the leaves. By measuring water-vapor loss from the leaves of a girdled plant, the researchers could determine whether the *main* pathway of water movement was through the bark or through tissues interior to the bark which were not destroyed by girdling. As did Malpighi, Mason and Maskell observed that the bark swelled only above the girdle. But their study determined that girdling did not change the rate of water-vapor loss from the plant's leaves. Based on these results, Mason and Maskell concluded that the bulk of water must move up through tissues interior to the bark. To explain the swelling above the girdle, the researchers concluded that other materials move *down* through the bark itself. By the 1920s, botanists had named the distinctive tissue within the bark the *phloem* and the tissue interior to the bark the *xylem*. Since most of the water that moves through a plant is released as water vapor from leaves, Mason and Maskel concluded that water circulates mostly upward through the xylem.

During the 1940s, a new technique was developed that enabled researchers to label organic compounds with radioactive carbon and, consequently, to trace the movement of these molecules through an organism. To trace the movement of sugars produced during photosynthesis, for example, plants were exposed to radioactive carbon dioxide, which was absorbed and incorporated into photosynthesized sugars. Scientists then traced the movement of the labeled sugars and discovered that they were concentrated completely in the phloem tissues (in the bark) rather than in the xylem. The predominant direction of movement was downward through the phloem from the leaves—the centers of photosynthesis—to all remaining plant tissues. Later studies using labeled compounds also revealed that virtually all organic molecules produced by a plant circulate through the phloem.

Nearly 300 years since Malpighi's experiments, scientists had finally proven that plants—like most animals—have a circulatory system. But in contrast to animals, water and minerals are transported in plants through tissues that are separate from those through which organic molecules are transported.

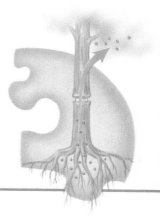

flow down through the bark, while water streams up through internal tissues.

Sometimes the English language gets in the way of our understanding biology. For example, on the one hand, being "animated" means being energetic and lively, like an animal. On the other hand, "vegetating" means being sluggish, dull, and passive —in other words, plantlike. But as we will discover in this and subsequent chapters, plants are not inactive at all. Hidden behind their immobility and stillness is not only the biochemical bustle of myriad metabolic reactions but also a number of surprisingly "animated" behaviors:

- *Competition* Some plants, such as mints and manzanitas, engage in battle with others, competing for space and environmental resources. And, like some animals, many of these plants use biochemical warfare tactics to defeat their rivals.

- *Movement* Although their responses are often slower than are those of most animals, some plants, like the Venus flytrap, move fast enough to catch living, motile prey.

- *Body temperature regulation* Like warm-blooded animals, some plants, such as a snow plant, store the heat produced from their own metabolism, melting their snowy habitats so that water is easily available.

- *Periods of dormancy* Like hibernating animals, many plants (sycamores, walnuts, and aspens, for example) become dormant during cold or dry periods; these plants monitor environmental conditions and resume growth and reproduction when favorable conditions return.

- *Circulation* Like many animals, most plants contain a network of cells that forms a "circulatory system" that distributes materials throughout the plant body. Unlike animals, however, plants have no muscles or heart to pump materials through their circulatory system. Instead, plants rely on the physical properties of water and concentration gradients to transport needed materials, the subject of this chapter.

▼ ▼ ▼

XYLEM: WATER AND MINERAL TRANSPORT

Plants take in and transport far greater amounts of water than do most animals. A single sunflower plant, for example, will take in more than 17 times as much water as you will in 1 day. Plants demand such large amounts of water because many lose up to 98 percent of the water they absorb to the air as water vapor. Water vapor is lost when plants expose moist leaf cells to the air to take up gaseous carbon dioxide for photosynthesis. When water is exposed to air, it evaporates; the drier the air, the faster the rate of evaporation.

The loss of water vapor from plant surfaces is called **transpiration.** Most water vapor is transpired from leaf surfaces through open stomates; the cuticle prevents evaporation from the surfaces of epidermal cells (page 366). Unless the plant's roots absorb enough water to replace the amount lost during transpiration, the plant will wilt and eventually die from dehydration. Recall from Chapter 8 that some plants have evolved biochemical and anatomic adaptations to hot, dry environments (C_4 and CAM synthesis) which help them reduce water-vapor loss. Transpiration will be discussed in more detail later in the chapter.

ABSORPTION OF WATER AND MINERALS

In the previous chapter, we learned that plants absorb water and minerals through root hairs found near root tips. Since water and minerals are usually widely dispersed in the soil, plants often develop extensive root systems with many root tips to mine the soil for these scant resources. In addition, more than 90 percent of vascular plants form root "partnerships" with soil fungi. In these fungus-root associations, called **mycorrhizae** (*myco* = fungus, *rhiza* = root), fungal filaments often form a tangled sheath around the root, with some filaments penetrating into the root's cortex cells (Figure 19-1a).

Mycorrhizal associations increase a plant's ability to extract water and minerals from soil because fungal filaments extend further into the surrounding soil than does the plant's root system. Some mycorrhizal alliances are so critical that young seedlings that fail to develop mycorrhizae perish from dehydration or nutrient deficiency (see Bioline: Mycorrhizae and Our Fragile Deserts). Furthermore, experiments have shown that plants for whom mycorrhizae are not a matter of life and death (citrus, some pines, and ornamental plants, such as eucalyptus and sugar maple) grow more efficiently with mycorrhizae than without them. Mycorrhizal partnerships are mutually beneficial: The plant gains greater supplies of water and minerals than it could absorb on its own, and the heterotrophic fungus absorbs food stored in the plant's root. Some studies reveal that plants with mycorrhizae may also be less susceptible to certain diseases caused by soil-borne pathogens. Furthermore, there is evidence that mycorrhizae enable some plants to grow better in soils that could not otherwise support favorable growth.

Mycorrhizal hyphae can extend up to 8 meters (25 feet) out from the root. Sometimes the same fungus penetrates the roots of different plants, forming linkages through which materials are exchanged, even among unrelated plants. In addition to water and minerals, carbohydrates and hormones can be exchanged in this manner. Some field biologists are currently investigating the extent of mycor-

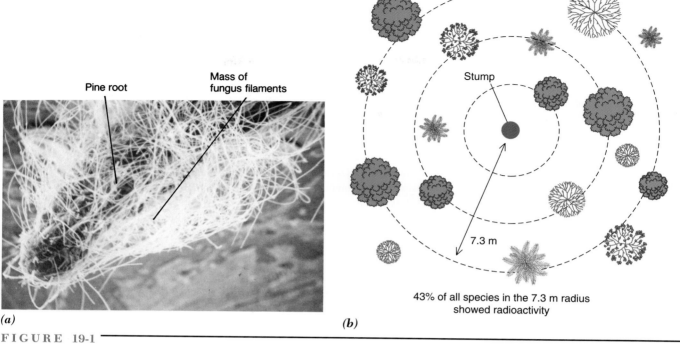

Pine root

Mass of
fungus filaments

Stump

7.3 m

43% of all species in the 7.3 m radius
showed radioactivity

(a) *(b)*

FIGURE 19-1

Mycorrhizae: An intimate association with mutual benefits. The filaments of a soil fungus join with the roots of plants to form mycorrhizae. The fungus absorbs some of the food (carbohydrates, amino acids) stored in the root, and the plant receives water and minerals absorbed by widespread fungal filaments. *(a)* In most mycorrhizae, a mass of fungal filaments forms a tangled sheath around the root tip, with some of the filaments penetrating root cells. *(b)* Scientists discovered that the roots of many plants in a North Carolina forest were interlinked by mycorrhizal filaments that extended from plant to plant. In this study, researchers added radioactive calcium and phosphorus to a freshly cut stump of a red maple tree. The radioactive materials quickly spread through 43 percent of the species within a radius of 7.3 meters, indicating that, because of mycorrhizae linkages, the roots of these forest plants form a single, widespread functional unit for absorbing water and minerals over very broad areas.

rhizal interconnections in ecosystems. For example, F. Woods and K. Brock added radioactive calcium and phosphorus to a freshly cut maple tree stump; within 8 days, 19 other types of trees and shrubs within a radius of 7.3 meters contained the labeled elements (Figure 19-1b). Woods and Brock suggested that because mycorrhizae interconnect root systems, the root mass of a forest may act as a single functional unit. This possibility is considered further in Chapter 41.

Avenues of Water Uptake

Water enters a plant's roots by osmosis, whereas dissolved minerals are actively transported into root cells through the cell membrane. Once inside the root, water travels through the root cortex in one of two ways: along cell walls or from cell to cell via connecting passages called **plasmodesmata** (Figure 19-2). Once water reaches the endodermis, however, aligned, waxy Casparian strips prevent further movement of water between or along cell walls. Water then

moves through the endodermal cells as minerals are actively transported into interior cells; that is, water flows along an osmotic gradient. To help visualize this phenomenon, consider what happens when you place a limp celery stalk in a bowl of water to make it crisp. Water from the bowl flows into the celery stalk along an osmotic gradient; that is, the concentration of solutes in the stalk is greater than that in the bowl so water flows into the stalk. Water continues to follow the inbound path of actively transported minerals into a conducting cell in the xylem. Once inside the xylem, water and minerals become part of a stream that conducts them through the plant.

Mineral Absorption and Requirements for Normal Growth

In the nineteenth century, two independent researchers, J. Sachs and W. Knop, grew plants by immersing their roots in a solution of inorganic salts rather than soil (a technique now referred to as **hydroponics**). The studies proved for

◁ B I O L I N E ▷
Mycorrhizae and Our Fragile Deserts

Hot, dry, and desolate. That is the image most of us have of the desert. In fact, *Webster's New World Dictionary* defines a desert as "arid, barren land . . . incapable of supporting any considerable population without an artificial water supply." Although this description may aptly characterize some areas of the Sahara Desert in Africa, it certainly does not describe the deserts of North America. In the Colorado Desert of southeastern California, for example, biologists have identified more than 1,200 plant species, 200 species of vertebrate animals, and numerous insects and invertebrates.

Some people believe that because desert plants and animals can withstand unusually harsh conditions, they are "tough" enough to withstand virtually any kind of abuse, from hordes of off-road vehicles and motorcycles to the many disturbances incurred from the construction of roads and power-transmission lines. Unfortunately, this perception of "toughness" is a false one. In reality, the desert is easily damaged and is very slow to recover from even small disturbances; 1,000-year-old Indian trails and campsites are still clearly visible, proving that virtually no natural recovery has taken place after thousands of years.

The California deserts are a valuable recreational, industrial, and biological re-

source. More than 100 of the plant and animal species found in these deserts are listed as rare, endangered, or threatened with extinction. With continued fast-paced development and heavy abuse of the deserts, these species cannot be saved without care, habitat protection, and efforts to rehabilitate damaged habitats.

In 1970, a local electrical company built a power-transmission line in a portion of the Mojave Desert in California, cutting through one of the few remaining populations of the Mojave ground squirrel, an endangered species. The wide power line road and path of disturbance destroyed the surrounding plants, preventing the ground squirrels on one side of the line from crossing over and breeding with squirrels on the other side. (Ground squirrels scamper from plant to plant not only to find food but also to hide from predators. When a ground squirrel reaches an area without plants, it turns around.) This segregation quickly lowered the squirrels' reproduction rate, causing the number of Mojave ground squirrels to plummet. In an effort to reverse this trend, field biologists tried to reseed the disturbed area with desert plants a number of times, but all efforts failed. Even when seeds and seedlings of the desert plants were repeatedly planted and irrigated, none survived more than a few weeks. Investigators eventually discov-

ered that the damaged soils lacked the fungus that naturally forms mycorrhizae with desert plant roots. Without mycorrhizae, the young plants were incapable of forming a root system large enough to gather sufficient water supplies, and they quickly died of dehydration. When the researchers added mycorrhizal fungi to the disturbed soils, however, many of the young plants lived. Unfortunately, the costs of injecting several square miles of disturbed soil with mycorrhizal fungi is so astronomical that the remaining population of Mojave ground squirrels will simply have to survive on its own.

A number of important lessons were learned from the Mojave ground squirrel studies. The electrical company vowed not to scrape the plants and desert soil to one side when constructing a power line, removing critical habitats and essential microorganisms (like the mycorrhizal fungus) in the soil. The company also pledged to adopt new construction practices that cause fewer environmental disturbances. And, perhaps most importantly, the power company agreed to be much more careful in the future in planning new power line corridors or construction sites in order to avoid tampering with the habitats of any endangered species.

the first time that plants could receive all their needs from sunlight and inorganic elements. Since that time, researchers have used hydroponics to grow plants in solutions in which only one element is missing at a time. From these studies, botanists have identified 16 **essential nutrients**—elements that are absolutely necessary for plants to complete their life cycle and for normal growth and development to occur (Table 19-1). Nine of these essential nutrients are needed in very large amounts; as such, they are called **macronutrients.** The seven remaining **micronu-**

trients are needed only in small amounts. Among their many roles in plant metabolism, essential nutrients are used to construct organic compounds, to activate enzymes, and to contribute to turgor pressure in plant cells. A plant deficient in an essential nutrient displays characteristic symptoms, some of which are described in Table 19-1.

Minerals are actively transported into roots, even when mineral concentration is lower in the root than it is in the soil or surrounding solution. Once inside the root, dissolved minerals are actively transported through cortical paren-

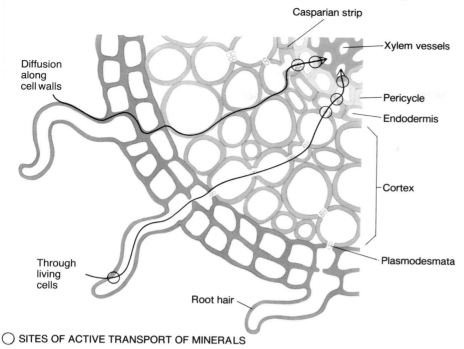

Casparian strip

Xylem vessels

Diffusion along cell walls

Pericycle

Endodermis

Cortex

Plasmodesmata

Through living cells

Root hair

○ SITES OF ACTIVE TRANSPORT OF MINERALS

FIGURE 19-2
Roadmap to the xylem. Water enters roots by osmosis, whereas most dissolved mineral ions are actively transported into root cells. Water moves either through living cortex cells or along cortex cell walls but is prevented from moving between endodermal cells by adjoining, waxy Casparian strips. Minerals are actively transported through the endodermis, the pericycle, the procambium, and into xylem conducting cells, creating an osmotic gradient. Water flows along this gradient into the xylem.

chyma cells, the endodermis, the pericycle, and, finally, into the xylem (Figure 19-2). Because of this one-way active transport path, minerals become concentrated in the xylem.

As minerals travel through the xylem, some diffuse out into surrounding root and stem tissues; the rest continue to flow within the xylem stream to the leaves. Minerals not used in the leaf are discharged into the descending phloem stream. Some minerals travel back to the root, where they are either used by root cells or reenter the ascending xylem stream.

▮▶ Some plants, particularly legumes (peas, beans, clover, alfalfa) form **root nodules**—swellings that house nitrogen-fixing (capable of converting nitrogen gas to ammonia or nitrate) *Rhizobium* bacteria (Figure 19-3). As in the case of mycorrhizal associations, both organisms benefit from this association: The plant receives a supply of nitrogen (an

FIGURE 19-3
Root nodules. Legumes, such as this snow pea *(a)* commonly form root nodules, *(b)*. The lumplike root nodules house nitrogen-fixing bacteria. The bacteria benefit from this alliance by harvesting some of the plant's carbohydrate stores, and the plant benefits by receiving nitrogen from the bacteria in a form it can use.

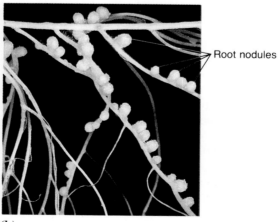

Root nodules

(a)   *(b)*

TABLE 19-1

PRIMARY USES OF THE 16 ESSENTIAL PLANT NUTRIENTS

Nutrient	Function	Percentage Dry Weight[a]	Deficiency Symptoms
Macronutrient			
Carbon (C)	Major component of organic molecules	45.0	Severely impaired growth
Oxygen (O)	Major component of organic molecules	45.0	Severely impaired growth
Hydrogen (H)	Major component of organic molecules	6.0	Severely impaired growth
Nitrogen (N)	Component of amino acids, nucleic acids, chlorophyll, and coenzymes	1.0–4.0	Yellowing of older leaves; plant spindly
Potassium (K)	Component of proteins and enzymes; helps regulate opening and closing of stomata	1.0	Mottled, yellow older leaves with die back at tips and margins
Calcium (Ca)	Component of middle lamella; enzyme cofactor; influences cell permeability	0.5	Youngest leaves die back at tip and margins; deformed leaves
Phosphorus (P)	Component of ATP, ADP, proteins, nucleic acids, and phospholipids	0.2	Youngest leaves marked with purple
Magnesium (Mg)	Component of chlorophyll; enzyme activator; amino acid and vitamin formation	0.2	Yellowing between leaf veins of older leaves; leaves may turn orange
Sulfur (S)	Component of proteins and coenzyme A	0.1	Yellowing of young leaves; veins bright red; stunted
Micronutrient			
Iron (Fe)	Catalyst for chlorophyll synthesis; component of cytochromes	0.01	Yellowing between veins of youngest leaves
Chlorine (Cl)	Osmosis and photosynthesis	0.01	Wilting of leaf tips; yellow leaves with bronze color
Copper (Cu)	Enzyme activator and component; chlorophyll synthesis	0.0006	Dark green leaves with black spots at tips and margins
Manganese (Mn)	Enzyme activator; component of chlorophyll	0.005	Black spots on veins in youngest leaves
Zinc (Zn)	Enzyme activator; influences synthesis of hormone, chloroplasts, and starch	0.002	Reduced internodes; leaf margins puckered
Molybdenum (Mo)	Essential for nitrogen fixation	0.0001	Yellowing between veins of older leaves with black spots
Boron (B)	Important to flowering, fruiting, cell division, water relations, hormone movement, and nitrogen metabolism	0.002	Black spots on young leaves and buds

[a] Percentage Dry Weight: Proportion of weight after all water has been removed.
Note: To help remember the macronutrients, use the phrase "C. HOPK(*i*)NS Car is an MG";
to help remember the micronutrients, use the phrase "*A Festive MoB comes in* (= CuMnZn) C*lapping*"

essential macronutrient), and the bacteria receive food, water, and living space. Mutually beneficial interactions like mycorrhizae and root nodules impart a survival advantage to both participants; as a result, such relationships are a common outcome of evolution even between very different kinds of organisms (this principle of coevolution is discussed in Chapter 34).

Nitrogen can mean the difference between success or disaster of agricultural crops. Nitrogen is a key element in many vital organic compounds, including nucleic acids and proteins; of the nine macronutrients, only carbon, oxygen,

and hydrogen are needed in greater quantities than is nitrogen (Table 19-1). Plants absorb nitrogen only in the form of nitrate (NO_3) or ammonia (NH_3) but not as nitrogen gas (N_2), the form of nitrogen found in our atmosphere.

⟲ As we mentioned in the previous paragraph, nitrogen gas is converted to ammonia or nitrate during **nitrogen fixation,** 90 percent of which is accomplished by nitrogen-fixing bacteria (the other 10 percent is performed by lightning). Nitrogen-fixing bacteria are found in the soil or in root nodules, where the plant receives a direct supply of fixed nitrogen. Root-nodule bacteria fix more nitrogen than the plant can actually use. Excess ammonia is then released into the soil, a natural means of enriching croplands. In fact, many farmers have discovered that they can improve productivity by rotating crops every other year—one year they plant clover or alfalfa, which produces root nodules that enrich the soil with nitrogen; the next year they plant corn, a plant that does not form root nodules. The corn grows vigorously in the nitrogen-rich soil. If corn were planted year after year, it would quickly deplete nitrogen supplies, and productivity would plummet.

Root Pressure and Guttation

As root cells pump minerals into the xylem by active transport, the concentration of dissolved particles increases, compared to the concentration in surrounding cells. This increase causes water to be drawn into the xylem by osmosis. The continuous influx of water can produce a positive **root pressure** that pushes water and dissolved minerals up a xylem column. Under certain conditions, root pressure builds high enough to force water and minerals completely out the tips of leaves, a process called **guttation** (Figure 19-4). Guttation only happens at night when there is no transpiration; as soon as transpiration begins in the morning, root pressure plunges and guttation stops. Guttation also occurs only in small plants growing in soils saturated with water. A drop of water on the tip of a leaf of a small plant in the early morning may be the result of either guttation or dew formation. A dew drop forms when water from the air condenses on a cool leaf tip, whereas a guttation drop forms as internal water is forced out of a plant.

Researchers have demonstrated that water cannot be pushed up a column to a height greater than 10 meters (about 30 feet, the height of a medium-size tree). Thus, root pressure is inadequate for moving water and minerals up plants that are taller than 10 meters and does not explain how water and minerals move through a plant during the daytime.

TRANSPIRATION

Why is it that when you put a cut flower in a vase of water, sometimes the flower still wilts? Doesn't the water in the vase simply replace the water being lost from leaves, stems, and petals through transpiration? It depends. In order for the water in the vase to replace transpired water, there must be a continuous connection between the water molecules in the xylem and the water in the vase; that is, water molecules in the xylem must be *cohesively bonded* to water molecules in the vase. If the flower was cut off the plant on a warm afternoon when transpiration was high, or if you waited some time before putting the flower in a vase, cohesive bonding between the xylem water and the water in the vase has been prevented.

If you could have observed the xylem under a microscope as you cut the flower from the plant, you would have seen that the tension of the water in the xylem channel from high transpiration caused the water to "snap up" the minute the stem was cut, similar to the way a stretched rubber band retracts in two directions if it is cut. The snapped water column creates an air pocket in the xylem channel. (Air pockets will also form in the xylem if you wait a while before putting the flower in a vase because the flower continues to lose water through transpiration.) When the flower is put in the vase, the air pocket separates the xylem water from the water in the vase, preventing cohesive bonding between them. The flower continues to lose water through transpiration. When it eventually uses up all the water in the xylem, the flower wilts and dies, even though there is plenty of water in the vase. The solution, as many florists will tell you, is to recut the stem of the flower under water before putting the flower in a vase. This removes the section of the stem with the air pocket and allows cohesive bonds to form between liquid water molecules in the xylem and the vase. In this case, as the flower loses water vapor, transpired water is replaced with the water from the xylem, which is replaced with water from the vase.

Transpiration Pull Mechanism for Transport

Plant physiologists have now shown that the primary mechanism for water and mineral translocation is **transpiration**

FIGURE 19-4 ———————

Forcing the issue. Glistening droplets of water and dissolved minerals are forced out of leaf tips by positive root pressure during guttation.

pull—a process triggered by water-vapor loss. Transpiration pull results from a chain of events that starts when leaves begin absorbing solar radiation in the morning. Sunlight heats the plant's leaves, causing more water to evaporate from moist cell walls. The evaporated water is immediately replaced with water from inside the cell, which is replaced with water from neighboring cells deeper in the leaf, which, in turn, is replaced with water in the xylem (Figure 19-5).

Recall from Chapter 4 that liquid water molecules are unique in their ability to cling to one another by strong cohesive bonds. In the xylem, liquid water molecules are cohesively bonded to one another, forming an unbroken water chain that extends all the way from the leaf vein, through the stem, and down into the root. Because of these strong bonds, as water molecules exit the xylem, they pull adjacent water molecules up, which, in turn, pull more

adjacent water molecules up, and so on, down the xylem column to the roots below. In other words, cohesive bonds pull water up the xylem. This process is transpiration pull.

☯ The constant pull between cohesively bonded water molecules creates a tension in the xylem that is transmitted down the entire water column. As transpiration increases during the day, the tension in the xylem builds, as more and more water molecules exit the xylem to replace the transpired water. But why don't the cohesive bonds simply snap from the tension, pulling liquid water molecules apart and breaking the continuous water column? The reason has to do with the strength of the cohesive bond *and* the narrow width of xylem tracheids and vessels (page 370). In narrow xylem cells, many water molecules are both cohesively bonded to other water molecules *and* adhesively bonded to xylem cell walls. This combination of cohesive and adhesive bonds enables a column of water to withstand tension ten times greater than that needed to pull water to the top of the tallest tree.

Peak periods of transpiration can build such tremendous tension in the xylem that the diameter of a large tree trunk may actually shrink! Because some water molecules in the xylem adhere to xylem cell walls as well as cohere to other water molecules, high tension pulls tracheid and vessel walls inward, enough to reduce the diameter of a tree measureably as the day progresses.

Regulating Transpiration Rates

The difference in concentration of water vapor between the air spaces *inside* a leaf and that found in the air *outside* a leaf determines the rate of transpiration; that is, the speed at which water vapor will be lost. Simply put, the greater the difference in concentration, the faster the rate of transpiration; the smaller the difference, the slower the rate (Figure 19-6). Since the air inside a leaf remains saturated with water vapor, there is almost always a difference between the air inside and the air outside a leaf. The only time there would not be a difference is when it is raining, foggy, or very humid outside.

Under normal daytime conditions, a leaf could transpire its entire water content in about 1 hour. Unless transpired water is replaced just as quickly, the leaf (and plant) will wilt and die. Thus, for plants growing in dry environments, the ability to control transpiration is a matter of life and death.

▶ Because shoot surfaces are covered with a waxy cuticle, less than 10 percent of transpired water is lost directly from the surface of epidermal cells, unless something damages the cuticle layer. One reason biologists are so concerned about acid rain is that it dissolves the cuticle, destroying one of the plant's natural adaptations for controlling water-vapor loss. Acid rain also kills soil bacteria (which supply plants with needed nutrients) and increases the rate of leaching of nutrients from the soil—two more reasons why it is so important to control acid rain.

Most water-vapor loss occurs through open stomates—the pores formed between adjacent guard cells. You will

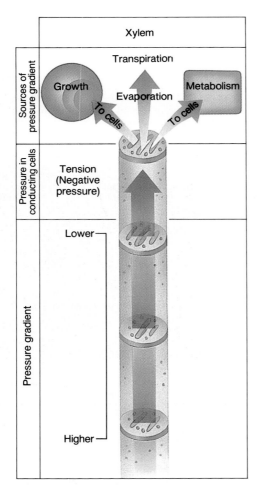

FIGURE 19-5
Xylem transport. Tension (negative pressure) builds in the xylem as water evaporates during transpiration or is used by cells for metabolism and growth. As water exits the xylem, cohesively bonded water molecules are pulled up.

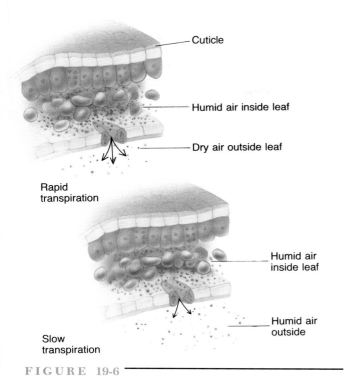

Rapid
transcription

Slow
transpiration

F I G U R E 19-6

The rate of transpiration.   The greater the difference in concentration of water-vapor molecules between the air spaces inside a leaf and that of the surrounding outside air, the greater the rate of transpiration. Water vapor is always high inside the leaf (unless it is wilted); changes in the humidity of the surrounding air affects the rate of water-vapor loss.

recall from Chapter 18 that stomates open to allow carbon dioxide to diffuse into the leaf for photosynthesis; when stomates open, water vapor usually flows out. Guard cells are able to change the size of a stomatal pore, thereby changing the rate of transpiration. The ability of guard cells to narrow, or even close, a stomatal pore, can save a plant from wilting or dying from dehydration. Plants have evolved other adaptations for regulating transpiration rate, three of which are illustrated in Figure 19-7.

Water and dissolved minerals are transported through the xylem. Since plants often transpire large quantities of water vapor from leaf surfaces, water and minerals flow up through the xylem, from the roots to the leaves. (See CTQ #2.)

PHLOEM: FOOD TRANSPORT

Substances are continually moving through the vascular tissues of a plant. The rate at which food moves through the phloem can reach speeds up to 100 cm/hour, or roughly 0.0013 miles/hour—about the speed of a tired snail. This is about one-fifth the rate at which water and minerals move through the xylem. Although phloem transport is slower than xylem transport, it is still much faster than cytoplasmic

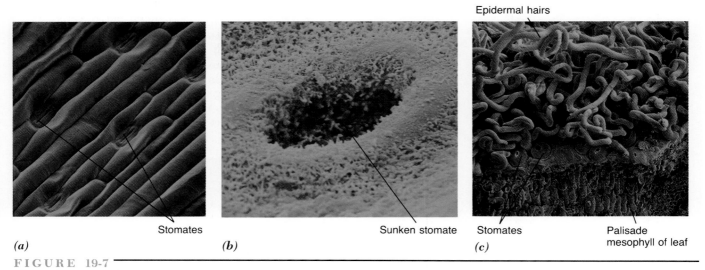

(a) Stomates   *(b)* Sunken stomate   *(c)* Stomates   Palisade mesophyll of leaf

F I G U R E 19-7

Plant adaptations that control transpiration rates.   *(a)* **Number of stomates.** Some plants have many thousands of stomates per each square centimeter of leaf surface. Plants adapted to growing in moist habitats have stomates on both the top and bottom surfaces; plants adapted to arid areas have fewer stomates, which are restricted to the cooler, lower surfaces where the vapor gradient between the inside and outside of the leaf is less. *(b)* **Position of stomates.** Recessed stomates help some plants adapted to arid habitats conserve water. This placement doesn't affect carbon dioxide diffusion but retards water-vapor loss by forming pockets of high humidity. Sunken stomates also protect the plant from the drying effects of wind. *(c)* **Epidermal hairs.** Dense epidermal hairs not only reflect light, reducing the leaf's heat load, but they also slow down transpiration by shielding stomate openings from drying winds.

streaming or diffusion of substances from cell to cell. Such rates would be too slow to distribute adequate supplies of nutrients throughout a plant.

TRANSPORT OF ORGANIC SUBSTANCES

It wasn't until 40 years ago that investigators actually identified which substances were being transported through the phloem. Before then, no matter how carefully researchers tried to extract the sap from phloem, the sieve tubes collapsed and plugged up, preventing researchers from getting a pure sample of phloem sap. Oddly enough, a common plant pest called *aphids* provided scientists with the "tool" they needed to collect pure phloem sap. Not only did the aphids enable botanists to verify what they had suspected all along—that food and other organic substances are transported through the phloem—but the pest also allowed researchers to measure how fast these substances move through phloem tubes.

▪▪▶ Aphids are insects that gather on the growing stem tips and leaves of many plants in early spring. Their hollow, needlelike mouthparts (stylets) enable aphids to bore directly into the phloem and feed on the plant's juices. The moment the aphid's stylet penetrates a sieve tube, its body balloons out, as sap gushes from the phloem transport cells; sometimes the pressure in the sieve tube is so high that it forces sap completely through the insect's digestive tract and out its anus, forming small droplets of honeydew (Fig-

ure 19-8*a*). But researchers did not want to analyze the contents of honeydew drops because the drops would be "contaminated" with materials from the aphids' digestive system. To collect "uncontaminated" phloem sap, investigators simply allowed aphids to tap into the phloem, then quickly anesthetized the insects and cut off their bodies, leaving the stylet still tapped into the phloem. The pure sap that exuded from the remaining stylets (Figure 19-8*b*) contained only materials flowing through the phloem—10 to 25 percent sugar (mainly sucrose), small amounts of amino acids, other nitrogen-containing substances, and plant hormones—all molecules synthesized by the plant. A few dissolved minerals were found in the sap as well. After nearly 350 years, the puzzle of what materials flow through the phloem was finally solved.

TRANSLOCATION THROUGH THE PHLOEM

Phloem sap always flow from *sources* to *sinks* (Figure 19-9*a*). Sources are those areas of a plant where sugars and other organic substances are synthesized (such as a photosynthesizing leaf cell) or stored (such as in parenchyma tissues). Sinks are plant tissues, such as actively growing meristems, developing fruits and seeds, or storage tissues, that use organic molecules. (Notice that storage tissues can be both sources and sinks. They are sources when they export their stored molecules, and they are sinks when they import molecules to build reserves.)

Stylet   Honeydew drop

(a)

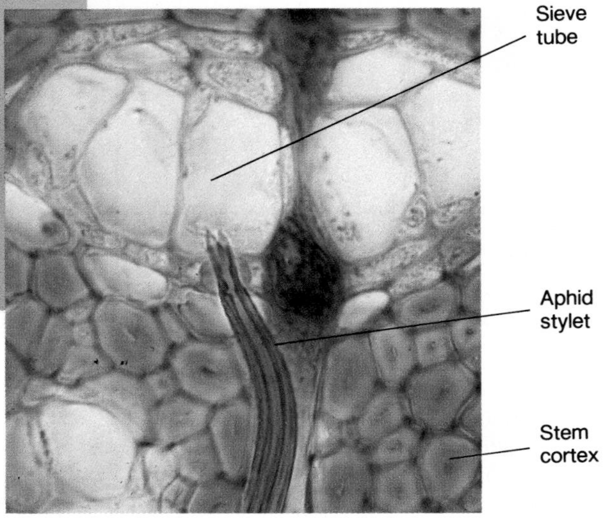

FIGURE 19-8

Aphids and phloem research. *(a)* The stylet mouthparts of aphids enable these insects to penetrate phloem cells without damaging delicate conducting cells. Once the phloem is tapped *(b)*, the aphid feeds on the plant's food supply. Pressure in the phloem forces phloem sap through the digestive tract of an overfed aphid, forming a honeydew drop. Aphids provided investigators with the perfect tool for studying phloem content and transport. When the aphid's body is removed, phloem sap flows out the insect's stylet for analysis.

Sieve tube

Aphid stylet

Stem cortex

(b)

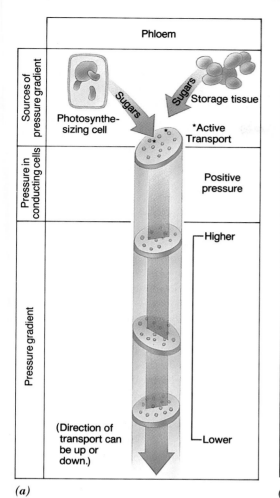

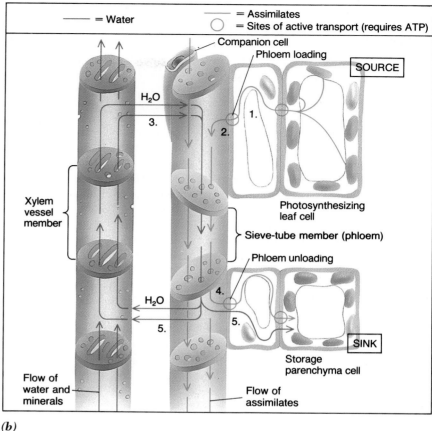

(a)   *(b)*

FIGURE 19-9

Phloem transport by pressure flow. *(a)* Sugars and organic molecules are actively loaded into sieve elements, creating an osmotic gradient that draws water into phloem transport cells, thereby increasing internal pressure at the point of active loading. At the other end of a phloem tube, pressure falls as molecules are actively unloaded and water flows out of the phloem. This pressure gradient pushes materials through the phloem from sources to sinks. *(b)* Phloem sap flows from sources to sinks. (1) A photosynthesizing leaf cell is a source because it produces sugars. (2) Sugars are actively transported into a sieve tube during phloem loading. (3) As sugar concentration increases in the sieve tube, water is drawn in from nearby xylem vessels. This influx of water raises the pressure in the sieve tube at the source. (4) At sinks (in this example, a storage parenchyma cell is a sink because it stores sugars), sugars are actively unloaded, lowering sugar concentration in the sieve tube and (5) causing water to exit the phloem tube and be recycled back to the xylem or drawn into surrounding cells. The pressure differences between sieve-tube members at sources (high pressure) and sinks (low pressure) initiate the flow of an assimilate stream.

Because phloem is so delicate and therefore difficult for researchers to work with, there is still some debate over exactly how materials are translocated through the phloem. Biologists have proposed and tested many hypotheses. So far, the greatest evidence supports the "pressure-flow hypothesis" of transport, which we discuss below. But the quest for understanding organic-molecule transport has not been abandoned. In fact, some studies suggest there may be more than one mechanism involved in phloem transport, illustrating the ongoing nature of scientific investigations.

PRESSURE-FLOW MECHANISM FOR PHLOEM TRANSPORT

To initiate **pressure flow,** higher pressures are created at sources when organic molecules are actively transported into sieve-tube members; this is called **phloem loading** (Figure 19-9*b*). Phloem loading increases the concentration of solutes inside the sieve tube and creates an osmotic gradient that causes water to flow in from nearby xylem cells. This process builds water pressure in the sieve tube at the source. At the sink end of the sieve tube, molecules are actively transported out of the phloem. This **phloem un-**

◁ THE HUMAN PERSPECTIVE ▷
Leaf Nodules and World Hunger: An Unforeseen Connection

Roots are not the only plant organ that form associations with microorganisms. Several hundred species of tropical plants develop bacteria leaf nodules, including an important crop plant, yams (*Dioscorea*). Because leaf nodules occur on less than 0.07 percent of plants, they have been mostly overlooked by investigators. In fact, it still isn't clear which bacteria are involved or whether these bacteria play a beneficial role for the plant. The work that has been done has made one thing certain: Leaf nodule bacteria, unlike root nodule bacteria, are not nitrogen-fixers. If leaf nodules are so rare and have no clear benefits, why study them at all?

The bacteria in leaf nodules are unique in that they are passed from one plant generation to the next. As the ovule develops into a seed, bacteria from the parent plant are transported to the embryo. When the embryo germinates, the transported bacteria immediately infect the leaves of the growing seedling. The ability of these plants to coexist permanently with resident leaf bacteria and to transfer bacteria from generation to generation may hold a key to solving some critical human problems. For example, yams are cultivated in many parts of the world for their edible tubers (fleshy roots). If leaf nodule bacteria on yams were genetically engineered to fix nitrogen or to deliver systemic pesticides, they could be transferred from one year's crop to the next, increasing yam production while reducing the need

for fertilizers or pesticides, neither of which are available in many regions that depend on yams as a food staple. Of course, this cannot happen until more basic research is done on plant leaf nodules.

Sometimes studying rare, seemingly unimportant problems may sound like a useless and wasteful pursuit. We often hear criticisms of government-supported research projects that, on the surface, seem like a waste of our tax dollars. But studying rare phenomena may also provide new scientific breakthroughs that can solve important human problems, such as world hunger, in the case of leaf nodule bacteria. Shutting the door on conducting research also shuts the door on reaching new understanding of the world around us.

---

loading reduces solute concentration in the phloem, causing water to flow out of the phloem and into the sink or back to the xylem. As a result, water pressure drops at a sink. Since all sources and sinks are connected by columns of sieve-tube members, higher water pressure at sources pushes the phloem sap to lower water pressures at sinks (Figure 19-9*b*). In other words, pressure differences cause organic molecules to flow from sources to sinks. The ATP that is needed for phloem loading and unloading is not produced by the sieve-tube members because, as we discussed in Chapter 18, sieve-tube members do not contain any mitochondria. The ATP needed for loading and unloading is provided by adjacent companion cells—cells that

also supply all the other metabolic needs of the sieve-tube member. Only by working together do sieve-tube members and companion cells create conditions necessary to power assimilate flow.

Organic substances that are synthesized by a plant move through the phloem. Since plants usually synthesize more sugar through photosynthesis than they do any other organic molecule, and since photosynthesis takes place predominantly in leaves, sugars and other organic substances generally flow downward through the phloem, from leaves to stems and roots. (See CTQ #5.)

REEXAMINING THE THEMES

Relationship between Form and Function

▨ The structure of vascular plant tissues enables them to translocate materials from one region of the plant to an-

other. Water and minerals flow through hollow, lifeless xylem cells. Their thick walls are able to withstand the high tensions that build during transpiration, the driving force behind the movement of water and minerals up to the top of the tallest trees. In the phloem, both sieve-tube members and companion cells create the pressure necessary to cause assimilates to flow from a source to a sink. Because of their interdependence, sieve-tube members and companion cells must join together. Finally, the juxtaposition of xylem and phloem in the same vascular bundle allows water to circulate back and forth between xylem and phloem; water is the basic medium for both mineral and assimilate flow.

Unity within Diversity

All multicellular organisms that grow larger than a few layers of cells face a similar problem: how to distribute essential materials throughout all of the cells in the body. Although higher plants and animals may be very different, they have all evolved very similar adaptations for solving the problem of moving materials from one region to another. Both have specialized vascular tissues that form a network of tubes through the body for distributing materials, and both rely on water as the primary transport medium.

Evolution and Adaptation

Water supply is probably the most important environmental factor that affects the growth, distribution, and health of any organism. Plants grow in many kinds of habitats, from those that are soaking wet all of the time (a tropical rainforest), to those that receive intense radiation and scant amounts of rainfall (a desert). Plants in dry, hot environments must control water-vapor losses; otherwise, the plant can easily lose more water than is available from its environment. Guard-cell behavior, stomate number and placement, the presence of epidermal hairs, reflective surfaces, and photosynthetic pathways are all adaptations plants have evolved for regulating water-vapor loss.

SYNOPSIS

A vascular plant's circulatory system includes two sets of conduits. Water and dissolved minerals are conducted through the xylem; organic substances synthesized by the plant are translocated through the phloem.

Water and dissolved minerals are absorbed through root hairs or mycorrhizae fungal filaments. Interconnecting Casparian strips on endodermal cells in the root direct water through the endodermis to xylem vascular tissues and prevent water from leaking out of roots as the soil dries.

Most of the water that is absorbed evaporates from leaf cell surfaces and exits through stomates during transpiration. As water vapor is lost during transpiration, water from the xylem replaces the lost water vapor. This builds a strong tension in the xylem that *pulls* water and minerals through the columns of hollow conducting cells. The faster the rate of transpiration, the stronger the tension, and the more accelerated the flow of the xylem stream.

Products of biochemical reactions (sugars from photosynthesis, hormones, amino acids, and so on) are transported through living sieve tubes of the phloem. Organic substances are actively loaded into the phloem at sources and are unloaded at sinks, where they are used or stored. Loading at sources causes water from nearby xylem to flow into the sieve tube, generating pressure that pushes the phloem sap to sinks by pressure flow.

The roots of many plants establish mutually beneficial partnerships with soil microorganisms. Many plants form mycorrhizae associations with soil fungi, greatly improving their ability to absorb water and minerals. Some plants form lump-shaped root nodules housing nitrogen-fixing bacteria that directly supply roots with usable nitrogen.

Key Terms

transpiration (p. 386)
mycorrhizae (p. 386)
plasmodesmata (p. 387)
hydroponics (p. 387)
essential nutrient (p. 388)

macronutrient (p. 388)
micronutrient (p. 388)
root nodule (p. 389)
nitrogen fixation (p. 390)
root pressure (p. 391)

guttation (p. 391)
transpiration pull (p. 391)
pressure flow (p. 395)
phloem loading (p. 395)
phloem unloading (p. 395)

Review Questions

1. Using an "X" for xylem and a "P" for phloem, indicate where each of the following is found:

   transpiration pull   micronutrients
   calcium   sugars
   pressure flow   organic molecules
   sieve tubes   vessel members

2. Where appropriate, put an "R" for root nodule or an "M" for mycorrhize.

   _____ a. root and fungus partnership
   _____ b. supplies nitrogen to plants
   _____ c. root and bacteria partnership
   _____ d. creates large water-absorption surface area
   _____ e. is mutually beneficial

3. How can you determine whether a drop of water on the tip of a leaf is dew (condensation from the air) or water that has been forced out by guttation?

4. List the essential nutrients that are:
   a. components of plant organic molecules
   b. needed for chlorophyll formation
   c. used to construct plant cell walls

5. Why are roots considered one of the main "sinks" in an actively growing plant? List three sources and three sinks in a typical plant.

6. What happens to the water in sap that flows down a stem and into the root, the site of water and mineral absorption?

7. Why is flow rate of water and minerals through the xylem several times faster than is the flow rate of phloem sap?

Critical Thinking Questions

1. Why does a girdled tree eventually die? Why doesn't the tree die immediately?

2. (*a*). The table below shows the number of stomates for several species of land plants. What is the adaptive value of the differences between the numbers of stomates on the upper and lower leaf surfaces?

	Number of Stomates per Square NM	
Plant Species	*Upper Surface*	*Lower Surface*
sunflower	175	325
cottonwood	89	132
olive	0	625
birch	0	237

   (*b*). In some plants, stomates are found only in the bottom depressions of the epidermis. What advantage would this have?

3. Most plants will die if the water content of their cells drops below 60 percent. Discuss some of the possible causes of death from dehydration.

4. Transpiration rate is affected by many environmental conditions, including temperature, wind, light, relative humidity, and water availability in the soil. For the following environments, rank each of these factors, from most important to least important, in *increasing* the rate of transpiration: tropical rainforest, desert, and arctic tundra (refer to Chapter 40 for descriptions). Justify your rankings.

5. Girdling showed that the area just above the removed bark would accumulate sugars and swell up and that the loss of water from leaves above the girdling was not affected. Explain how these observations led scientists to a correct description of the functions of the xylem and phloem.

Additional Readings

Allen, Michael F. 1991. *The ecology of mycorrhizae.* New York: Cambridge University Press. (Intermediate)

Brum, G., R. Boyd, and S. Carter. 1983. Recovery rates and rehabilitation of powerline corridors. In *Environmental effects of off-road vehicles*, R. Webb and H. Wilshire (eds.). New York: Springer-Verlag. (Intermediate)

Cronshaw, J. 1981. Phloem structure and function. *Ann. Rev. Plant Physiol.* 32:465–484. (Intermediate)

Raven, P., R. Evert, and S. Eichhorn. 1992. *Biology of plants.* New York: Worth. (Introductory)

Taiz, L., and E. Zeiger. 1991. *Plant physiology.* Reading, MA: Benjamin/Cummings. (Advanced)

CHAPTER
◀ 20 ▶

Sexual Reproduction of Flowering Plants

STEPS
TO
DISCOVERY
What Triggers Flowering?

FLOWER STRUCTURE AND POLLINATION

Flower Diversity

FORMATION OF GAMETES

Production of Female Gametes
Production of Male Gametes

FERTILIZATION AND DEVELOPMENT

Embryo Development
Seed Development
Fruit Development

FRUIT AND SEED DISPERSAL

GERMINATION AND SEEDLING DEVELOPMENT

Seed Viability and Seed Dormancy
Seedling Development
Seedling Survival and Growth to Maturity

ASEXUAL REPRODUCTION

AGRICULTURAL APPLICATIONS

BIOLINE
The Odd Couples: Bizarre Flowers and Their Pollinators

THE HUMAN PERSPECTIVE
The Fruits of Civilization

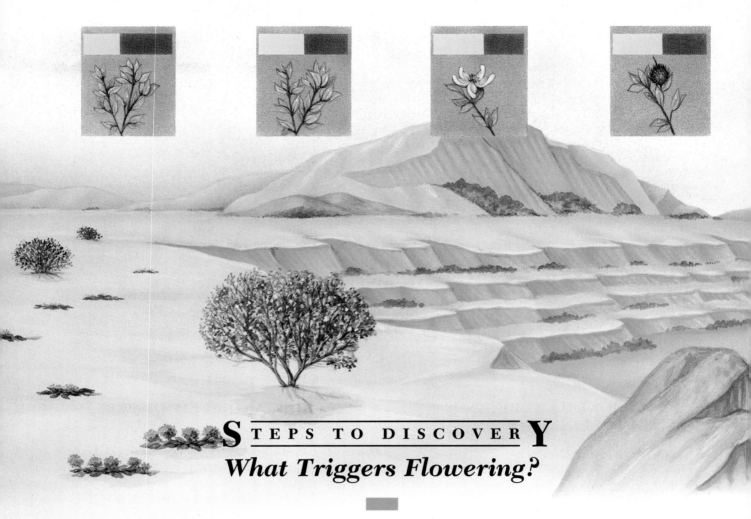

STEPS TO DISCOVERY
What Triggers Flowering?

During springtime, plants in the Mojave Desert of California begin to flourish following the cold winter season and brief, warm spring rains. After a period of rapid vegetative growth, during which plants produce new stems, leaves, and roots, the plants promptly shift to reproductive growth, triggering the development of a profusion of desert flowers. What activates this shift from vegetative to reproductive growth? Oddly enough, this dramatic transition is not set in motion by plant stems, where the flowers are produced. Instead, researchers have discovered that the internal trigger for flowering originates elsewhere in the plant.

By the 1930s, botanists had clearly demonstrated that, in many plants, flowering is regulated by seasonal changes in the photoperiod—the relative length of day to night during a 24-hour period. For example, plants such as spinach are called *long-day plants* because they flower only after being exposed to the lengthening days of spring. Conversely, other plants, like chrysanthemums, bloom only after exposure to the progressively shorter days of autumn; thus, they are called *short-day plants*. (*Day-neutral plants* are not affected by photoperiod.) Although it was easy for botanists to document that changes in photoperiod induced flowering, it was not as easy to reveal how plants were able to detect photoperiod changes or to determine which part of the plant actually detected the change.

Part of the puzzle was solved in 1938 through the work

Increasing daylength (the yellow bar in the boxes) triggers the leaves of desert creosote bush and sand verbena to synthesize a

of two botanists, Karl Hamner and James Bonner. The scientists reasoned that because leaves are well adapted to absorbing light energy for photosynthesis, perhaps they are also adapted to monitoring photoperiod. Hamner and Bonner chose a short-day plant, the cocklebur (*Xanthium*), to test their hypothesis because this plant blooms within only 2 weeks following exposure to daylight. They exposed only a single leaf of the plant to the correct photoperiod, while the rest of the plant's stems and leaves remained exposed to longer periods of light. By the end of the experiment, the entire plant had bloomed, even though only one leaf had been exposed to the proper day length. Hamner and Bonner concluded that the flowering "signal" had to have originated in the plant's leaves and that some impulse or transmissible substance had acted on the plant's stems to promote a shift from vegetative bud development to flower bud development. But whether the trigger was an impulse or a substance could not be determined from these studies.

Part of the answer on the nature of the trigger came from research done in 1958 by Jan A. Zeevaart in the Netherlands. Zeevaart removed the leaves of a short-day plant, exposed the detached leaves to the correct photoperiod, and then grafted a single leaf onto short-day plants that had been maintained on long days. A control group of short-day plants received no grafts and were maintained on long days. None of the plants in the control group bloomed, whereas all the plants that had received leaf grafts produced flowers. Like Hamner and Bonner, Zeevaart concluded that the flowering signal originates in the leaf and that the signal must be a chemical that is transmitted from the leaf to the stem, where flowers develop.

In 1968, R. W. King, L. T. Evans, and I. F. Wardlaw, three Australian plant physiologists, devised an experiment to establish the speed at which the flowering signal travels from the leaves to the stem. The scientists exposed the leaves of a series of plants to the correct photoperiod and then removed all of the leaves from the plants, each of which had retained the exposed leaves for differing lengths of time. Plants that had retained their leaves for only a very brief time following exposure to the correct photoperiod failed to flower, indicating that a minimum length of time was needed for the trigger to travel to the stem. But how much time? Since the experiment was designed so that there was a succession of plants, each of which had retained leaves for increasing periods of time, the investigators were able to determine the shortest time necessary for the trigger to travel to the stem by identifying the plant that had retained leaves for the shortest period yet still bloomed.

The results of this study revealed that the flowering signal traveled at a rate equal to that reported by other researchers for materials traveling through the plant's phloem tissues, an average of 100 centimeters/hour. King, Evans, and Wardlaw deduced that the flowering signal was an organic chemical that travels through the phloem from the leaf to the stem (recall from Chapter 19 that all organic molecules are transported through the phloem). When this chemical reaches the stem, it triggers axillary buds to shift from vegetative growth to flower development. The exact nature of the chemical is still being investigated.

chemical that travels to stems where it induces flowering in the Spring.

*I*t could be a scene from a horror movie. A victim is lured to a pool by a powerful force and then knocked into the water. Although the victim frantically attempts to climb out, the sides of the pool are so high and slippery it is impossible for him to get a firm hold. Eventually, he discovers a narrow passageway in the back of the pool, squeezes through it, and escapes. But the ordeal is far from over.

As the victim emerges, he discovers that two bulging sacs have been attached to his back. As he tries to flee the scene, the victim is lured to yet another pool, unable to resist the same powerful force that enticed him to the first one. The events that follow are the same: The victim is dunked into the pool; he makes several unsuccessful attempts to climb out the sides; and he eventually escapes through a narrow exit. This time, however, the two sacs are gone.

This is not a scene from a movie but a real-life drama between a male euglossine bee (the "victim") and a bucket orchid flower, *Coryanthus leucocorys* (the "captor"). During its ordeal, the bee inadvertently transfers the orchid's male gametes from one flower to another. Like other flowering plants, *Coryanthes* houses male gametes in **pollen grains**—the contents of the bulging sacs attached to the bee's back. The bucket orchid is a marvelously complicated flower that baits and then captures male orchid bees by dunking them into a pool of water (Figure 20-1). When the bee escapes through a narrow channel in the back of the flower, bags of pollen grains are attached to or removed from the insect's back, transferring gametes between flowers.

Male bees are attracted to the bucket orchid not only because of its alluring fragrance but also because the flower produces a waxy substance the male bee uses to make its own sexy "perfume," which he uses to attract female euglossine bees. Thus, the sexual reproduction of both the orchid and the bee are inseparably intertwined; each organism depends on the other to form a critical link in a chain that leads to the sexual reproduction of its species.

▼ ▼ ▼

FLOWER STRUCTURE AND POLLINATION

The flower is the center of sexual activity for all flowering plants. Flowers come in an array of sizes, shapes, colors, color patterns, and arrangements, each representing an adaptation that promotes pollen transfer for sexual reproduction. The transfer of pollen grains between flowers is called **pollination.** Many plants, like *Coryanthes*, rely on insects for pollination; others are adapted to utilizing wind, water,

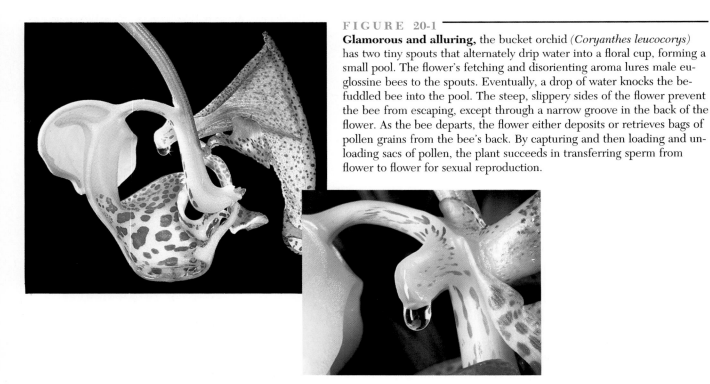

FIGURE 20-1

Glamorous and alluring, the bucket orchid (*Coryanthes leucocorys*) has two tiny spouts that alternately drip water into a floral cup, forming a small pool. The flower's fetching and disorienting aroma lures male euglossine bees to the spouts. Eventually, a drop of water knocks the befuddled bee into the pool. The steep, slippery sides of the flower prevent the bee from escaping, except through a narrow groove in the back of the flower. As the bee departs, the flower either deposits or retrieves bags of pollen grains from the bee's back. By capturing and then loading and unloading sacs of pollen, the plant succeeds in transferring sperm from flower to flower for sexual reproduction.

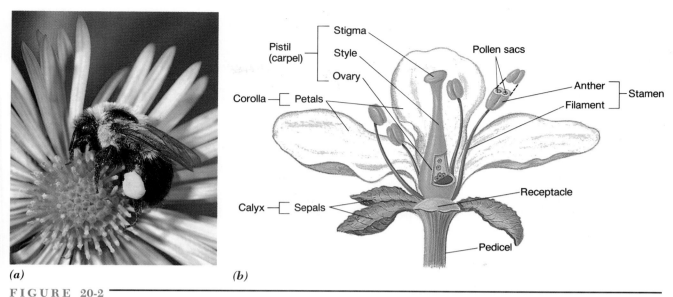

(a)   *(b)*

F I G U R E 20-2

Structures of a typical flower. *(a)* Flowers produce the plant's gametes for sexual reproduction and attract pollinators that transport male gametes from flower to flower. *(b)* The anthers of the stamen form pollen, which eventually produce the sperm gametes. The pistil is the female part of a flower. Within the pistil, the stigma collects pollen; the style provides a channel that directs the growth of a pollen tube to the egg (which is produced in the ovary) and, following fertilization, the ovary develops into a fruit that protects, nourishes, and helps disperse developing embryos contained in its seeds.

or other types of animals, such as birds and even a few mammals. Most animal-pollinated flowers produce food (usually nectar and pollen) that attracts the animal to the flower. However, a few plants employ some rather unusual tactics to accomplish pollination (see Bioline: The Odd Couples).

To understand how flowers accomplish sexual reproduction, we need to investigate the structure of the flower itself. A flower, such as the one illustrated in Figure 20-2, is a cluster of highly modified leaves attached to a shortened stem, called the **pedicel.** The flower parts emerge from a **receptacle**—the widened end of the pedicel which forms the flower base. A typical flower is made up of four groups of modified leaves—*sepals, petals, stamens,* and a *pistil*—each functioning in some way to achieve sexual reproduction.

Generally, the **sepals** surround and protect the flower bud as it develops. The collection of individual sepals make up the plant's calyx. To discourage hungry animals, the sepals of some flowers contain a distasteful latex, a milky white fluid that quickly deters animals from biting into the nutrient-rich flower bud. Bright or patterned **petals** distinguish the flower from its surroundings, catching the attention of pollen-transferring animals. The petals collectively make up the flower's corolla. Although some petal colors and markings are invisible to humans, they are readily spotted by the animals they attract (Figure 20-3). Petals are not

the only structures that attract animals; sometimes, the sepals, clumps of flowers, and even brightly colored leaves attract animals for pollination (Figure 20-4).

Although it is easy to see that sepals and petals are really modified leaves (they are leaf-shaped and have prominent veins), the stamens and pistil of the flower do not appear leaflike at all, but they too are modified leaves. These structures are the product of extensive evolutionary modifications of spore-producing leaves in the ancestors of flowering plants—modifications that enabled these leaves to manufacture and house the plant's gametes. The **stamens** are the flower's male reproductive structures; they produce pollen grains, which contain sperm. Most stamens are made up of a slender stalk—the **filament**—with a swollen end, called the **anther.** Pollen grains are produced inside the anther. When pollen matures, the anthers split open, releasing the pollen.

The **pistil** consists of one or more **carpels**—modified leaves that produce egg gametes. Each pistil has three parts:

* the **stigma,** or sticky area to which pollen adheres;

* the **style,** or tube that connects the stigma to the ovary; and

* the **ovary,** or enlarged bottom part of the pistil, where eggs are produced and embryos and seeds develop.

In plants adapted to utilizing wind for pollination, the stigma is often enlarged and feathery, creating a large sur-

◁ B I O L I N E ▷
The Odd Couples: Bizarre Flowers and Their Pollinators

Many plants enlist an army of "middle-men" of various sizes, shapes, and behavior for pollination. Usually, the plant entices pollinators by providing "rewards," such as nectar, pollen, oils, or plant tissue. But some plants seem to resort to evil and manipulative means to reach this end, as in the following cases:

CASE 1: THE IMPOSTOR

The dull-red, wrinkled flowers of the South African milkweed, *Stapelia gigan-*

tea, look and smell like putrefying flesh, deceiving female blowflies into laying her eggs on them. In doing so, the duped female blowfly carries pollen from flower to flower. But the floral hoax has grave consequences for the blowfly. All of the fly's maggots perish from lack of proper food because blowfly maggots can live only on animal flesh, not flower tissue—no matter how foul it is.

CASE 2: OPEN HOUSE

The elongated petal of the Dutchman's pipe (*Aristolochia clematitis*) is covered with millions of tiny grains of wax. Gaining a footing on such a surface is like trying to walk down a mountain covered with greased ball bearings. Even with hooks and suckers, the gnats and flies attracted to the flower cannot gain a foothold on this slick surface. As soon as an insect lands, it slips down the flower tube, past a forest of downward-pointing hairs, and finally

crashes into a cluster of stigmas (part of the female reproductive structure) at the bottom of the chamber. "Visitors" are prevented from leaving this hairy forest for at least 2 or 3 days. During that time, the stamens (the male reproductive structures) mature and shower the insects with pollen. Then the flower tilts over, and the retaining hairs shrivel. The insects walk out, usually only to be quickly lured into captivity by yet another Dutchman's pipe flower, completing the pollination cycle.

(a)

(b)

CASE 3: THE CASE OF MISTAKEN IDENTITY

The flowers of the orchid *Ophrys speculum* take advantage of the powerful sex drive and energetic nature of male wasps. To a male wasp, the orchid looks and smells like a female of his own species. The illusion is perfect; as the male wasp "copulates" with a number of orchids, it trans-ports pollen among them. The flower is just as successful in arousing the amorous attention of the male wasp as is the genuine female of his own species!

CASE 4: R.I.P.—MURDER AMONG THE LILIES

The South African waterlily attracts hundreds of eager hoverflies, bees, and beetles to its flowers, using an abundance of nutritious pollen as bait. The insects soon learn that waterlilies are a rich source of food. Lily flowers first go through a fe-male stage when no pollen is offered and then go into the male stage. But by the time pollen is offered, the mortal deed has al-ready been done, for hidden below the bountiful stamens float the remains of previous visitors.

The deceased visitors happened upon the lily flower on the first day the flower opened, a day in which the flower was in its female stage. In their feeding frenzy, the insects carrying pollen from other lily flow-ers in their male stage failed to notice there was no pollen in this flower. The immature stamens form a ring of slick columns surrounding a watery pool. Once the in-sects land and start crawling over the sta-mens in search of pollen, it is too late. Un-able to get a grip, the insects helplessly slide down the slippery stamens into the pool and drown. At night, the flower closes, and the pollen collected on the insects from visits to other lily flowers washes off the corpses and settles to the bottom of the pool, where the stigma is located. Pollina-tion is completed. By morning, the murky burial pool is hidden by a blanket of mature stamens, the flower now luring insects with a bounty of pollen. In its male stage, the lily flower is safe, but insects cannot tell the difference between flowers that offer pol-len and those that offer certain death, a fact on which the lily has staked its sexual re-production.

◀ FIGURE 20-3 ━━━━━━━━━━━

A bee's-eye view. Humans and bees see different wave-lengths of light, as these two photos of a marsh marigold (*Caltha palustris*) illustrate. *(a)* To humans, this marsh mari-gold is almost uniformly yellow. *(b)* To bees, the petals have distinct lines that converge in the dark center where the nectar (and stamens and pistils) are found. These nectar guides absorb ultraviolet (UV) wavelengths, while the remaining parts reflect UV wavelengths (humans do not see UV wavelengths, but bees do).

face that gleans pollen from the air (Figure 20-5*a*). In flow-ers that utilize animals for pollination, the stigma is shaped in such a way that it contacts the pollen-laden areas of the animal's body. For example, the stigma of the monkey flower depicted in Figure 20-5*b* remains wide open until a bee deposits pollen on it. The stigma then closes, reducing inbreeding by preventing pollen produced from the same flower from being deposited on the stigma surface. In most cases, pollen must physically and chemically fit the stigma before a pollen tube will grow down the style and into the ovary to deliver sperm to the egg. This "recognition" of pollen helps ensure that only sperm from members of the same species will fertilize the egg.

☀ Many flowers have **nectaries**—glands that secrete sugary nectar that attracts animals for pollination. Some animals collect both nectar and protein-rich pollen. As an animal laps up nectar and collects pollen, some of the pollen will adhere to its body. As the animal moves from flower to

(a)   *(b)*

(c)

FIGURE 20-4

Flower power. In most cases, only the petals of a flower are brightly colored to help attract animal pollinators, but some plant species do not produce flowers with petals or have very inconspicuous flowers. In these cases, other structures have become modified to attract pollinators. *(a)* Although they look like petals, flannel bushes *(Fremontedendron)* have brightly colored sepals that attract animals. *(b)* An African daisy is not a single flower but a miniature bouquet of tiny, densely packed flowers that forms a *composite*. Each of the outer *ray flowers* has long petals that attract animals. The petals of the inner, inconspicuous *disk flowers* are barely visible. *(c)* In poinsettias, it isn't the flower that attracts animal pollinators but a cluster of colored leaves located just below the inconspicuous flowers.

Stigma

(a)   *(b)*

FIGURE 20-5

Improving the odds. *(a)* Wind-pollinated plants, like this grass, produce tremendous quantities of pollen and form large feathery stigmas. Together, these characteristics improve the chances of pollination. *(b)* This two-lipped monkey flower stigma *(Mimulus)* "licks" pollen off the heads of entering bees and then quickly closes to prevent pollen from its own stamens from entering as the bee exits.

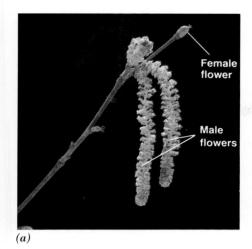

(a)

FIGURE 20-6
Separating the sexes. *(a)* Monoecious plants, such as this oak, have both male and female flowers (the male flowers are in a pendulous, or suspended, group, and the female flower is at the tip of the stem). Dioecious plants, such as meadow rue *(Thalictricum)*, produce male flowers on one plant *(b)* and female flowers on another *(c)*. Separate-sexed flowers reduce the chances of fertilization between sperm and eggs from the same individual, thereby increasing genetic variability.

(b)

(c)

flower, pollen is carried from the animal's body to the flower's stigma, completing pollination and setting the stage for fertilization.

FLOWER DIVERSITY

A flower that contains all four groups of modified leaves—sepals, petals, stamens, and pistil—is a **complete flower,** as opposed to an **incomplete flower,** which lacks one or more of these four basic parts. Roses and daisies are examples of complete flowers, whereas dogwood and palms produce incomplete flowers. Some plants produce flowers with only stamens (male flowers) or only pistils (female flowers). A flower that lacks either stamens or pistils, or both, is called an **imperfect flower,** as opposed to a **perfect flower,** which has both stamens and pistils. If separate male and female flowers are produced on one plant, the plant is said to be **monoecious** (*mono* = one, *ecious* = house) (Figure 20-6a), whereas **dioecious** (*di* = two) plants produce male and female flowers on separate individuals (Figure 20-6b). All monecious and dioecious plants produce imperfect flowers. The tassels of male flowers and the separate cobs of female flowers make corn a monoecious plant, whereas eggplant and other squashes are dioecious plants because

each individual produces either male or female flowers. Producing different-sexed flowers on different plants guarantees that eggs cannot be fertilized by sperm from the same individual, thus promoting genetic variability.

A flower is a multipurpose structure for sexual reproduction. The flower produces sexual gametes; facilitates pollination; and serves as the site of fertilization and embryo, seed, and fruit development. (See CTQ #1.)

FORMATION OF GAMETES

Gamete formation always begins with meiosis (Chapter 11). In plants, special diploid mother cells inside the anthers and ovaries of flowers divide by meiosis to form haploid *spores.* Mother cells in the anthers divide to form **microspores,** whereas mother cells in the ovary produce **megaspores.** (The prefixes "micro" and "mega" do not necessarily refer to the size of the spores but to the size of the gametes produced; the male sperm is smaller than the female egg.)

Before gametes are formed, both microspores and megaspores divide by mitosis to form a haploid, multicellular gametophyte, which produces the gametes for sexual reproduction. Megaspores develop into **embryo sacs,** the female gametophytes; microspores develop into pollen grains, the male gametophytes.

PRODUCTION OF FEMALE GAMETES

Female gametophytes form within **ovules**—round masses of cells that are attached by a thin stalk to the inner surfaces of the flower's ovary. The outer cells of the ovule form one or two protective layers (the integuments) that leave a small opening, the *micropyle*, through which the pollen tube grows, to deliver sperm.

As illustrated in Figure 20-7, the production of the egg gamete begins when a diploid mother cell inside the ovule undergoes meiosis. Three of the four resulting megaspores degenerate. The nucleus of the surviving megaspore divides three times by mitosis to produce the female gametophyte, which contains eight haploid nuclei. (The first mitotic division produces two nuclei; the second, four; and the third, eight.) The eight nuclei become partitioned into seven cells, one of which contains two nuclei; only one of the cells— that nearest the micropyle—becomes the egg. The cell that contains two nuclei is called an **endosperm mother cell,** which later proliferates into the **endosperm,** an important tissue that nourishes the embryo and the growing seedling.

PRODUCTION OF MALE GAMETES

As embryo sacs develop in the ovary, pollen grains form in the anthers of stamens (Figure 20-8). Mother cells in the anther divide by meiosis to produce four microspores, each of which divides once by mitosis, forming a two-celled pollen grain. One of the cells—the *generative cell*—is enclosed within the other—the *tube cell*; the generative cell will divide by mitosis to produce two sperm. Pollen grains are surrounded by a tough wall that protects the male gametophyte while being transported to a stigma. The design of the outer walls of some pollen grains helps them adhere to animals (Figure 20-9*a*) or remain airborne during pollination (Figure 20-9*b*).

The production of plant gametes begins with meiosis and the formation of four haploid spores. Each spore divides by mitosis to form a many-celled structure that produces either an egg or sperm gamete. (See CTQ #2.)

FERTILIZATION AND DEVELOPMENT

Even after pollination is completed, the plant's male and female gametes remain physically separated. The pollen grain lies on top of the stigma, while the egg is buried deep within the ovary. But when pollen lands on a stigma, the

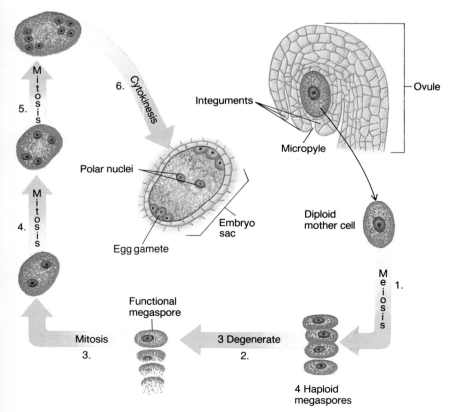

FIGURE 20-7
Formation of the female gametophyte is a six-step process that takes place inside an ovule. (1) A diploid mother cell divides by meiosis to produce four haploid megaspores. (2) Three megaspores degenerate, leaving a single functional megaspore. (3–5) The nucleus of the surviving megaspore divides by mitosis three times to form an eight-nucleate gametophyte. (6) Cytokinesis (page 206) partitions the nuclei into seven cells, producing the mature embryo sac with a single egg gamete.

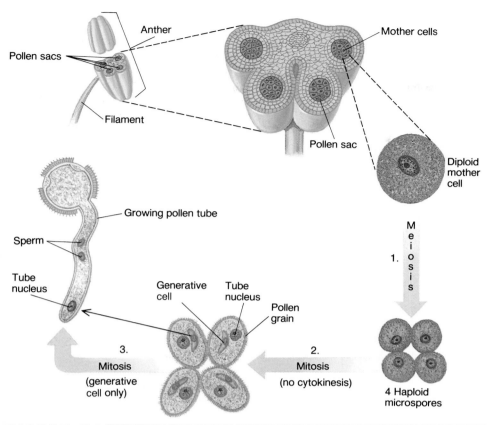

FIGURE 20-8

Formation of the male gametophyte is a three-step process that takes place inside the anthers of stamens. (1) A mother cell divides by meiosis to produce four haploid microspores. (2) Each microspore divides by mitosis to a pollen grain containing a tube cell and a generative cell. (3) When deposited on a stigma, the tube cell forms a pollen tube, while the generative cell divides to form two sperm.

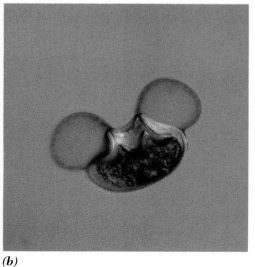

FIGURE 20-9

Pollen form and function. *(a)* Sculpturing of cotton pollen helps them adhere to insects. *(b)* Bladderlike wings of pine pollen *(Pinus nigra)* help keep them airborne.

(a)   *(b)*

tube cell of the microgametophyte forms a pollen tube that grows down through the stigma, style, and ovary tissues (Figure 20-10), eventually worming its way through the micropyle of an ovule. Two sperm are discharged; one fuses with the egg to produce a diploid zygote, while the other fuses with the diploid endosperm mother cell to form a triploid (3N) *primary endosperm cell*. Because both sperm from a pollen grain fuse with cells in the embryo sac, this is sometimes called "double fertilization," although technically there is only one fertilization—the fusion of egg and sperm to form the zygote.

Fertilization sparks dramatic transformations in the flower, including the following:

- the zygote develops into the embryo;
- the primary endosperm cell forms nourishing endosperm tissue;
- the ovule is transformed into a seed; and
- the ovary develops into a fruit.

☀ Together, the endosperm, seed, and fruit nourish and protect the embryo as it develops (Figure 20-11) and during its journey to a new habitat.

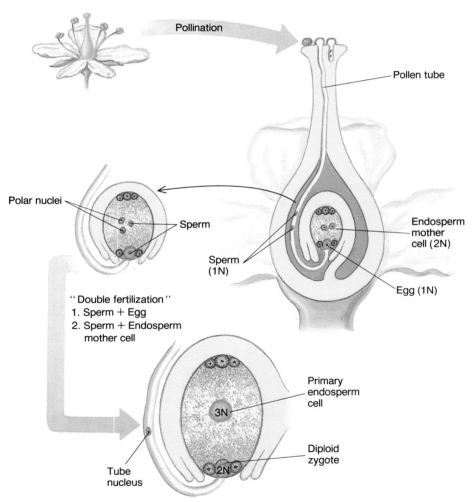

FIGURE 20-10

Pollination and fertilization. The beginnings of a new generation. Pollen is transferred from anthers to a stigma during pollination. The pollen grain grows a pollen tube down the style, into the ovary, and through the micropyle of an ovule. The generative cell divides to produce two sperm; one fertilizes the egg to form a zygote, the other fuses with the two nuclei of the endosperm mother cell to form a triploid primary endosperm cell.

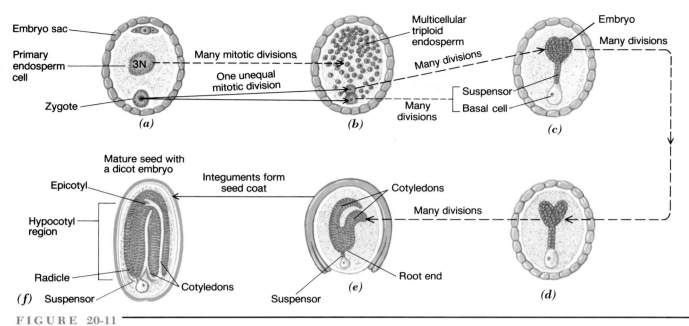

FIGURE 20-11

Development of a dicot embryo, *Capsella*. *(a)* A zygote and primary endosperm cell form following fertilization. *(b)* The triploid (3N) primary endosperm cell divides many times by mitosis to produce nutritive endosperm. The first mitotic division of the zygote produces different-sized daughter cells. *(c)* The smaller cell divides many times to produce the plant embryo. Divisions of the larger, basal cell produce the suspensor. *(d–f)* Continued mitotic divisions leads to the formation of a mature dicot embryo.

EMBRYO DEVELOPMENT

A newly formed zygote divides by mitosis to produce two different-sized daughter cells (Figure 20-11b). The smaller cell eventually grows into the plant embryo, while the larger, basal cell produces the *suspensor*—a column of cells that conducts nutrients to the growing embryo (Figure 20-11c).

The plant embryo is a relatively simple structure, consisting of only three parts: one or two cotyledons, an epi-

cotyl, and a hypocotyl. The **cotyledons,** or seed leaves, are the first structures to begin photosynthesis when the seedling emerges from the soil. In most dicots, the endosperm (food) is located within the cotyledons and is part of the embryo; in most monocots, the endosperm surrounds the embryo and is part of the seed. As their names indicate, dicot embryos have two cotyledons, whereas monocot embryos have only one (Figure 20-12). The **epicotyl** is the portion of the embryo above the cotyledons (*epi* = above, *cotyl* = cotyledon); it develops into the shoot system of a plant. The **hypocotyl** is the portion of the embryo below the cotyledons (*hypo* = below). At the tip of the hypocotyl is the **radicle,** which eventually develops into most or all of the plant's root system.

SEED DEVELOPMENT

Following fertilization, the ovule develops into a seed, which contains the embryo, endosperm, and a seed coat (Figure 20-13). The seed coat forms from the outer cell layers of the ovule. The integuments become tough and hard, protecting the embryo from abrasion or other dangers as it is dispersed to new habitats. Some seed coats have projections that help absorb water, attach the seed to the fur of an animal for dispersal, or protect the embryo from hungry animals.

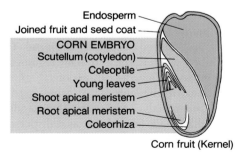

FIGURE 20-12

Corn (*Zea mays*) fruit (kernel) with mature embryo. A corn kernel is a combination of fruit and seed. The ovary of the pistil forms an outer "fruit + seed" coat. The single cotyledon digests and absorbs endosperm stored outside the embryo.

Bean fruit with seeds

BEAN SEED

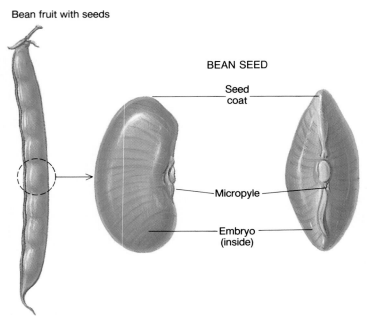

Seed coat

Micropyle

Embryo (inside)

FIGURE 20-13

Seed. The seed is a life-sustaining capsule that houses, nourishes, and protects the embryo after separation from its parent and during dispersal.

FRUIT DEVELOPMENT

With sexual reproduction completed, those parts of the flower that were involved in pollination (sepals, petals, stamens, stigma, and style) wither. As embryos and seeds develop, the ovary of the pistil matures into the *fruit*. Many animal species, including humans, rely on plant fruits for nutrition. In fact, fruits have had a profound effect on human life, perhaps even sparking the development of human civilization (see The Human Perspective: The Fruits of Civilization).

Whether a fruit is green or brightly colored, juicy or dry, soft or hard, all (or part) of it formed from a ripened ovary. Some so-called vegetables, such as cucumbers, tomatoes, squash, string beans, and grains, technically are fruits because they develop from an ovary. There are three categories of fruits (Figure 20-14)—simple, aggregate, and multiple—depending on whether a fruit develops from the ovary of one pistil of a single flower (**simple fruits**), from many pistils in a single flower (**aggregate fruits**), or from the pistils of a number of flowers (**multiple fruits**).

Flower structure and function change following pollination and fertilization. The parts of the plant involved in pollination (petals, sepals, nectaries, and stamens) wither as the pistil undergoes a dramatic transformation during embryo, seed, and fruit development. (See CTQ #3.)

FRUIT AND SEED DISPERSAL

▮▶ In just one flowering season, a single plant can produce hundreds, thousands, and even millions of seeds, each containing an embryo capable of growing into a new plant. If a parent plant simply dropped all of its seeds onto the ground directly beneath its stems and branches, the resulting seedlings would compete with each other and with the parent plant for limited space, light, water, and nutrients as they grew. Such intense competition would lead to widespread malnourishment and massive numbers of deaths. *Seed dispersal* reduces competition by separating seeds. Dispersal also scatters seeds to new areas where conditions may be more favorable for growth and development.

With such clear advantages, it is not surprising that natural selection has resulted in the proliferation of seeds and fruits that are adapted to enhance dispersal. The fruits of wild cucumber, witch hazel, and many legumes literally explode, flinging seeds in various directions (Figure 20-15). However, the majority of plants rely on some agent—wind, water currents, or animals—to disperse their seeds (Figure 20-16). Plants that utilize animals for seed dispersal recruit a broad range of species, from worms to bears. In plants that develop juicy fruits, the seeds are often consumed along with the fruit but pass undamaged through the animal's digestive system and are scattered along with the animal's feces. For example, the seeds of the giant Saguaro cactus are dispersed by birds that eat the fruit and seeds simulta-

Simple fruits

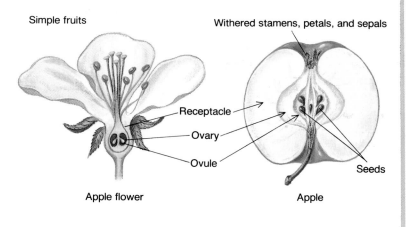

Withered stamens, petals, and sepals

Receptacle
Ovary
Ovule
Seeds

Apple flower
Apple

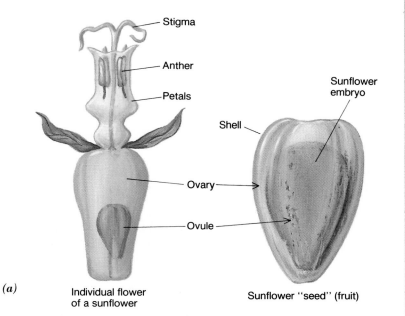

Stigma
Anther
Petals
Sunflower embryo
Shell
Ovary
Ovule

(a)
Individual flower of a sunflower
Sunflower ''seed'' (fruit)

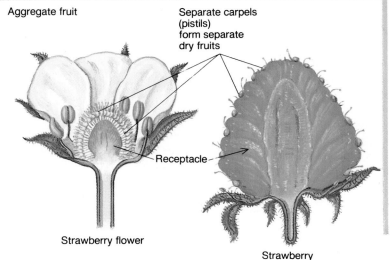

Aggregate fruit
Separate carpels (pistils) form separate dry fruits

Receptacle

Strawberry flower
Strawberry

(b)

Multiple fruits

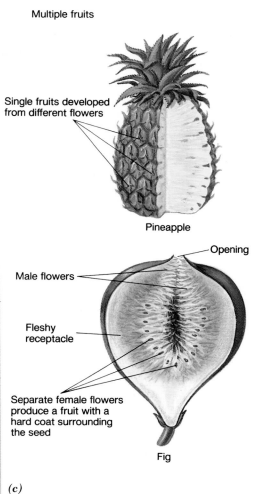

Single fruits developed from different flowers

Pineapple

Opening
Male flowers
Fleshy receptacle
Separate female flowers produce a fruit with a hard coat surrounding the seed

Fig

(c)

FIGURE 20-14

Development of three main fruit types. *(a)* Whether fleshy (apple) or dry (sunflower), simple fruits develop from one pistil. *(b)* Aggregate fruits, such as strawberries, are formed from separate pistils in the same flower. *(c)* Multiple fruits develop as the ovaries of several, often densely packed flowers mature together. Multiple fruits form a single fruit (such as a pineapple) or develop from many flowers that are physically joined by other flower tissues. A fig, for example, is an enlarged receptacle filled with many tiny simple fruits, each of which developed from a separate flower.

◁ THE HUMAN PERSPECTIVE ▷
The Fruits of Civilization

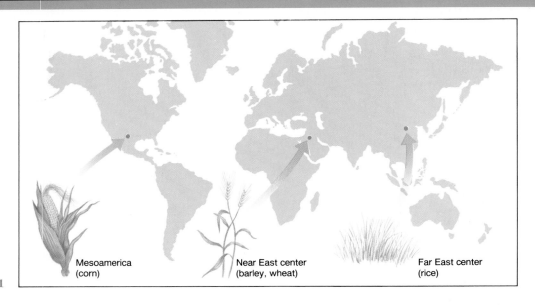

Mesoamerica
(corn)

Near East center
(barley, wheat)

Far East center
(rice)

FIGURE 1

As humans, our existence depends on fruits and seeds, but our dependency goes beyond the obvious apple, orange, or peanut butter sandwich. Flour for making bread is ground grain, the fruit of grasses; we feed domestic livestock fruits and grains for meat production. In fact, most of our food comes directly or indirectly from fruits and seeds.

The same was true for our early ancestors. Early humans traveled in nomadic tribes, forced to follow seasonal changes in plant growth in order to harvest enough food to feed their members. The perpetual search for food occupied most of the time and energy of every member of the tribe. There was no time for contemplation, no time or energy to develop art or science or even to begin the first scratchings of a written language. So how did these intellectual and civilized pursuits begin? Human civilization may owe its existence to one of the simplest kinds of fruits—grains.

Historians have identified three geographic regions where human civilization flourished and then radiated out to other areas of the world (Figure 1). Each of these centers of civilization had at least one naturally growing grain that likely provided the nutritional foundation for the development of civilization. The Near East Center had wheat and barley, the Far East Center had rice, and the Middle America Center had corn. Because of the abundance of these grasses and their high nutritional content, a few members of each tribe could gather, tend, and cultivate enough grain to feed the entire group. In addition, grass grains could be stored for long periods without spoiling, eliminating the need to travel from place to place to find a fresh supply of food. Some anthropologists speculate that as more and more members of the tribe were freed from the oppression of an endless, exhausting search for food, they had time to contemplate their surroundings and to develop the innovations that allowed civilization to progress. The development of plant and animal husbandry (production and care of domestic plants and animals) practices freed even more people to specialize in other activities. Rice terraces, such as these in the Phillipines (Figure 2) greatly expand food production in mountainous regions. Eventually, language, arts, and sciences began to flourish, leading to the social and technological advances we enjoy today.

FIGURE 2

FIGURE 20-15

Exploding fruits disperse their own seeds. This common vetch (*Vicia sativa*) produces a power-packed legume that flings seeds 3 meters (10 feet) from the plant.

(a)   *(b)*

FIGURE 20-16

Wind dispersal. Airborne seeds and fruits often have wings or plumes that help them remain aloft, enabling them to travel further. *(a)* Plumes of milkweed, *(b)* wings of swamp maple.

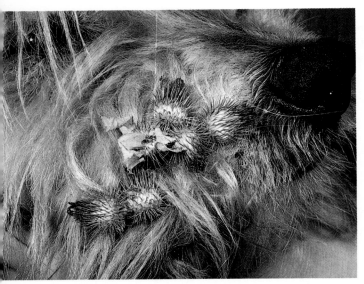

FIGURE 20-17
Hitchhiking fruits, such as the burrs of common Burdock (*Arctium minus*) have hooks, spines, or barbs to "bum a ride" to new habitats by attaching to the skin or fur of animals.

neously. Saguaro seeds pass undamaged through the bird's digestive system as it flies about and are released along with the bird's guano (excrement). In some cases, rodents gather and stockpile nuts and seeds in hiding places; the uneaten seeds grow into new plants.

 Seeds can be dispersed over very long distances — across oceans, to other continents or to distant islands — carried on a bird's feathers or in mud adhering to the feet of migrating birds. (Some of the original plants that grew on the Hawaiian Islands were dispersed there in this way.) Barbs, hooks, spines, or sticky surfaces that temporarily fasten fruits or seeds to an animal's fur aid dispersal of some plants to new locations (Figure 20-17).

Fruits and seeds protect the plant embryo and aid in the dispersal of offspring to new habitats as well as reduce competition among offspring or competition between offspring and the parent plant. (See CTQ #4.)

GERMINATION AND SEEDLING DEVELOPMENT

When environmental conditions are favorable (proper temperature, sufficient water, and adequate oxygen, for example), a seed germinates and a new generation of plants begin to grow and develop. During **germination,** the radicle of the hypocotyl is usually the first structure to thrust through the seed coat. The radicle then forms a root that anchors the new plant and begins absorbing water and minerals from the soil. As the root system develops, the epicotyl begins growing into the shoot system; photosynthesis begins when the shoot system emerges from the soil surface.

SEED VIABILITY AND SEED DORMANCY

The embryos of different plant species remain viable (capable of germinating) for different periods of time. The *seed viability* of some species, such as willows, is only a few days, whereas the seeds of other plants may remain viable for months or even years. For example, oriental-lotus seeds that were recovered from ancient tombs germinated after 3,000 years!

 A seed may fail to germinate for one of two reasons: (1) the embryo is dead (that is, the seed is not viable), or (2) the embryo is "resting," a condition known as **seed dormancy.** Seeds may remain dormant if there is a lack of adequate water; the embryo is underdeveloped; hormones are blocking germination; the seed requires a cold period; or the radicle is unable to break through a tough seed coat.

SEEDLING DEVELOPMENT

A young plant's (seedling's) battle for survival begins instantly. Producing a root system to obtain water is more important to survival than is breaking through the surface for photosynthesis. Nutrients stored in the seed can usually last for several days after germination, but without an immediate supply of moisture the seedling will quickly die of dehydration. Most plants establish mycorrhizae associations with soil fungi during the initial growth of the root system (page 386), improving their ability to absorb water and minerals.

Seedling development is slightly different for monocots than for dicots. To illustrate these differences, we discuss a representative example of each: a dicot bean seedling (*Phaseolus*), and a monocot corn seedling (*Zea*).

Bean Seedling Development

During bean germination, the radicle emerges and grows downward, producing many lateral roots from a single main root (Figure 20-18). Unequal cell growth causes the bean hypocotyl to bend, forming a hook that nudges its way up through the soil, carving a channel so that the delicate apical meristem, cotyledons, and young leaves can be drawn along without injury. Once the hypocotyl breaks the soil surface, it straightens, exposing the cotyledons and epicotyl to sunlight.

 Up to this point, the seedling has depended on the nutrients stored in the endosperm of its cotyledons. Once the cotyledons and young leaves are exposed to light, however, the plant manufactures chlorophyll and other pigments, and photosynthesis begins. As food reserves are used up, the cotyledons wither and fall.

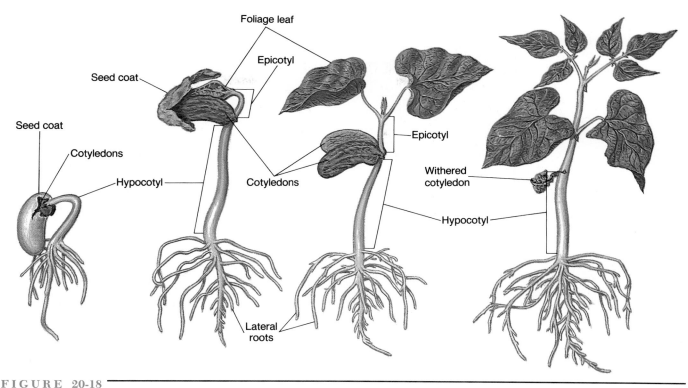

FIGURE 20-18

Seed germination and seedling development of a dicot, bean *(Phaseolus vulgaris).*

Corn Seedling Development

Food reserves usually aren't stored in the single cotyledon of a monocot embryo. Instead, the cotyledon absorbs food from surrounding endosperm tissues, which fuels germination and initial seedling growth.

When a corn kernel germinates, the root tip breaks through the surrounding tissues, as the epicotyl forces its way through a protective sheath (the *coleoptile*) and then up through the soil (Figure 20-19). As in dicots, once young monocot leaves are exposed to sunlight, photosynthesis begins, and the young plant begins its life of autotrophic independence.

SEEDLING SURVIVAL AND GROWTH TO MATURITY

A seedling's life does not depend solely on its ability to absorb water and minerals or to photosynthesize enough food to support its own growth and development. Often, the seedling must also be able to outcompete other plants for nutrients and light as well as avoid the many hazards of life, including continuous onslaughts of plant-eating animals, attacks by hundreds of disease-causing microbes, fluctuations in temperature, slashing winds, and a host of other life-threatening environmental conditions. The few seed-

lings that survive establish a new generation of plants. The ability of a plant to survive depends on a number of growth and development strategies and behaviors that are regulated by plant hormones, the subject of the next chapter.

A new plant begins its independent life after it germinates; it develops a root system for absorbing water and minerals and a shoot system for photosynthesis. (See CTQ #5.)

ASEXUAL REPRODUCTION

Flowering plants reproduce new individuals both sexually and asexually (without the fusion of gametes and resulting embryo, seed, and fruit development). Asexual, or *vegetative reproduction,* results from mitosis and consequently produces offspring that are always genetically identical to the parent plant.

Some plants whose stems grow over the soil surface, such as strawberries and spider plants, form asexual offspring from horizontal stems—*runners* (or *stolons*). If a plant's stems grow at or below the soil surface (as in the case of crab grass and bamboo), asexual offspring are formed

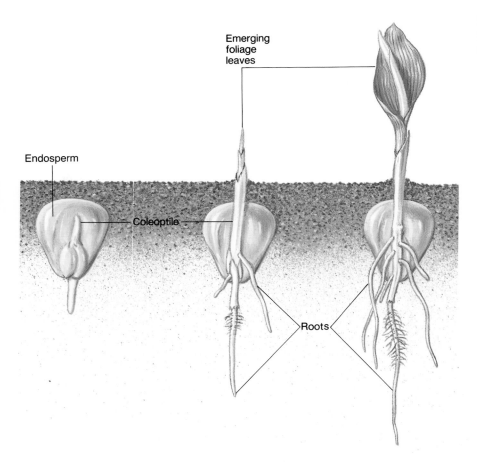

Emerging
foliage
leaves

Endosperm

Coleoptile

Roots

FIGURE 20-19
Seed germination and seedling development of a monocot, corn (*Zea mays*).

from *rhizomes*. Other plants develop offshoots from vertical stems (African violets), leaves (piggyback plants), bulbs (garlic, daffodils, tulips), or tubers (potato, Jerusalem artichoke, tuberous begonias). Botanists are learning to take advantage of the ability of plants to form asexual individuals to improve crop production by employing propagation techniques, such as stem, leaf, and root cuttings, budding and grafting, and tissue culture.

In a few plants, fruits may develop without fertilization, a process called *parthenocarpy* (*partheno* = virgin, *carpy* = carpel). Parthenocarpic fruits contain no embryos and no seeds. Some cultivated plants (bananas, pineapples, seedless grapes, seedless watermelons, seedless cucumbers) lack seeds because their fruits developed parthenocarpically, often aided by sprayed artificial plant hormones that induce the ovary to mature.

Most flowering plants have the ability to reproduce both sexually and asexually. Asexual offspring develop as a result of mitotic divisions and are therefore genetically identical to their parent. (See CTQ #6.)

AGRICULTURAL APPLICATIONS

In the early 1900s, agriculturists used their knowledge of how plants reproduce to improve crops by crossing plants that produced offspring with higher yields and other desirable characteristics. For example, through selective breeding, the sugar content in sugar beets increased from 7 to 20 percent; strawberry yields increased from 100 to 200 percent; and corn production increased from 19 to 37 percent. Despite these dramatic increases, many people around the world still go to bed hungry because crop increases are not enough to keep up with high human population growth rates.

The problem is further compounded because many crops, including bananas and potatoes, do not lend themselves to breeding programs. Many agricultural scientists are now using biotechnological techniques to engineer superior crop plants from somatic mutations (mutations that occur in cells not involved in sexual reproduction). In a technique called *protoplast manipulation*, a piece of tissue is removed from a crop plant, and the cells are separated from one another. The cell walls are dissolved, leaving the protoplast, which can then be examined for desirable traits, such as improved growth rate or increased disease resistance. Once identified, the cells with the desirable mutation can be grown using tissue culture techniques, thereby producing entire plants. Using this same technique, scientists are attempting to fuse protoplasts of different species to produce entirely "new" plants with combinations of desirable traits.

> **Like all other animals, humans depend on photosynthetic plants for food and nutrients, most of which come from flowering plants. Understanding plant reproduction, growth, and development has led to a number of agricultural innovations that greatly improve crop production and help feed the world's population. (See CTQ #7.)**

REEXAMINING THE THEMES

Relationship between Form and Function

The shape, position, size, color, and many other characteristics of flower parts affect the frequency of successful pollination. For example, the filaments of stamens help position anthers so that pollen will be deposited on specific areas of animal pollinators; long, pendulous filaments dangle anthers away from other plant parts in wind-pollinated plants. Similarly, the styles of the pistils position stigmas so that pollen is picked up from animal pollinators; in wind-pollinated plants, long, divided styles cause stigmas to protrude into the air, where airborne pollen grains are more likely to be passing by.

Acquiring and Using Energy

All animals are heterotrophs and ultimately acquire food (energy and nutrients) from autotrophic organisms. This dependency forms the basis for pollination and seed dispersal by animals. While foraging for food in flowers, animals transport pollen from the anthers to a stigma, positioning the plant's male gametophytes for eventual fertilization. A proportion of the acquired energy the plant used to synthesize its own organic molecules is used to manufacture nectar, bright pigments, and a strong fragrance—resources that are necessary for attracting and "rewarding" pollinators.

Unity within Diversity

All organisms have the potential to reproduce. Offspring that are produced asexually are genetically identical to their parent, and as a result, develop virtually identical traits. Since most of the earth's habitats are not uniform from one location to another, however, and since all habitats eventually change over time (the recurring ice ages, for instance), a parent who is ill-equipped for change will produce asexual offspring that are equally ill-equipped. In contrast, offspring produced by sexual reproduction have slightly different genetic compositions, producing individuals with slightly different traits. Such variability improves the chances that some individuals will possess characteristics that enable them to survive and reproduce in a new habitat or following a change in the environment.

Evolution and Adaptation

The interdependency and complementary structures found in plants and their animal pollinators and in plants and animals that disperse the plants' seeds illustrate coevolution—the reciprocal evolution of adaptations in different species based on mutual advantages. Part of the great success of flowering plants compared to other plant groups is attributed to the fact that most flowering plants are pollinated by animals—pollen carriers that substantially increase successful crossbreeding over wind or water currents. Increased pollination increases sexual reproduction, enabling flowering plants to proliferate in virtually all the earth's habitats.

SYNOPSIS

Plants utilize wind, water, or animals to transfer pollen grains for sexual reproduction and to disperse seeds to new habitats. Most animals are attracted to plants for food (nectar, pollen, and edible flower and fruit tissues) and inadvertently transport pollen or seeds.

The flower is the site of sexual reproduction, seed formation, and eventual fruit development. Variations in flower design are adaptations that promote successful dissemination and collection of pollen for sexual reproduction.

Meiosis leads to the formation of gametes for sexual reproduction. In plants, the daughter cells resulting from meiosis form spores. The spores produced in the anthers develop into the male gametophyte (pollen grain), whereas those formed in the ovary of a pistil develop into the female gametophyte (embryo sac) contained within an ovule in the ovary of the pistil.

Following the transfer of pollen during pollination, fertilization occurs in the ovary of a pistil. Two sperm cells are delivered to each ovule, even though there is only a single egg. One sperm fertilizes the egg, while the second fuses with the endosperm mother cell to begin building nutrient supplies for the offspring.

Following fertilization, the zygote develops into a plant embryo; the ovule transforms into the seed; and the ovary (and sometimes other flower parts) matures into the fruit. Fruit and seed adaptations help disperse seeds to new habitats and reduce competition between offspring and the parent plant.

Favorable environmental conditions break seed dormancy and trigger seed germination. Various adaptations regulate germination so that young seedlings do not grow in unfavorable conditions.

Key Terms

pollen grain (p. 402)
pollination (p. 402)
pedicel (p. 403)
receptacle (p. 403)
sepal (p. 403)
petal (p. 403)
stamen (p. 403)
filament (p. 403)
anther (p. 403)
pistil (p. 403)
carpel (p. 403)

stigma (p. 403)
style (p. 403)
ovary (p. 403)
nectary (p. 405)
complete flower (p. 407)
incomplete flower (p. 407)
imperfect flower (p. 407)
perfect flower (p. 407)
monoecious (p. 407)
dioecious (p. 407)
microspore (p. 407)

megaspore (p. 407)
embryo sac (p. 408)
ovule (p. 408)
endosperm mother cell (p. 408)
endosperm (p. 408)
cotyledon (p. 411)
epicotyl (p. 411)
hypocotyl (p. 411)
radicle (p. 411)
seed dormancy (p. 416)

Review Questions

1. Arrange the following events in the sexual reproduction of a flowering plant in their correct order:

   a. pollen dispersal
   b. seed germination
   c. pollination
   d. fertilization
   e. seed dispersal
   f. fruit development
   g. pollen-tube growth
   h. attraction of animal pollinator

2. Match the following sexual reproductive structures to the structure in which it is located:

   _____1. megaspore
   _____2. ovule
   _____3. micropyle
   _____4. generative cell
   _____5. microspore

   a. ovary of pistil
   b. pollen grain
   c. ovule
   d. anthers
   e. pollen tube

3. How does pollination differ from fertilization? Are each of these processes necessary for the other to occur, or is only one required before the other can happen?

4. Compare monocot and dicot seedling development (Figures 20-18 and 20-19). List three similarities and three differences in their development.

5. Discuss the basic function(s) of fruits.

6. Discuss the relationship among seed dormancy, viability, and germination. Is each necessary for the others to occur? Does one always occur before or after another?

7. Match the structure to its function:

   _____1. embryo sac
   _____2. stigma
   _____3. style
   _____4. nectaries
   _____5. petals

   a. attract animal pollinator
   b. site of fertilization
   c. pollen deposition
   d. produce pollen grains
   e. connects stigma and ovary; provides a channel for pollen-tube growth

8. Explain how complete flowers are always perfect flowers yet incomplete flowers are not always imperfect.

9. List five adaptations plants have evolved to reduce the possibility of inbreeding. (Recall that inbreeding reduces genetic variability and increases recessive trait expression, both of which reduce the frequency of adaptive traits.)

10. Identify the three main components of a flowering plant embryo and state what each will develop into in a young seedling.

Critical Thinking Questions

1. One of the principles of biology is that form (structure) and function are complementary, i.e., that form determines function and that function shapes form. Select three examples from flowers and explain how each illustrates this principle.

2. Using simple diagrams or drawings, illustrate the place of meiosis, mitosis, and fertilization in the life cycle of a flowering plant (from seed to mature plant to new generation of seed).

3. Pollination efficiency is often based on a comparison of the number of pollen grains a plant produces for each egg (ovule) available for fertilization. Pollination efficiency ranges from several million pollen grains per egg to one pollen grain per egg—the most efficient pollination can be. Which pollination system (flower + pollinating agent) do you suppose has the least efficiency? Which has the highest? Explain your answers.

4. Interpret the graph below which compares the number of seeds and the distance they were dispersed from the parent plant. Describe the graph in terms of the relationship between number of seeds and distance traveled. What is the dispersal agent—wind, water, bird (carried internally), or animal (carried externally)? Why did you choose this vector? How tall do you think the

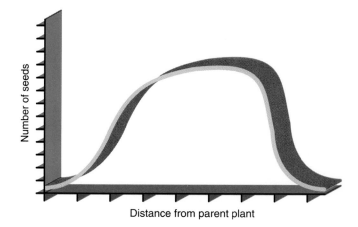

Distance from parent plant

parent plant is? How large would you suppose the seeds were, and what do you think their relative weight was —extremely light, light, moderate, heavy, or very heavy?

5. Although light is necessary for photosynthesis, it is not required for germination. Explain why this is so.

6. What are some of the advantages to people of the ability of some plants to reproduce asexually? What are some of the disadvantages? [Hint: Review the story of the grape disaster in Europe.]

7. Most of the food for the people of the world comes from flowering plants, the most important of which are the grains, such as wheat, rye, barley, oats, and rice. Why do few people think of these as "flowering plants"? Why are they, in fact, classified as flowering plants?

Additional Readings

Corbet, S. 1987. More bees make better crops. *New Sci.* 23:40–43. (Intermediate)

Hamner, K., and J. Bonner. 1938. Photoperiodism in relation to hormones as factors in floral initiation and development. *Botanical Gazette* 101:658–687. (Intermediate)

Meeuse, B., and S. Morris. 1984. *The sex life of flowers.* New York: Facts on File Publications. (Introductory)

Richards, A. 1986. *Plant breeding systems.* Winchester, MA: Allen & Unwin. (Intermediate)

Stiles, E. 1984. Fruit for all seasons. *Natural History* August: 43–53. (Introductory)

CHAPTER
‹ 21 ›

How Plants Grow and Develop

**STEPS
TO
DISCOVERY**
Plants Have Hormones

**LEVELS OF CONTROL OF GROWTH
AND DEVELOPMENT**

**PLANT HORMONES—
COORDINATING GROWTH AND
DEVELOPMENT**

Auxins

Gibberellins

Cytokinins

Ethylene Gas

Abscisic Acid

Other Plant Hormones

**TIMING GROWTH AND
DEVELOPMENT**

Temperature and Growth

Light and Growth

Biological Clocks

PLANT TROPISMS

Phototropism

Gravitropism

Thigmotropism

THE HUMAN PERSPECTIVE

Saving Tropical Rain Forests

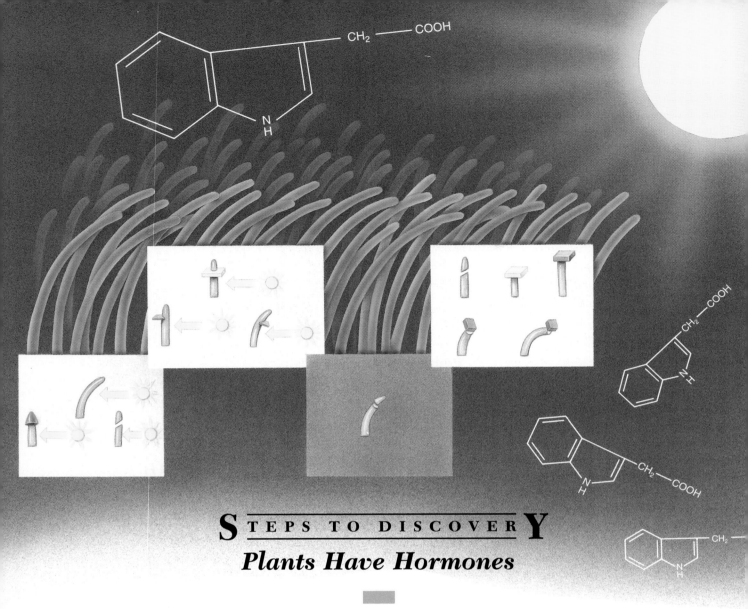

Plants Have Hormones

*A*lthough we are likely to associate Charles Darwin only with the theory of evolution and natural selection, Darwin actually completed a number of experiments that also added to the knowledge of plant growth and development; some of his findings were published in 1881 in a book entitled *The Power of Movement in Plants*. Many of Darwin's experiments focused on the growth and movements of plants in response to gravity, touch, chemicals, and light, helping to confirm that plants—like animals—manufacture hormones.

In the 1880s, Darwin collaborated with his son Francis in a study on the bending of plants toward light. The Darwins tested the growth responses of young canary-grass seedlings to light shining from one direction. The Darwins demonstrated that light coming from only one direction

caused the young shoot to bend toward the light. Like many other grasses, canary grass seedlings form a protective sheath (coleoptile) that covers the plant's young leaves and apical meristem. If the tip of the coleoptile was removed or covered with foil, however, the shoot did not bend. The Darwins concluded that the coleoptile tip somehow affected cell growth, causing the shoot to bend toward the light. The Darwins knew that most plant growth was the result of cell elongation, which takes place several millimeters below the coleoptile tip. In order to explain how the coleoptile tip influences elongation several millimeters away, the Darwins suggested that a growth signal produced in the coleoptile tip traveled down the shoot, triggering cells to enlarge differentially. The differential elongation of cells then caused the young shoot to bend toward light. The

By varying experimental conditions depicted in the boxes, investigators discovered that a growth hormone produced at the

identity of the signal and the way it affected cell elongation was beyond the scope of the Darwins' experiments, however.

In 1910, Peter Boysen-Jensen, a professor of plant physiology at the University of Copenhagen, designed studies that not only confirmed the Darwins' findings but added new information on the nature of the growth signal. In two sets of experiments, Boysen-Jensen demonstrated that the growth signal was most likely a chemical stimulus that caused cells on the dark side of the illuminated shoot to elongate more than those on the lighted side. In his experiment, Boysen-Jensen removed a plant's coleoptile tips, positioned a block of water-permeable gelatin on the cut shoot, and then placed the coleoptile tips on top of the gelatin. When the seedlings were exposed to light, the shoots curved toward the direction of the light, demonstrating that the growth signal had traveled through the gelatin (otherwise, there would not have been curved growth). Because dissolved chemicals travel through gelatin, Boysen-Jensen concluded that the growth stimulus was probably a chemical.

In another experiment, Boysen-Jensen inserted pieces of water-insoluble mica (a thin piece of quartz or other silicate mineral) part way into either the lighted side or the dark side of the shoot. Mica blocks the movement of all dissolved substances. Only those shoots in which the mica was inserted on the lighted side bent toward the direction of light. Because none of the shoots bent when mica was inserted on the dark side, Boysen-Jensen concluded that the mica blocked the downward movement of the growth stimulus. Under normal circumstances, the growth-stimulating substance traveled down the dark side of the shoot. Therefore, stem bending was the result of the greater concentration of growth stimulus on the dark side than on the illuminated side. That is, a higher concentration of growth stimulus on the dark side triggered greater cell elongation so the shoot bent toward the light.

In 1918, A. Paal in Hungary conducted experiments that provided further evidence that the growth-stimulating substance in the coleoptile tip was responsible for shoot bending. Paal removed coleoptile tips and then displaced them to one side of the shoot or the other. All the plants in Paal's experiment were kept in the dark. It didn't matter on which side of the shoot Paal placed the severed tip; the

shoot always bent in the opposite direction. Paal's study demonstrated that the growth stimulus originated from the coleoptile tip and promoted elongation of the cells immediately below the displaced tip, causing the shoot to bend.

The proof that the growth stimulus was a chemical was finally demonstrated in experiments conducted by Fritz Went in 1926. Went's first experiments, like those of the Darwins, demonstrated that without the coleoptile tip, shoots would not bend toward light. In his next experiments, Went placed severed coleoptile tips onto agar (a gelatinous substance extracted from seaweed), allowing the growth substance ("juice") to diffuse into the agar and then discarded the tips. He then placed a small block of the agar with juice on a shoot that had had its coleoptile tip removed. As a control, he also placed plain agar blocks (no coleoptile juice) on decapitated shoots. None of the stems with plain agar elongated, while all of the stems with coleoptile juice in the agar grew straight up. When Went offset the agar blocks with coleoptile juice on the decapitated shoots, the shoots bent in the opposite direction. Went's explanation was that the growth substance had diffused into the agar, stimulating cell growth; if all the shoot's cells were stimulated, the shoot grew straight up; if cells on only one side of the shoot were stimulated, the shoot bended.

Went also developed a technique whereby he could quantify the effect of the growth stimulus on stem bending. Went varied the amount of coleoptile juice in the agar by allowing severed tips to remain on the agar for longer periods of time, demonstrating that stem curvature was proportional to the amount of growth-promoting juice found in the agar. Went named the growth-promoting chemical in the juice *auxin*. Based on his work and on that of previous researchers, Went concluded that auxin promotes cell elongation on the side away from the light source. Since these cells elongate more than do those on the lighted side, the shoot bends toward the light. The amount of bending is proportional to the concentration of the auxin present.

tips of stems causes plants to bend toward light.

Without plants, we would quickly die. Like all other animals, we depend on plants for the food and oxygen they produce during photosynthesis. But our dependency on plants goes beyond these basic necessities:

- We use plants to medicate ourselves against disease. More than half our medicines come directly from plants, most of which are from tropical rain forests (see The Human Perspective: Saving Tropical Rain Forests).

- We shelter ourselves with plants (lumber for houses).

- We clothe ourselves in plant fibers (cotton and linen).

- We brighten our world with plant extracts (coloring and dyes).

- We spice up our food with plants (vanilla, paprika, ginger, coffee, cocoa, and teas).

- We scent our bodies with plant oils (perfumes and colognes).

- We brighten our lives with indoor and outdoor plants and gardens.

Our dependency on plants puts us in a rather vulnerable position. Soaring world population, widespread habitat destruction, rapidly diminishing natural resources, and an alarming extinction rate for the world's plant species are fast depleting and destroying the very things that supply us with so many necessities and pleasures. Now, more than ever, our understanding of plant growth and development will not only affect the quality of our lives but may even determine the number of people that can live on earth.

▼ ▼ ▼

LEVELS OF CONTROL OF GROWTH AND DEVELOPMENT

How does a tiny saguaro embryo the size of a grain of sand grow into the world's tallest cactus? What makes a root grow down, while stems and branches reach up toward light? Why do some plants bloom in the spring, whereas others flower in summer, fall, or even winter? And why, when farmers sow the best-quality seed, do their crops sometimes yield only a fraction of their potential or fail entirely?

Plant growth and development are the products of a complex interplay among three levels of control:

(a)

(b)

FIGURE 21-1 ────────────

Nature versus nurture. The environment profoundly influences how a plant grows. *(a)* Persistent ocean winds have produced this picturesque Monterey cypress growing near the shoreline. *(b)* Continuously frigid winds kill axillary buds on the windward side of stems so that branches grow only from leeward axillary buds, giving high-elevation plants, such as this Alpine fir, a "flag-form" appearance.

- *Intracellular level* (within a cell) — certain genes in the cells operate at precise times to code for specific characteristics and activities.

- *Intercellular level* — hormones (potent chemicals that coordinate growth and development in different parts of the plant) trigger reactions in target cells, including determining which genes are expressed.

- *Environmental level* — environmental controls include the direction or intensity of light; amount of moisture; rain acidity; extremes in temperatures; availability of minerals in the soil; and many more (Figure 21-1). The environment plays a large role in molding plant growth and development.

Plant growth and development are regulated by internal genetic controls and hormone production, which, in turn, are influenced by the surrounding environment. (See CTQ #2.)

PLANT HORMONES: COORDINATING GROWTH AND DEVELOPMENT

In multicellular plants, as in multicellular animals, coordinating the growth and activities of the hundreds or thousands of cells requires hormones. But because plants are stationary organisms, their growth and development must also be adjusted to accommodate the peculiarities of their immediate surroundings. As a result, the production and distribution of plant hormones are very sensitive to environmental conditions.

▸ Plant hormones, like those of animals, are chemicals that are synthesized in a particular part of the plant which stimulate or inhibit specific responses in target tissues.

Plant hormones are effective in extremely minute amounts; some researchers suggest that only one molecule of a hormone may be needed to trigger a response. Furthermore, the effects of plant hormones vary, depending on the time of year and the developmental stage of the plant. Plant hormones also often work in combination to trigger developments changes; combinations of hormones evoke responses different from those each hormone would produce alone.

To date, botanists have identified three groups of plant substances (auxins, gibberellins, and cytokinins), and two specific molecules (ethylene gas and abscisic acid) that influence plant growth and development; all are plant hormones (Table 21-1).

AUXINS

Auxins stimulate plant growth by promoting both cell division and, most often, cell elongation. In fact, the name comes from the Greek "auxien," meaning "to increase" or "augment." Auxins stimulate cell elongation by softening cell walls, allowing internal turgor pressure to stretch the wall.

Auxins affect plant growth and development in many ways. They:

- Cause stems to bend, which orients the plant more favorably to its environment. For example, bending allows the stems of newly germinated seedlings to grow up through the soil (Figure 21-2) and enables shaded plants to grow out from under another plant, toward light (Figure 21-3).

- Stimulate root development, increasing water and mineral absorption.

- Stimulate cell division in the vascular cambium, increasing production of xylem and phloem tissues which, in turn, increases translocation capability.

TABLE 21-1

KNOWN PLANT HORMONES

Hormone	Effects
Auxins	Promotes cell elongation; stimulates cell division of vascular cambium; inhibits axillary bud development; promotes flowering and fruiting
Gibberellins	Stimulates cell division and elongation; mobilizes food reserves in seeds; inhibits seed formation; stimulates flowering and pollen-tube growth; increases fruit size; ends dormancy
Cytokinins	Promotes cell division in stems; facilitates healing of wounds; prolongs life of leaf, flowers, and fruits
Ethylene gas	Regulates cell metabolism; promotes fruit ripening; stimulates stems to thicken; regulates shedding of leaves and flowers
Abscisic acid	Reduces cell division and elongation; inhibits growth; promotes dormancy and stomate closure

◁ THE HUMAN PERSPECTIVE ▷
Saving Tropical Rain Forests

FIGURE 1

Satellite photographs reveal a perpetual cloud of smoke covering the forests of Brazil—a cloud the size of the state of Texas. Beneath the smoke, one of the earth's most critical life-sustaining habitats is being consumed. Every minute, 100 acres of tropical rain forest disappear, burned or hacked away by people seeking new land for cultivation, by subsistence farmers whose lives depend on the crops grown on small plots of land (FIGURE 1). But the livelihoods of these farmers are doomed to a few years of marginal success; they then must clear more forest for their crops.

Nowhere is life more abundant or diverse than in the tropical rain forests, where more than half the world's species of plants and animals reside. But the abundance of life in the natural rain forest does not translate into lush crop growth once the forest has been destroyed. In fact, cleared rain forest land is paradoxically unproductive. Because the forest abounds with so much life, litter does not accumulate. Dead organisms are quickly consumed by other organisms. Consequently, little or no humus collects in the soil to condition it, and precious little nitrogen or phosphorous seeps into the ground to fertilize it. In other words, the topsoil is thin and poor. With the rain forest cleared away, the area can no longer support a rich abundance of life, and what few nutrients are present in the soil rapidly erode away with each rainfall. Within a few years of clearing, the land is abandoned by its cultivators, who then burn off more rain forest in their need to grow food for survival (FIGURE 2).

Deforestation in the tropics has rapidly accelerated since World War II. New

FIGURE 2

highways penetrate into the deep forests, providing easy access by timber harvesters, land developers, cattle ranchers, cash-crop planters, and oil and mineral exploration teams. Each year, 27 million acres of tropical forest disappear, along with the habitats that support the most diverse range of species on earth. One of the most distressing aspects of tropical rain forest destruction is the extinction of thousands of species. Although extinction is part of the natural history of life, deforestation has accelerated the rate by 10,000 times, compared to the extinction rate that existed before humans. Extinction claims the lives of three to four species *every day*!

The tropics provide humans with most of our edible plant species (a typical breakfast of bananas, orange juice, cornflakes and sugar, coffee or hot chocolate, and hashed-brown potatoes consists entirely of tropical plants). More than 40 percent of our medicines, from pain-relieving

aspirin to treatments for malaria, have come from tropical plants. For example, the rain forest's rosy periwinkle provides a life-saving drug used for treating otherwise fatal cases of lymphocytic leukemia. Cures for cancer and AIDS may be hiding in plants on the brink of extinction, plants that could disappear before they are ever discovered.

Destruction of the tropical rain forests also alters the earth's climate. Ecologists refer to tropical rain forests as "rain machines" because they recycle water back into the atmosphere (roots draw up the ground water, which evaporates from leaves, returning about 50 percent of rainwater to the atmosphere). Deforested areas receive much lower amounts of rainfall; nearly 100 percent of the water runs off into streams, most of it flowing all the way to the the ocean. Some lush tropical areas have become desertlike following deforestation.

Deforestation also plays a very important role in accelerating global warming. The forests are critical to the balance of carbon dioxide on earth. The plants of the forest remove carbon dioxide from the atmosphere for photosynthesis. In burning the rain forests, we are not only destroying our major carbon-dioxide "sink," we are also turning the trees into carbon dioxide (about 1 billion tons of carbon dioxide are ejected into the atmosphere by deforestation). Because carbon dioxide in the atmosphere traps heat, this ecological "combination punch" will likely hasten global warming which, in turn, will trigger unexpected and unwelcome environmental changes.

Already half the rain forests have disappeared; at the current rate of destruction, virtually all the earth's rain forests may be eliminated (along with more than half the earth's species) within the next 50 years. What will the world be like then?

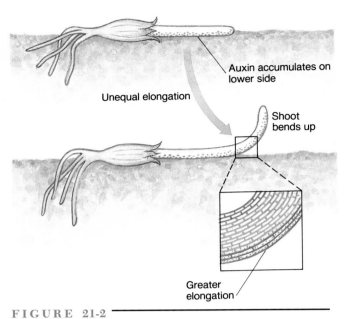

FIGURE 21-2

Which way is up? When a seedling is placed on its side, auxin accumulates on the lower side of the shoot, increasing cell elongation there and causing the shoot to grow up.

- Induce cells to differentiate into secondary xylem, thereby increasing water and mineral transport and adding reinforcement for greater support.

- Promote the development of flowers and fruits.

The first (and most common) auxin to be isolated was **indoleacetic acid (IAA)**. As you will recall from the chapter opening, researchers knew long ago that the center of auxin production in grasses was in the tips of their coleoptiles. It took the collection of more than 10,000 oat coleoptile tips to obtain a single microgram of the hormone, confirming the fact that plant hormones are indeed active in *extremely* small amounts.

In addition to their growth-promoting functions, auxins sometimes inhibit development. For example, auxins inhibit the development of axillary buds, producing a familiar growth pattern, known as **apical dominance** (Figure 21-4). As horticulturists and gardeners will tell you, pinching off the tips of a plant's main stems stimulates the production of more leafy branches. Removing the tips of stems also removes the apical meristem, a center of IAA production. Normally, IAA moves down the stem and inhibits axillary buds from growing into new leafy stems. By removing the center of IAA production, you allow axillary buds to begin to develop into new stems with more leaves and more axillary buds.

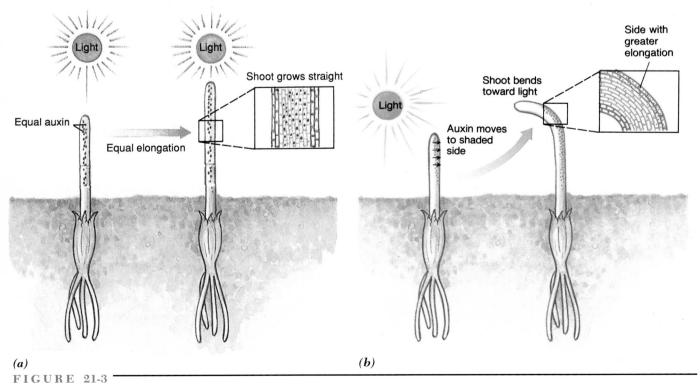

FIGURE 21-3

Light and stem growth. *(a)* Shoots grow straight when light comes from above because auxin is evenly distributed, resulting in equal cell elongation on all sides. *(b)* Shoots grow toward light because auxin is transported to the shaded side, where it increases cell elongation.

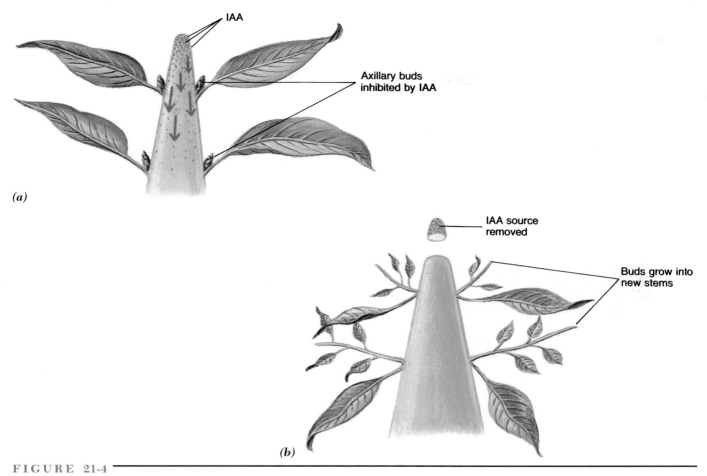

FIGURE 21-4

Apical dominance. *(a)* Produced by apical meristems, IAA moves down the stem and inhibits axillary bud growth. *(b)* When the shoot tip is clipped, the source of IAA is removed, and axillary buds grow into new stems.

Over the past 40 years, hundreds of synthetic auxins have been manufactured in chemical laboratories. Synthetic auxins (in fact, synthetic versions of all plant hormones) are called *growth regulators*. The active ingredient in Agent Orange, the defoliant (herbicide) that was used extensively in Vietnam, is a potent growth regulator that disrupts normal plant growth. Agent Orange was intended to defoliate densely populated jungles that provided enemy cover. Unfortunately, the herbicide also contains the contaminant dioxin, which is now known to be one of the most toxic substances for humans and has been linked to leukemia, miscarriages, birth defects, and liver and lung disorders.

Experiments confirm that other plant hormones play a role in many of the effects produced by auxins. As more research is completed, it becomes increasingly clear that hormonal control of plant growth and development is not simply the result of a single hormone acting alone but of a combination of hormones that triggers specific responses.

GIBBERELLINS

At the same time that researchers were attempting to isolate IAA during the 1920s, several Japanese scientists were investigating the cause of a fungal disease found in rice. Rice infected with a fungus (*Gibberella fujikuroi*) grew taller than did noninfected plants. As a result, the plants' stems weakened, and the plants collapsed. In 1926, E. Kurosawa purified the chemical in the fungus which stimulated cell elongation and division in infected rice; he named the substance **gibberellin** after the genus name of the fungus. During the 1940s and 1950s, intensive studies in Japan, the United States, and England revealed a number of growth-promoting chemicals, all of which had similar molecular structures to gibberellin. Today, more than 70 compounds are classified as gibberellins.

When gibberellins are applied to dwarf plants, normal stem growth is restored (Figure 21-5), suggesting that dwarf plants lack the gene needed to manufacture their own sup-

FIGURE 21-5
Gibberellin-enhanced growth. Applications of gibberellins to a plant cause skyrocketing vertical growth due to rapid enlargement of internode cells.

ply of gibberellins. Applications of gibberellins stimulate both cell division and cell elongation. But normal stem growth is not regulated by gibberellins alone; auxins and cytokinins are also involved.

▮▶ In addition to affecting stem growth, gibberellins are critical to embryo and seedling development. In all monocot seeds (corn, orchids, and grasses), gibberellins stimulate mobilization of food reserves stored in the endosperm. Without such mobilization, stored energy and nutrients would not be available to the embryo; without energy and nutrients, the embryo cannot grow, the seed cannot germinate, and a seedling cannot grow into a new plant. Gibberellins also inhibit seed formation, stimulate pollen-tube

growth, increase fruit size, initiate flowering, and end periods of dormancy in seeds and axillary buds.

CYTOKININS

In the 1950s and 1960s, researchers discovered that rapid growth could be stimulated in plant embryos and plant tissue cultures by adding coconut milk and yeast extract to the culture growth solution. Later studies isolated the active ingredients, which were named **cytokinins** because they stimulated rapid growth by promoting cell division (cytokinesis). Injuries to plants (such as a cut or a torn branch) are quickly healed as a result of the production of cytokinins.

In addition to repairing damaged tissues, cytokinins also play a role in regulating normal stem and root growth. Investigators have found that applying cytokinins stimulates cell division in stems and inhibits cell division in roots. These opposite effects on cell division in stems and roots is apparently related to the presence of auxins and gibberellins; but just how varying levels of these hormones control promotion or inhibition of cell division is still not known.

Cytokinins also delay **senescence**—the aging and eventual death of an organism, organ, or tissue. In leaves, cytokinins slow senescence by delaying the breakdown of chlorophyll. In flowers and fruits, cytokinins delay senescence by increasing nutrient transport.

ETHYLENE GAS

Some enterprising entrepreneurs have made a great deal of money from an unusual plant hormone, **ethylene gas**. The entrepreneurs' advertisement claims that when green fruits are placed in a "specially designed bowl," they will ripen two or three times faster than normal. This claim is true, but it would also be true for any closed container, including an ordinary paper bag. A covered bowl or closed bag traps ethylene gas that is released from ripening fruit. Ethylene gas hastens fruit ripening, and as fruits ripen, they produce more and more ethylene gas. Enclosing fruits sets up a positive feedback system (ethylene breeds more ethylene), causing fruits to ripen faster and faster. Production of ethylene gas by overripe fruit explains why "one rotten apple can spoil the whole barrel."

The discovery that ethylene gas promotes fruit ripening has dramatically lowered the cost of shipping fruits to markets. Tomatoes, bananas, apples, oranges, and many other fruits are usually picked when they are still green (reducing the chance of spoilage), transported in ventilated crates (to prevent ethylene gas buildup), and then gassed with synthetic ethylene at distribution centers to promote last-minute ripening.

In addition to its role in fruit ripening, ethylene gas stimulates stems to grow thicker in response to physical stress (Figure 21-6). Trees that have been staked while growing (and thus remain protected from the bending

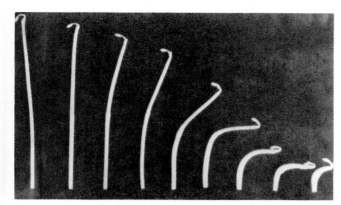

FIGURE 21-6

Ethylene gas and the emergence of seedlings. Stresses from pushing through the soil stimulate ethylene gas production in young seedlings. The bigger the obstacle, the greater the stress, and the more ethylene gas produced. This photo shows that increasing level of ethylene (from left to right) causes the stems of pea seedlings to thicken and curve. Thickening and curving of stems aid a seedling as it forces its way around obstacles and through the soil in its struggle to emerge.

stresses caused by winds) usually have weaker trunks and stems than do plants allowed to bend in the wind. The production of ethylene gas in response to physical stresses would be favored by natural selection since it increases survivorship under natural conditions, which, in turn, increases the plant's chances of living long enough to reproduce.

High concentrations of ethylene gas cause abnormal stem bending and leaf curling, discolor flowers, and inhibit root growth. Ethylene gas interacts with other plant hormones to control leaf fall (ethylene gas + abscisic acid), to stimulate flower production (ethylene gas + auxins), and to control the ratio of male to female flowers on some monoecious plants (ethylene gas + gibberellins).

ABSCISIC ACID

Not all plant hormones promote growth. **Abscisic acid (ABA)** almost always *inhibits* growth by either reducing the rate of cell division and elongation or halting the processes altogether. Abscisic acid was given its name because early investigators believed that ABA was involved in **abscission** —the process that causes flowers, fruits, and leaves to fall. While originally thought to be associated with both dormancy and abscission, more recently, abscisic acid has been shown to have little or no role in either process.

⊘ ABA-suspended growth can lead to plant dormancy, which can save a plant's life during periods of severe environmental conditions. Repressed growth better enables dormant plants to withstand the stresses of dry, hot periods

or periods of freezing temperatures. During fall and winter, abscisic acid levels remain high in the axillary buds of deciduous trees—plants that drop their leaves during unfavorable growth periods. High abscisic acid levels counteract growth-promoting auxins and gibberellins, preventing cell division and elongation at a time when delicate, growing tissues would normally be exposed to lethal temperatures. ABA concentration drops at the onset of favorable growing conditions, releasing buds from their dormancy and causing plants to erupt with new growth (Figure 21-7). Alternating periods of dormancy and active growth which are keyed to environmental changes are examples of order and regulation in plants.

Under drought conditions, the concentration of ABA in guard cells may increase up to 40 times. Increased ABA concentration causes stomates to close, thereby reducing water loss and averting wilting and dehydration. Because ABA is active during periods of stress (triggering stomate closure during periods of water stress or promoting plant dormancy during periods of temperature stress), it is sometimes referred to as "the stress hormone."

OTHER PLANT HORMONES

Experimental evidence suggests there may be more than just five groups of plant hormones. For example, some studies suggest the presence of a flower-stimulating hor-

FIGURE 21-7

Deciduous plants burst out with new growth after a period of dormancy, when conditions become favorable for growth.

mone. In one experiment, two plants were grafted together; both were kept completely in the dark, except for the tip of one leaf on one plant. When this leaf tip was exposed to the critical photoperiod, both plants bloomed, suggesting that a hormone may have been involved. The hormone was transported to stems via the phloem, where it triggered flower-bud development. Since both experimental plants produced flowers, the hormone must have traveled to the stems of the grafted plant as well, inducing flower development. This elusive flowering chemical has yet to be isolated, although it already has been named **florigen.**

In addition to the possibility of a flowering hormone, some investigations suggest the existence of a root-growth hormone. Like the flowering hormone, the root-forming hormone has not yet been isolated.

Recently, scientists have discovered that some plants produce chemicals that act like the growth hormones of insects. These plant chemicals mimic the insect's hormones so closely that they disrupt the insect's normal growth and development, preventing sexual reproduction of the insect.

Plant growth and development is regulated by five groups of hormones—auxins, gibberellins, cytokinins, ethylene gas, and abscisic acid—although more hormones may be discovered in the future. Each hormone affects the plant differently; however, many growth and development responses result from the combined affects of two or more hormones. (See CTQ #3.)

TIMING GROWTH AND DEVELOPMENT

In many parts of the world, spring triggers a lavish surge of plant growth and development. Seeds are jolted out of dormancy; ghostly deciduous trees erupt with new leaves; and branches become crowded with masses of blossoms. But not all plants bloom or germinate in early spring. Flowering or germination is delayed in some plants until late spring, summer, or even fall. Clearly, plants are able to synchronize growth and development with the changing seasons; some, with great precision. In fact, Indian tribes in Arizona based their annual calender on the punctual flowering and fruiting of the saguaro cactus; a new year was marked by the appearance of the first saguaro fruits.

In addition to seasonal cycles, plants also have daily cycles. For example, the flowers of some plants open only at certain times of the day—morning glory (*Ipomoea purpurea*) at 4:00 A.M., fireweed (*Epilobium angustifolium*) at 6:00 P.M., nightshades (*Datura*) at 7:00 P.M., and the queen of the night (*Selenicereus grandiflora*) at 9:00 P.M. In the mid 1700s, Carolus Linnaeus, the Swedish scientist who introduced the binomial system for naming organisms (see page 13) established a "flower clock" in the Botanical Gardens of Uppsala, using plants that bloomed throughout the day and night to indicate the time of day.

Just how do plants monitor the time of day or the season of the year? Plants are able to detect and respond to seasonal environmental changes by monitoring changes in temperature and **photoperiod**—the relative length of daylight to dark in a 24-hour period. Plants use an internal *biological clock* to measure the time of day. Together, temperature, photoperiod, and an internal biological clock regulate cell metabolism and hormone production so that plant growth and development remain synchronized with environmental changes, enabling plants to live in some of the earth's harshest environments.

TEMPERATURE AND GROWTH

Plants grow in virtually all parts of the world, including Death Valley, California, where the world's highest air temperature was recorded at 57.8°C (136°F), and Siberia, where the world's lowest recorded temperature was −87°C (−126°F).

Each plant species has a particular range of temperatures it can tolerate, including an optimum temperature at which it grows best, and maximum and minimum temperatures it can tolerate for only limited periods time. Thus, where a plant grows is partly determined by the prevailing temperatures and the duration of temperature extremes. For example, saguaro cactus seedlings cannot survive more than 19 hours of freezing temperatures, so the saguaro stops growing at the Haulapi Mountains of Arizona. Any further north, freezing temperatures persist for longer than 19 hours, and the saguaro could not survive.

Because plant responses are sensitive to temperature, wide temperature fluctuations or prolonged exposure to extreme temperature can threaten a plant's life. Because a dormant plant's metabolism is significantly reduced, however, the plant is able to withstand greater temperature extremes for longer periods. On the days that the highest and lowest air temperatures were recorded in Death Valley and Siberia, most of the plants were dormant. Had they been actively growing, vulnerable meristems and primary tissues would have been killed, jeopardizing the life of the entire plant.

⏩ Dormancy requires a number of preparations, including building up stores of food, water, and other nutrients and growing protective structures (such as axillary bud scales) to shield delicate tissues. Plants must be able to detect the approach of unfavorable growing conditions to prepare for dormancy as well as to detect the end of unfavorable growing conditions so that they can resume growth as favorable conditions return. Because temperature is closely tied to other weather factors (such as relative humidity, rainfall, and snowfall), gradual temperature changes act as a predictor of upcoming weather. By responding to temperature trends, plants synchronize growth and development to upcoming weather.

Temperature pattern plays another role in synchronizing plant activity. Some plants must be exposed to a certain period of cold before they will flower or before their seeds will germinate. The promotion of flowering by cold is called *vernalization*. Many stone fruit trees—plants that produce fruit that have pits (peaches, apricots, plums)—and apple trees fail to flower (and, therefore, fail to produce fruit) unless they are exposed to cold temperatures for precise numbers of hours (ranging from 250 to more than 1,200 hours, depending on the species and the latitude and altitude of the environment). Similarly, many seeds germinate only after they are exposed to near-freezing temperatures for several weeks. Plant sensitivity to cold is an adaptation favored by natural selection. A requisite, minimum cold period is an indicator that bleak winter conditions have passed and is not just a brief cold spell. If a plant were to resume growth following *any* length of cold, it might do so soon after at the onset of winter, freezing all new tissues and threatening the plant's life.

LIGHT AND GROWTH

In addition to responding to temperature cues, many plants react to variations in photoperiod. In the Northern Hemisphere, the period of daylight gradually lengthens from December until late June and then gradually shortens from late June until early December. Photoperiods are a very reliable indicator. On any given day of the year, the length of daylight and darkness is nearly constant from one year to the next in a particular region. This constancy explains the precise blooming of the saguaro cactus each year in the southwest deserts of North America.

The ability of organisms to respond to day length is known as **photoperiodism**. Photoperiodism enables plants to induce flowering at times when pollinators are active; to initiate dormancy as conditions gradually become less favorable; to correlate seed germination with periods favorable for seedling growth; and to control stem and root development of an emerging seedling. But how do plants measure photoperiod?

Phytochrome

In the 1950s, a group of plant scientists at the U.S. Department of Agriculture in Beltsville, Maryland, conducted a series of studies that exposed seeds and other plant parts to specific wavelengths of light and measured the plant responses. The Beltsville group observed that red light (660 to 680 nanometers) induced certain responses, while far-red light (700 to 740 nanometers) inhibited or nullified responses (Figure 21-8). These observations, together with results from studies on the affects of length of darkness on

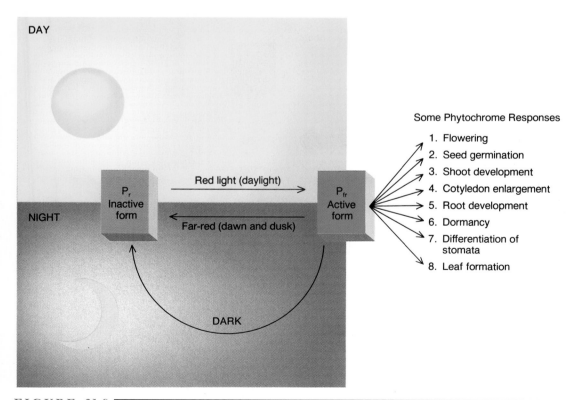

FIGURE 21-8

Phytochrome and plant development. P_{fr}, the active form of phytochrome, is formed in daylight when P_r absorbs red light and converts to P_{fr}. P_{fr} reverts back to inactive P_r at dawn, dusk, and in the dark and is gradually destroyed by cell enzymes. The length of the dark period largely determines how much active P_{fr} remains to stimulate or inhibit a number of plant growth responses.

plant responses, led the Beltsville researchers to propose that plants contained a blue-green pigment that exists in two, interconvertible forms. They named this alternating pigment **phytochrome** (*phyto* = plant, *chrome* = color).

After many studies, scientists verified that phytochrome contributes to the plant's ability to monitor photoperiod. One form of phytochrome absorbs red light at 660-nanometer wavelengths and is called P_r (phytochrome *red*). The other form of phytochrome absorbs far-red light at 730 nanometers and is called P_{fr} (phytochrome *far-red*). P_{fr} is the active form of phytochrome; it promotes or inhibits photoperiodic responses.

Phytochrome flips back and forth between its two forms as it absorbs light: When P_r absorbs red light, it immediately converts to the P_{fr} form; when P_{fr} absorbs far-red light, it immediately converts to P_r. P_{fr} also gradually converts into P_r in the dark (refer to Figure 21-9). These back-and-forth conversions between the two forms of phytochrome enables the plant to measure the length of daylight and darkness during a 24-hour period.

At sunrise, red light predominates, triggering the conversion of phytochrome to the P_{fr} form. Thus, the first appearance of P_{fr} marks the beginning of a new day. Since red light predominates throughout the day, P_{fr} accumulates. At sunset, far-red light predominates, so phytochrome is converted to the P_r form; the appearance of P_r denotes the end of daylight. In other words, P_{fr} accumulates during daylight, and P_r accumulates in the dark. The proportion of P_{fr} to P_r is the means by which plants chemically measure the length of daylight and darkness. But, as researchers have discovered during their experiments, even a brief flash of light during the dark hours converts P_r back to P_{fr}, thereby changing the P_{fr} to P_r ratio, and, ultimately, the photoperiodic response.

The means by which phytochrome stimulates or blocks photoperiodic responses are still being investigated. Because phytochrome is not transported from cell to cell it is

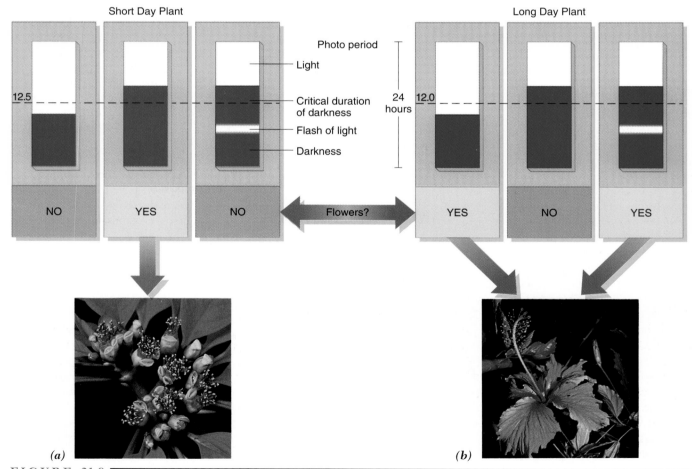

(a)   *(b)*

FIGURE 21-9

Photoperiod and flowering. Short-day and long-day plants flower in response to changes in photoperiod. *(a)* The poinsettia is a short-day plant that flowers when the day length becomes *less* than 12.5 hours of daylight (critical day length). *(b)* The hibiscus is a long-day plant that flowers when the day length becomes *greater* than 12 hours (critical day length).

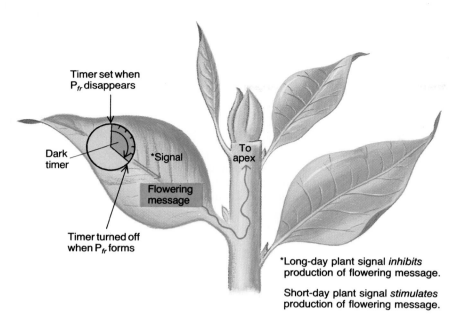

Timer set when P$_{fr}$ disappears

Dark timer

*Signal

To apex

Flowering message

Timer turned off when P$_{fr}$ forms

*Long-day plant signal *inhibits* production of flowering message.

Short-day plant signal *stimulates* production of flowering message.

FIGURE 21-10

Regulating flowering in long-day and short-day plants. Leaf cells have a "dark timer" that measures the length of darkness. When the night reaches a critical length, the signal either inhibits (in long-day plants) or stimulates (in short-day plants) the production of one or more hormones that convey a flowering message to the stem, promoting flower bud development.

not a hormone, although evidence suggests that phytochrome probably controls the activity of plant hormones. For example, studies suggest that phytochrome is embedded in cell membranes and controls the passage of hormones into the cell, thereby altering cell activity.

Photoperiod and Flowering

◆ It would be advantageous for plants to time flowering so that the process coincides with pollinator activity as well as with favorable growing conditions that last long enough for flowers, fruits, and seeds to mature before severe weather returns. Natural selection favors those plants that have adaptations that enable them to monitor photoperiod so that reproduction is optimized.

Plants are classified into three groups, depending on how changes in day length affect their flowering. **Short-day plants**, such as poinsettias, rice, morning glory, chrysanthemum, and ragweed, flower when the length of daylight grows shorter than some critical photoperiod, usually in late summer or fall (Figure 21-9a). **Long-day plants** flower when the length of daylight becomes longer than some critical photoperiod, usually in the spring and early summer (Figure 21-9b). Larkspurs, spinach, wheat, lettuce, mustard, and petunias are examples of long-day plants. **Day-neutral plants** flower independently of day length. Most day-neutral plants flower only when they are mature enough to do so, although other factors, such as temperature and water supply, may also affect flowering. Day-neutral plants include carnations, roses, snapdragons, sunflowers, and, as most gardeners know, many common weeds. Many agriculturally important plants, such as corn, beans, and tomatoes, are also day-neutral plants.

It is easy to assume mistakenly that short-day plants will always bloom when days are short, and long-day plants will only bloom when days are long. In fact, early researchers

developed this terminology by noting that some plants required shorter days to bloom, while others required either longer days or no specific day length. But further experimentation revealed that the difference between short- and long-day plants depended on whether the daylength was shorter or longer than some *critical photoperiod* (Figure 21-10). In long-day plants, the critical photoperiod represents a *minimum* day length; in short-day plants the critical photoperiod represents a *maximum* day length. For example, hibiscus (*Hibiscus syriacus*) and rice (*Oryza sativa*) are plants with an identical critical day length of 12 hours. However, the hibiscus is a long-day plant because it must be exposed to a day length *longer* than 12 hours in order to flower (12 hours is the *minimum* day length), whereas rice is a short-day plant because it must be exposed to a day length *shorter* than 12 hours in order to flower (12 hours is the *maximum* day length).

↻ As you might expect, phytochrome plays a role in inducing flowering in both short-day and long-day plants. But after a great deal of research, investigators discovered there was more to photoperiod and flowering than just phytochrome conversions. Experiments suggest that the formation and disappearance of P$_r$ and P$_{fr}$ apparently sets a natural timing system within plant cells. But instead of measuring the length of daylight, as the terms *long-day* and *short-day* suggested, the timer actually measured the length of darkness.

The cell's "dark timer" works like a stopwatch. The timer starts running when P$_{fr}$ begins to disappear at sunset and stops when P$_{fr}$ begins forming at sunrise. If the dark timer has run for the critical length of time, then a message (perhaps an unknown hormone) is sent from the leaves to the shoot apex, stimulating axillary buds to develop into flowers (Figure 21-10). Long-day and short-day plants differ in whether the timer triggers (short-day plants) or inhibits (long-day plants) the formation of this flowering message.

Phytochrome and Shoot Development

The moment a seedling nudges its way through the soil surface, light begins affecting the way the seedling grows and develops (for example, auxins cause shoots to bend). As a shoot emerges, daylight induces P_{fr} conversion. The appearance of P_{fr} stimulates hormonal changes that cause the shoot to straighten; stimulate leaf enlargement, and begin chlorophyll synthesis for photosynthesis—normal shoot growth and development for a seedling emerging from the soil.

But even before a seedling nudges through the soil surface, the *absence* of light has already had an impact on shoot development. In buried seeds, phytochrome remains in the P_r form. The active P_{fr} form that triggers normal growth regulators is lacking, so the shoot elongates rapidly; leaves remain small and underdeveloped; the shoot hook (if present) remains bent; and the plant remains colorless, devoid of chlorophyll. These features are associated with a condition called **etiolation**. Plants kept in the dark, such as the grass under a board, will also become etiolated because of an absence of P_{fr}.

Shade affects stem growth similarly. A plant growing in the shade receives mostly far-red light, red light having been absorbed by the leaves of surrounding taller plants. Far-red light triggers the conversion of phytochrome to P_r, reducing the amount of P_{fr} present. A lower P_{fr} to P_r ratio boosts stem elongation, increasing a plant's chances of growing out from under the shade.

Phytochrome and Seed Germination

Some seeds do not germinate in the dark, which means that they will not germinate under the ground either. When these seeds are exposed to light, however, they germinate. This is why you must continually pull weeds from your garden: The seeds of many weedy plants germinate only after exposure to light, which is just what they get when you turn the soil. Experiments demonstrate that even a short flash of red light induces light-requiring seeds to germinate. This is a clue that phytochrome is involved. The red light shifts P_r to P_{fr}, removing the germination block.

BIOLOGICAL CLOCKS

Plant growth and development are controlled not only by changes in temperature and light but also by internal biological clocks. Recall that the "dark timer" we mentioned earlier measures the length of night, launching flowering in some plants. Other plant activities are also controlled by biological clocks. For example, the leaves of the silk tree (*Albizzia*), as well as those of other legumes, are oriented horizontally during daylight but fall into a more compact vertical "sleeping" position at night. Even when kept in continuous light and at constant temperature, the leaves fall into the "sleeping" position at exactly the same time each day. This rhythm is said to be **endogenous** (controlled internally), rather than **exogenous** (controlled by outside environmental changes). The sleep movements of leaves follow a circadian (*circa* = about, *diem* = day) cycle, recurring at 24-hour intervals. Other endogenous circadian plant activities include flower opening and closing and nectar and flower odor production.

By monitoring changes in temperature and light, plants adjust their growth and development to coincide with favorable environmental periods. (See CTQ #4.)

PLANT TROPISMS

Several environmental factors change the rate of production or distribution of growth-regulating hormones, thereby changing the direction of plant growth. Such responses are called **tropisms**. Tropisms are stimulated by light, gravity, contact with objects, chemicals, temperature gradients, wounding, and the presence of water. The three most common tropisms are phototropism, gravitropism, and thigmotropism.

PHOTOTROPISM

A familiar tropism occurs in many houseplants. If left in one spot, houseplants gradually turn their leaf surfaces toward a light source, often a nearby window. Since light is the environmental stimulus, the response is called a **phototropism** (*photo* = light). Shoot bending, which we discussed in the chapter opener, is another example of phototropism. Experiments reveal that a light-absorbing, yellow pigment—a flavoprotein—promotes the movement of auxins away from the light source, increasing cell elongation in the cells on the shaded side (Figure 21-1 and 21-4).

GRAVITROPISM

When the pull of gravity causes changes in a plant's growth, the response is called **gravitropism**. Gravity indicates which way is down and which way is up, helping to direct growth of roots and shoots in the right direction. Shoot growth away from the gravitational force is called **negative gravitropism**, whereas root growth toward the gravitational force is called **positive gravitropism**.

Researchers believe that plants detect the direction of gravity by small particles inside their cells which are pulled to the bottom of the cell by gravity. The change in position of these particles apparently redirects the transport of growth hormones, generating the gravitropic response. In one experiment, a growing shoot was placed on its side. Within 15 minutes, the lower side of the shoot contained twice as much auxin than did the upper side. This unequal distribution caused the stem to bend upward (see Figure 21-3). Once upright, the particles settled into their original position inside the cell; auxin concentration became equally distributed on all sides; and the shoot grows straight up.

The root cap also plays an important role in controlling the direction and rate of root growth. If the root cap is experimentally removed, the root grows faster than normal but not downward. This suggests that the root cap produces a growth inhibitor that slows the production and elongation of certain root-tip cells so that growth is always downward. When a root is placed horizontally, the inhibitor accumulates in the cells on the lower side, reducing their growth relative to cells on the top side and producing downward bending (Figure 21-11). Unlike stems, root bending is probably not controlled by auxin; some experimental evidence suggests that abscisic acid may in fact be the growth inhibitor in roots.

THIGMOTROPISM

A change in plant growth that is stimulated by contact with another object is called a **thigmotropism**. For example, many climbing plants have stems and branches that are too weak to support their own weight so they rely on other objects to keep them upright. Using tendrils, some plants either wrap around a supporting structure (Figure 21-12*a*) or produce pads at the tips of branches that fasten the plant to a wall, fence post, or some other solid object for support (Figure 21-12*b*).

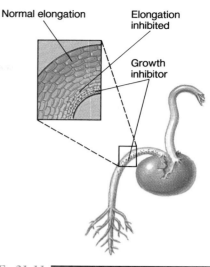

FIGURE 21-11
Which way is down? Gravity causes a growth-inhibiting hormone to accumulate on the lower side of a root, slowing cell elongation relative to cells on the upper side. This differential elongation causes the root to turn downward.

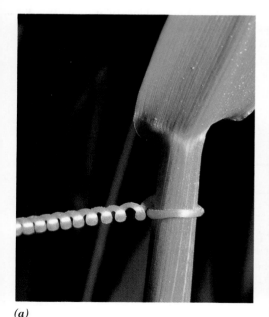

(a)   *(b)*

FIGURE 21-12
Some climbing vines have very weak stems and rely on a sturdier plant or some other solid object for support. The tendril of a pea plant grows around a corn stem for support *(a)*, while expanded pads of a Virginia creeper fasten its growing stems to a wall *(b)*.

Considered together—the diversity of plant tropisms, the mechanisms by which plants monitor and respond to changes in environmental conditions, and the production of hormones that coordinate growth and development—all help dispell a common misconception that plants are a static, comparatively unresponsive life form. The diversity of plant growth and development adaptations, the complexity by which plants are regulated, and the precision by which plants operate, reveal just the opposite: Plants are dynamic organisms; they continually respond and adjust to their surroundings in ways that promote survival and reproduction.

Tropisms are the result of changes in the size or shape of many cells or of changes in internal water pressure within a few strategically placed cells. Changes in a plant's internal water pressure bring about rapid responses, whereas long-term adjustments require permanent modifications in cell structure. (See CTQ #6.)

REEXAMINING THE THEMES

Relationship between Form and Function

Hormones control and coordinate plant growth and development. The production of hormones enables plants to adjust their form so that resulting structures will be better able to carry out their functions. Shoots growing upward, roots growing downward, stems and leaves growing toward light, flowering at certain times of the year, seedling stems increasing thickness and bending as they push through the soil, and tendrils and pad production to support stems are all examples of the modification of plant form as a result of hormone production to increase function.

Biological Order, Regulation, and Homeostasis

The regulation of growth and development in plants follows a hierarchy of controls, the combination of which enables plants to germinate, grow, develop, and produce flowers and fruits at the most optimum times of the year.

Genes are controlled by hormones. In turn, the production and distribution of hormones are affected by the environment.

Evolution and Adaptation

Plant seeds can find their way to any number of habitats, each with slightly different environmental circumstances. For example, a seed may be dispersed to the side of a mountain or on flat ground. To survive, a developing seedling would have to grow differently in each of these habitats. On a mountainside, the shoot cannot grow perpendicular to the surface, as it would on flat ground, and the root could not grow directly down. A plant disperses its embryos when they are in a very underdeveloped form, which enables them to adjust to particular environmental circumstances.

SYNOPSIS

Plant growth and development are the result of three interacting levels of control. The intracellular level contains fixed genetic controls, which are regulated by intercellular hormonal controls, which, in turn, are affected by various aspects of the environment. Environmental factors greatly influence plant growth and development, helping adjust a plant's form to its immediate surroundings.

Effects of plant hormones vary with hormone concentration, target tissue, time of year, developmental stage of the plant, and type of plant. Most growth and development responses are the result of combinations of plant hormones.

In general, four of the five groups of plant hormones (auxins, gibberellins, cytokinins, and ethylene gas) promote growth, whereas the fifth (abscisic acid) inhibits growth. Recent research suggests that there may be other plant hormones besides these five.

Many plants synchronize dormancy and periods of active growth with changing seasons by responding to variations in temperature and day length and by using an internal clock. Dormancy is often critical for a plant to survive periods of unfavorable growing conditions.

Interconvertible forms of light-absorbing phytochrome enable plants to monitor changes in photoperiod. Monitoring photoperiod helps plants synchronize flowering, seed germination, and renewed vegetative growth to favorable periods of environmental conditions.

Key Terms

auxin (p. 427)
indoleacetic acid (IAA) (p. 430)
apical dominance (p. 430)
gibberellin (p. 431)
cytokinin (p. 432)
senescence (p. 432)
ethylene gas (p. 432)
abscisic acid (ABA) (p. 433)
abscission (p. 433)

florigen (p. 434)
photoperiod (p. 434)
photoperiodism (p. 435)
phytochrome (p. 436)
short-day plant (p. 437)
long-day plant (p. 437)
day-neutral plant (p. 437)
etiolation (p. 438)
endogenous (p. 438)

exogenous (p. 438)
tropism (p. 438)
phototropism (p. 438)
gravitropism (p. 438)
negative gravitropism (p. 438)
positive gravitropism (p. 438)
thigmotropism (p. 439)

Review Questions

1. List five reasons why it has been so difficult for plant researchers to work with and determine the functions of plant hormones.

2. How do the characteristics of etiolated seedlings improve their chances of emerging from the soil?

3. Match the hormone with its function (multiple matches are possible).

_____auxins		a.	fruit ripening
_____cytokinins		b.	apical dominance
_____gibberellins		c.	senescence
_____ethylene gas		d.	stimulate stem elongation
_____abscisic acid		e.	seed dormancy

4. Growth can be the result of cell division or cell enlargement (mainly elongation). In plants, most growth results from cell elongation; a mature plant cell may be ten times larger than a cell produced by the apical meristem. Since all plant cells have a surrounding cell wall, what must happen in order for plant cells to be able to elongate? What force is involved?

5. Consider a young vine struggling to grow in a dense tropical rain forest. Which tropisms are likely be involved if the vine is to grow successfully and develop to maturity?

6. Are plant tropisms irreversible?

7. List three hormones that *promote* growth. What is different about the way the hormones increase growth?

8. Phytochrome is a plant pigment, yet it differs from those involved in photosynthesis, such as chlorophyll. How is phytochrome similar to photosynthetic pigments, and how is it different?

9. If you remove the apex of a stem, will growth in length of that stem stop altogether? Explain.

10. Explain how photoperiodism works. What is the adaptive advantage to photoperiodism for plants?

Critical Thinking Questions

1. Explaining the results of their experiments on light and stem bending, the Darwins reasoned that a growth signal was produced in the coleoptile tip. According to the Darwins, the signal traveled down the shoot and triggered the cells below to enlarge differentially which, in turn, caused the young shoot to bend toward light. Describe two ways in which cells could "enlarge differentially" and cause a stem to bend toward light. How can you determine which way is correct?

2. One molecule of a plant hormone can cause 150,000 glucose molecules to be converted to cellulose. What does this tell you about how plant hormones work? How are they similar to enzymes? Different from enzymes?

3. A botanist gave wheat seedlings a dose of X-rays sufficient to prevent cell division but which left the seedlings able to function normally otherwise. He gave half of the irradiated seedlings gibberellin, but not the other half. The length of the seedlings is plotted in the graph below. Which of the following conclusions is supported by the experiment, and why? (1) gibberellins stimulate cell elongation only; (2) gibberellins stimulate cell division only; (3) gibberellins stimulate both cell elongation and cell division.

4. The seeds of some plants do not germinate in the spring, even though temperatures are generally warm and water is plentiful—two conditions that are favorable for growth. Suggest what might be preventing these seeds from germinating. Would there be any advantage to delaying germination past spring?

5. Discuss some of the ways you can use your new knowledge of plant growth and development to improve the growth of plants in your garden.

6. The discovery of auxin began with the study of tropisms by Darwin and others. Knowledge of auxins has helped scientists understand tropisms. Summarize the significant events in this history by means of a diagram that illustrates the interaction between these two lines of research.

7. Colonizing space presents many difficult challenges, not the least of which is how plants can be grown for food and for providing a quality environment for the human inhabitants. Discuss some of the problems you would anticipate in trying to grow plants in a space station. For each problem, try to suggest a solution.

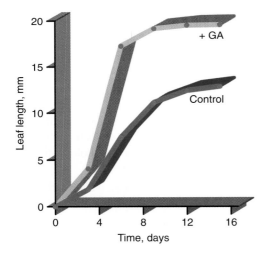

Additional Readings

Darwin, C. 1881. *The power of movement in plants*. New York: Appleton-Century-Crofts. (Intermediate)

Davies, P. 1987. *Plant hormones and their role in plant growth and development*. Dordrecht, Netherlands: Martinus Nijhoff. (Advanced)

Gregory, P. J., J. V. Lake, and D. A. Rose (eds.). 1987. *Root development and function*. Society for Experimental Biology. Press Syndicate of the Univ. of Cambridge, Cambridge, England. (Advanced)

Halstad, T., and D. Dilcher. 1987. Plants in space. *Ann. Rev. Plant Physiol.* 38: 317–345. (Intermediate)

Taiz, L., and E. Zeiger. 1991. *Plant physiology*. Reading, MA: Benjamin Cummings. (Advanced)

Torrey, J. 1985. The development of plant biotechnology. *Amer. Sci.* 73: 354–363. (Intermediate)

Van Overbeek, J. 1968. The control of plant growths. *Sci. Amer.* July: 75–81. (Introductory)

Zimmermann, M. H., and C. L. Brown. 1975. Trees: *Structure and Function*. Springer-Verlag, New York. (Intermediate)

Salisbury, F. B., and C. W. Ross. 1991. *Plant Physiology*, 4th ed. Wadsworth Publishing Co., Belmont, Ca. (Advanced)

Steeues, T. A., and I. M. Sussex. 1989. *Patterns in plant development*. 2nd ed. Prentice Hall, Inc. New Jersey. (Intermediate)

APPENDIX

◀ A ▶

Metric and Temperature Conversion Charts

Metric Unit (symbol)		Metric to English	English to Metric
Length			
kilometer (km)	= 1,000 (10^3) meters	1 km = 0.62 mile	1 mile = 1.609 km
meter (m)	= 100 centimeters	1 m = 1.09 yards	1 yard = 0.914 m
		= 3.28 feet	1 foot = 0.305 m
centimeter (cm)	= 0.01 (10^{-2}) meter	1 cm = 0.394 inch	1 inch = 2.54 cm
millimeter (mm)	= 0.001 (10^{-3}) meter	1 mm = 0.039 inch	1 inch = 25.4 mm
micrometer (μm)	= 0.000001 (10^{-6}) meter		
nanometer (nm)	= 0.000000001 (10^{-9}) meter		
angstrom (Å)	= 0.0000000001 (10^{-10}) meter		
Area			
square kilometer (km^2)	= 100 hectares	1 km^2 = 0.386 square mile	1 square mile = 2.590 km^2
hectare (ha)	= 10,000 square meters	1 ha = 2.471 acres	1 acre = 0.405 ha
square meter (m^2)	= 10,000 square centimeters	1 m^2 = 1.196 square yards	1 square yard = 0.836 m^2
		= 10.764 square feet	1 square foot = 0.093 m^2
square centimeter (cm^2)	= 100 square millimeters	1 cm^2 = 0.155 square inch	1 square inch = 6.452 cm^2
Mass			
metric ton (t)	= 1,000 kilograms	1 t = 1.103 tons	1 ton = 0.907 t
	= 1,000,000 grams		
kilogram (kg)	= 1,000 grams	1 kg = 2.205 pounds	1 pound = 0.454 kg
gram (g)	= 1,000 milligrams	1 g = 0.035 ounce	1 ounce = 28.35 g
milligram (mg)	= 0.001 gram		
microgram (μg)	= 0.000001 gram		
Volume Solids			
1 cubic meter (m^3)	= 1,000,000 cubic centimeters	1 m^3 = 1.308 cubic yards	1 cubic yard = 0.765 m^3
		= 35.315 cubic feet	1 cubic foot = 0.028 m^3
1 cubic centimeter (cm^3)	= 1,000 cubic millimeters	1 cm^3 = 0.061 cubic inch	1 cubic inch = 16.387 cm^3
Volume Liquids			
kiloliter (kl)	= 1,000 liters	1 kl = 264.17 gallons	
liter (l)	= 1,000 milliliters	1 l = 1.06 quarts	1 gal = 3.785 l
			1 qt = 0.94 l
			1 pt = 0.47 l
milliliter (ml)	= 0.001 liter	1 ml = 0.034 fluid ounce	1 fluid ounce = 29.57 ml
microliter (μl)	= 0.000001 liter		

TEMPERATURE

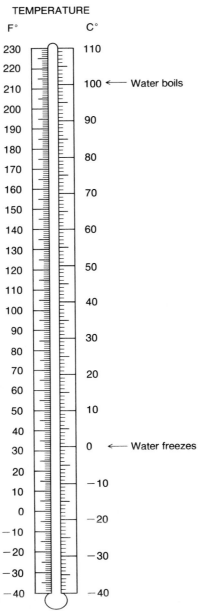

F°   C°

100 ←— Water boils

0 ←— Water freezes

Fahrenheit to Centigrade: °C = ⅝ (°F − 32)
Centigrade to Fahrenheit: °F = ⅝ (°C + 32)

APPENDIX
◄ B ►

Microscopes: Exploring the Details of Life

Microscopes are the instruments that have allowed biologists to visualize objects that are vastly smaller than anything visible with the naked eye. There are broadly two types of specimens viewed in a microscope: whole mounts which consist of an intact subject, such as a hair, a living cell, or even a DNA molecule, and thin sections of a specimen, such as a cell or piece of tissue.

THE LIGHT MICROSCOPE

A light microscope consists of a series of glass lenses that bend (refract) the light coming from an illuminated specimen so as to form a visual image of the specimen that is larger than the specimen itself (a). The specimen is often stained with a colored dye to increase its visibility. A special phase contrast light microscope is best suited for observing unstained, living cells because it converts differences in the density of cell organelles, which are normally invisible to the eye, into differences in light intensity which can be seen.

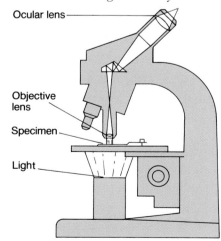

Ocular lens

Objective lens

Specimen

Light

(a)

All light microscopes have limited *resolving power*— the ability to distinguish two very close objects as being separate from each other. The resolving power of the light microscope is about 0.2 μm (about 1,000 times that of the naked eye), a property determined by the wave length of visible light. Consequently, objects closer to each other than 0.2 μm, which includes many of the smaller cell organ-

elles, will be seen as a single, blurred object through a light microscope.

THE TRANSMISSION ELECTRON MICROSCOPE

Appreciation of the wondrous complexity of cellular organization awaited the development of the transmission electron microscope (or TEM), which can deliver resolving powers 1000 times greater than the light microscope. Suddenly, biologists could see strange new structures, whose function was totally unknown—a breakthrough that has kept cell biologists busy for the past 50 years. The TEM (b) works by shooting a beam of electrons through very thinly sliced specimens that have been stained with heavy metals,

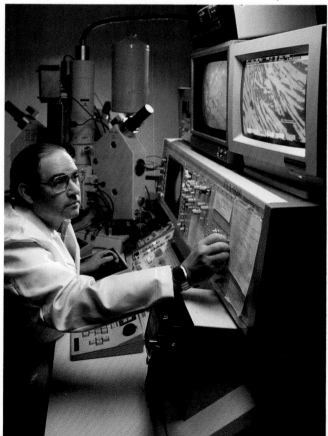

(b)

such as uranium, capable of deflecting electrons in the beam. The electrons that pass through the specimen undeflected are focused by powerful electromagnets (the lenses of a TEM) onto either a phosphorescent screen or high-contrast photographic film. The resolution of the TEM is so great—sufficient to allow us to see individual DNA molecules—because the wavelength of an electron beam is so small (about 0.0005 μm).

THE SCANNING ELECTRON MICROSCOPE

Specimens examined in the scanning electron microscope (SEM) are whole mounts whose surfaces have been coated with a thin layer of heavy metals. In the SEM, a fine beam of electrons scans back and forth across the specimen and the image is formed from electrons bouncing off the hills and valleys of its surface. The SEM produces a three-dimensional image of the surface of the specimen—which can

(c) *(d)*

range in size from a virus to an insect head (c,d)—with remarkable depth and clarity. The SEM produces black and white images; the colors seen in many of the micrographs in the text have been added to enhance their visual quality. Note that the insect head (d) is that of an antennatedia mutant as described on p 687.

APPENDIX

◀ **C** ▶

The Hardy-Weinberg Principle

If the allele for brown hair is dominant over that for blond hair, and curly hair is dominant over straight hair, then why don't all people by now have brown, curly hair? The **Hardy-Weinberg Principle** (developed independently by English mathematician G. H. Hardy and German physician W. Weinberg) demonstrates that the frequency of alleles remains the same from generation to generation unless influenced by outside factors. The outside factors that would cause allele frequencies to change are mutation, immigration and emigration (movement of individuals into and out of a breeding population, respectively), natural selection of particular traits, and breeding between members of a small population. In other words, unless one or more of these forces influence hair color and hair curl, the relative number of people with brown and curly hair will not increase over those with blond and straight hair.

To illustrate the Hardy-Weinberg Principle, consider a single gene locus with two alleles, A and a, in a breeding population. (If you wish, consider A to be the allele for brown hair and a to be the allele for blond hair.) Because there are only two alleles for the gene, the sum of the frequencies of A and a will equal 1.0. (By convention, allele frequencies are given in decimals instead of percentages.) Translating this into mathematical terms, if

p = the frequency of allele A, and
q = the frequency of allele a,

then $p + q = 1$.

If A represented 80 percent of the alleles in the breeding population ($p = 0.8$), then according to this formula the frequency of a must be 0.2 ($p + q = 0.8 + 0.2 = 1.0$).

After determining the allele frequency in a starting population, the predicted frequencies of alleles and genotypes in the next generation can be calculated. Setting up a Punnett square with starting allele frequencies of $p = 0.8$ and $q = 0.2$:

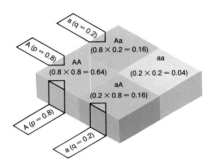

The chances of each offspring receiving any combination of the two alleles is the product of the probability of receiving one of the two alleles alone. In this example, the chances of an offspring receiving two A alleles is $p \times p = p^2$, or $0.8 \times 0.8 = 0.64$. A frequency of 0.64 means that 64 percent of the next generation will be homozygous dominant (AA). The chances of an offspring receiving two a alleles is $q^2 = 0.2 \times 0.2 = 0.04$, meaning 4 percent of the next generation is predicted to be aa. The predicted frequency of heterozygotes (Aa or aA) is 0.32 or $2pq$, the sum of the probability of an individual being $Aa(p \times q = 0.8 \times 0.2 = 0.16)$ plus the probability of an individual being $aA(q \times p = 0.2 \times 0.8 = 0.16)$. Just as all of the allele frequencies for a particular gene must add up to 1, so must all of the possible genotypes for a particular gene locus add up to 1. Thus, the Hardy-Weinberg Principle is

$$p^2 + 2pq + q^2 = 1$$
$$(0.64 + 0.32 + 0.04 = 1)$$

So after one generation, the frequency of possible genotypes is

$$AA = p^2 = 0.64$$
$$Aa = 2pq = 0.32$$
$$aa = q^2 = 0.04$$

Now let's determine the actual allele frequencies for *A* and *a* in the new generation. (Remember the original allele frequencies were 0.8 for allele *A* and 0.2 for allele *a*. If the Hardy-Weinberg Principle is right, there will be no change in the frequency of either allele.) To do this we sum the frequencies for each genotype containing the allele. Since heterozygotes carry both alleles, the genotype frequency must be divided in half to determine the frequency of each allele. (In our example, heterozygote *Aa* has a frequency of 0.32, 0.16 for allele *A*, plus 0.16 for allele *a*.) Summarizing then:

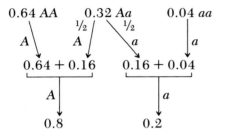

As predicted by the Hardy-Weinberg Principle, the frequency of allele *A* remained 0.8 and the frequency of allele *a* remained 0.2 in the new generation. Future generations can be calculated in exactly the same way, over and over again. As long as there are no mutations, no gene flow between populations, completely random mating, no natural selection, and no genetic drift, there will be no change in allele frequency, and therefore no evolution.

Population geneticists use the Hardy-Weinberg Principle to calculate a starting point allele frequency, a reference that can be compared to frequencies measured at some future time. The amount of deviation between observed allele frequencies and those predicted by the Hardy-Weinberg Principle indicates the degree of evolutionary change. Thus, this principle enables population geneticists to measure the rate of evolutionary change and identify the forces that cause changes in allele frequency.

APPENDIX

◄ D ►

Careers in Biology

Although many of you are enrolled in biology as a requirement for another major, some of you will become interested enough to investigate the career opportunities in life sciences. This interest in biology can grow into a satisfying livelihood. Here are some facts to consider:

- Biology is a field that offers a very wide range of possible science careers

- Biology offers high job security since many aspects of it deal with the most vital human needs: health and food

- Each year in the United States, nearly 40,000 people obtain bachelor's degrees in biology. But the number of newly created and vacated positions for biologists is increasing at a rate that exceeds the number of new graduates. Many of these jobs will be in the newer areas of biotechnology and bioservices.

Biologists not only enjoy job satisfaction, their work often changes the future for the better. Careers in medical biology help combat diseases and promote health. Biologists have been instrumental in preserving the earth's life-supporting capacity. Biotechnologists are engineering organisms that promise dramatic breakthroughs in medicine, food production, pest management, and environmental protection. Even the economic vitality of modern society will be increasingly linked to biology.

Biology also combines well with other fields of expertise. There is an increasing demand for people with backgrounds or majors in biology complexed with such areas as business, art, law, or engineering. Such a distinct blend of expertise gives a person a special advantage.

The average starting salary for all biologists with a Bachelor's degree is $22,000. A recent survey of California State University graduates in biology revealed that most were earning salaries between $20,000 and $50,000. But as important as salary is, most biologists stress job satisfaction, job security, work with sophisticated tools and scientific equipment, travel opportunities (either to the field or to scientific conferences), and opportunities to be creative in their job as the reasons they are happy in their career.

Here is a list of just a few of the careers for people with degrees in biology. For more resources, such as lists of current openings, career guides, and job banks, write to Biology Career Information, John Wiley and Sons, 605 Third Avenue, New York, NY 10158.

A SAMPLER OF JOBS THAT GRADUATES HAVE SECURED IN THE FIELD OF BIOLOGY°

Agricultural Biologist	Bioanalytical Chemist	Brain Function	Environmental Center
Agricultural Economist	Biochemical/Endocrine	Researcher	Director
Agricultural Extension	Toxicologist	Cancer Biologist	Environmental Engineer
Officer	Biochemical Engineer	Cardiovascular Biologist	Environmental Geographer
Agronomist	Pharmacology Distributor	Cardiovascular/Computer	Environmental Law Specialist
Amino-acid Analyst	Pharmacology Technician	Specialist	Farmer
Analytical Biochemist	Biochemist	Chemical Ecologist	Fetal Physiologist
Anatomist	Biogeochemist	Chromatographer	Flavorist
Animal Behavior	Biogeographer	Clinical Pharmacologist	Food Processing Technologist
Specialist	Biological Engineer	Coagulation Biochemist	Food Production Manager
Anticancer Drug Research	Biologist	Cognitive Neuroscientist	Food Quality Control
Technician	Biomedical	Computer Scientist	Inspector
Antiviral Therapist	Communication Biologist	Dental Assistant	Flower Grower
Arid Soils Technician	Biometerologist	Ecological Biochemist	Forest Ecologist
Audio-neurobiologist	Biophysicist	Electrophysiology/	Forest Economist
Author, Magazines & Books	Biotechnologist	Cardiovascular Technician	Forest Engineer
Behavioral Biologist	Blood Analyst	Energy Regulation Officer	Forest Geneticist
Bioanalyst	Botanist	Environmental Biochemist	Forest Manager

Forest Pathologist
Forest Plantation Manager
Forest Products Technologist
Forest Protection Expert
Forest Soils Analyst
Forester
Forestry Information Specialist
Freeze-Dry Engineer
Fresh Water Biologist
Grant Proposal Writer
Health Administrator
Health Inspector
Health Scientist
Hospital Administrator
Hydrologist
Illustrator
Immunochemist
Immunodiagnostic
    Assay Developer
Inflammation Technologist
Landscape Architect
Landscape Designer
Legislative Aid
Lepidopterist
Liaison Scientist,
    Library of Medicine
    Computer Biologist
Life Science Computer
    Technologist
Lipid Biochemist
Livestock Inspector
Lumber Inspector

Medical Assistant
Medical Imaging Technician
Medical Officer
Medical Products Developer
Medical Writer
Microbial Physiologist
Microbiologist
Mine Reclamation Scientist
Molecular Endocrinologist
Molecular Neurobiologist
Molecular Parasitologist
Molecular Toxicologist
Molecular Virologist
Morphologist
Natural Products Chemist
Natural Resources Manager
Nature Writer
Nematode Control Biologist
Nematode Specialist
Nematologist
Neuroanatomist
Neurobiologist
Neurophysiologist
Neuroscientist
Nucleic Acids Chemist
Nursing Aid
Nutritionist
Occupational Health Officer
Ornamental Horticulturist
Paleontologist
Paper Chemist
Parasitologist

Pathologist
Peptide Biochemist
Pharmaceutical Writer
Pharmaceutical Sales
Pharmacologist
Physiologist
Planning Consultant
Plant Pathologist
Plant Physiologist
Production Agronomist
Protein Biochemist
Protein Structure & Design
    Technician
Purification Biochemist
Quantitative Geneticist
Radiation Biologist
Radiological Scientist
Regional Planner
Regulatory Biologist
Renal Physiologist
Renal Toxicologist
Reproductive Toxicologist
Research and Development
    Director
Research Technician
Research Liaison Scientist
Research Products Designer
Research Proposal Writer
Safety Assessment Sanitarian
Scientific Illustrator
Scientific Photographer
Scientific Reference Librarian

Scientific Writer
Soil Microbiologist
Space Station Life Support
    Technician
Spectroscopist
Sports Product Designer
Steroid Health Assessor
Taxonomic Biologist
Teacher
Technical Analyst
Technical Science Project
    Writer
Textbook Editor
Theoretical Ecologist
Timber Harvester
Toxicologist
Toxic Waste Treatment
    Specialist
Urban Planner
Water Chemist
Water Resources Biologist
Wood Chemist
Wood Fuel Technician
Zoning and Planning
    Manager
Zoologist
Zoo Animal Breeder
Zoo Animal Behaviorist
Zoo Designer
Zoo Inspector

°Results of one survey of California State University graduates. Some careers may require advanced degrees

Glossary

◄ A ►

Abiotic Environment Components of ecosystems that include all nonliving factors. (41)

Abscisic Acid (ABA) A plant hormone that inhibits growth and causes stomata to close. ABA may not be commonly involved in leaf drop. (21)

Abscission Separation of leaves, fruit, and flowers from the stem. (21)

Acclimation A physiological adjustment to environmental stress. (28)

Acetyl CoA Acetyl coenzyme A. A complex formed when acetic acid binds to a carrier coenzyme forming a bridge between the end products of glycolysis and the Krebs cycle in respiration. (9)

Acetylcholine Neurotransmitter released by motor neurons at neuromuscular junctions and by some interneurons. (23)

Acid Rain Occurring in polluted air, rain that has a lower pH than rain from areas with unpolluted air. (40)

Acids Substances that release hydrogen ions (H^+) when dissolved in water. (3)

Acid Snow Occurring in polluted air, snow that has a lower pH than snow from areas with unpolluted air. (40)

Acoelomates Animals that lack a body cavity between the digestive cavity and body wall. (39)

Acquired Immune Deficiency Syndrome (AIDS) Disease caused by infection with HIV (Human Immunodeficiency Virus) that destroys the body's ability to mount an immune response due to destruction of its helper T cells. (30, 36)

Actin A contractile protein that makes up the major component of the thin filaments of a muscle cell and the microfilaments of nonmuscle cells. (26)

Action Potential A sudden, dramatic reversal of the voltage (potential difference) across the plasma membrane of a nerve or muscle cell due to the opening of the sodium channels. The basis of a nerve impulse. (23)

Activation Energy Energy required to initiate chemical reaction. (6)

Active Site Region on an enzyme that binds its specific substrates, making them more reactive. (6)

Active Transport Movement of substances into or out of cells against a concentration gradient, i.e., from a region of lower concentration to a region of higher concentration. The process requires an expenditure of energy by the cell. (7)

Adaptation A hereditary trait that improves an organism's chances of survival and/or reproduction. (33)

Adaptive Radiation The divergence of many species from a single ancestral line. (33)

Adenosine Triphosphate (ATP) The molecule present in all living organisms that provides energy for cellular reactions in its phosphate bonds. ATP is the universal energy currency of cells. (6)

Adenylate Cyclase An enzyme activated by hormones that converts ATP to cyclic AMP, a molecule that activates resting enzymes. (25)

Adrenal Cortex Outer layer of the adrenal glands. It secretes steroid hormones in response to ACTH. (25)

Adrenal Medulla An endocrine gland that controls metabolism, cardiovascular function, and stress responses. (25)

Adrenocorticotropic Hormone (ACTH) An anterior pituitary hormone that stimulates the cortex of the adrenal glands to secrete cortisol and other steroid hormones. (25)

Adventitious Root System Secondary roots that develop from stem or leaf tissues. (18)

Aerobe An organism that requires oxygen to release energy from food molecules. (9)

Aerobic Respiration Pathway by which glucose is completely oxidized to CO_2 and H_2O, requiring oxygen and an electron transport system. (9)

Afferent (Sensory) Neurons Neurons that conduct impulses from the sense organs to the central nervous system. (23)

Age-Sex Structure The number of individuals of a certain age and sex within a population. (43)

Aggregate Fruits Fruits that develop from many pistils in a single flower. (20)

AIDS See Acquired Immune Deficiency Syndrome.

Albinism A genetic condition characterized by an absence of epidermal pigmentation that can result from a deficiency of any of a variety of enzymes involved in pigment formation. (12)

Alcoholic Fermentation The process in which electrons removed during glycolysis are transferred from NADH to form alcohol as an end product. Used by yeast during the commercial process of ethyl alcohol production. (9)

Aldosterone A hormone secreted by the adrenal cortex that stimulates reabsorption of sodium from the distal tubules and collecting ducts of the kidneys. (23)

Algae Any unicellular or simple colonial photosynthetic eukaryote. (37,38)

Algin A substance produced by brown algae harvested for human application because of its ability to regulate texture and consistency of products. Found in ice cream, cosmetics, marshmellows, paints, and dozens of other products. (38)

Allantois Extraembryonic membrane that serves as a repository for nitrogenous wastes. In placental mammals, it helps form the vascular connections between mother and fetus. (32)

Allele Alternative form of a gene at a particular site, or locus, on the chromosome. (12)

Allele Frequency The relative occurrence of a certain allele in individuals of a population. (33)

Allelochemicals Chemicals released by some plants and animals that deter or kill a predator or competitor. (42)

Allelopathy A type of interaction in which one organism releases allelochemicals that harm another organism. (42)

Allergy An inappropriate response by the immune system to a harmless foreign substance leading to symptoms such as itchy eyes, runny nose, and congested airways. If the reaction occurs throughout the body (anaphylaxis) it can be life threatening. (30)

Allopatric Speciation Formation of new species when gene flow between parts of a population is stopped by geographic isolation. (33)

Alpha Helix Portion of a polypeptide chain organized into a defined spiral conformation. (4)

Alternation of Generations Sequential change during the life cycle of a plant in which a haploid (1N) multicellular stage (gametophyte) alternates with a diploid (2N) multicellular stage (sporophyte). (38)

Alternative Processing When a primary RNA transcript can be processed to form more than one mRNA depending on conditions. (15)

Altruism The performance of a behavior that benefits another member of the species at some cost to the one who does the deed. (44)

Alveolus A tiny pouch in the lung where gas is exchanged between the blood and the air; the functional unit of the lung where CO_2 and O_2 are exchanged. (29)

Alzheimer's Disease A degenerative disease of the human brain, particularly affecting acetylcholine-releasing neurons and the hippocampus, characterized by the presence of tangled fibrils within the cytoplasm of neurons and amyloid plaques outside the cells. (23)

Amino Acids Molecules containing an amino group (-NH_2) and a carboxyl group (-COOH) attached to a central carbon atom. Amino acids are the subunits from which proteins are constructed. (4)

Amniocentesis A procedure for obtaining fetal cells by withdrawing a sample of the fluid that surrounds a developing fetus (amniotic fluid) using a hypodermic needle and syringe. (17)

Amnion Extraembryonic membrane that envelops the young embryo and encloses the amniotic fluid that suspends and cushions it. (32)

Amoeba A protozoan that employs pseudopods for motility. (37)

Amphibia A vertebrate class grouped into three orders: Caudata (tailed amphibians); Anura (tail-less amphibians); Apoda (rare worm-like, burrowing amphibians). (39)

Anabolic Steroids Steroid hormones, such as testosterone, which promote biosynthesis (anabolism), especially protein synthesis. (25)

Anabolism Biosynthesis of complex molecules from simpler compounds. Anabolic pathways are endergonic, i.e., require energy. (6)

Anaerobe Organism that does not require oxygen to release energy from food molecules. (9)

Anaerobic Respiration Pathway by which glucose is completely oxidized, using an electron transport system but requiring a terminal electron acceptor other than oxygen. (Compare with fermentation.) (9)

Analogous Structures (Homoplasies) Structures that perform a similar function, such as wings in birds and insects, but did not originate from the same structure in a common ancestor. (33)

Anaphase Stage of mitosis when the kinetochores split and the sister chromatids (now termed chromosomes) move to opposite poles of the spindle. (10)

Anatomy Study of the structural characteristics of an organism. (18)

Angiosperm (Anthophyta) Any plant having its seeds surrounded by fruit tissue formed from the mature ovary of the flowers. (38)

Animal A mobile, heterotrophic, multicellular organism, classified in the Animal kingdom. (39)

Anion A negatively charged ion. (3)

Annelida The phylum which contains segmented worms (earthworms, leeches, and bristleworms). (39)

Annuals Plants that live for one year or less. (18)

Annulus A row of specialized cells encircling each sporangium on the fern frond; facilitates rupture of the sporangium and dispersal of spors. (38)

Antagonistic Muscles Pairs of muscles whose contraction bring about opposite actions as illustrated by the biceps and triceps, which bends or straightens the arm at the elbow, respectively. (26)

Antenna Pigments Components of photosystems that gather light energy of different wavelengths and then channel the absorbed energy to a reaction center. (8)

Anterior In anatomy, at or near the front of an animal; the opposite of posterior. (39)

Anterior Pituitary A true endocrine gland manufacturing and releasing six hormones

when stimulated by releasing factors from the hypothalamus. (25)

Anther The swollen end of the stamen (male reproductive organ) of a flowering plant. Pollen grains are produced inside the anther lobes in pollen sacs. (20)

Antibiotic A substance produced by a fungus or bacterium that is capable of preventing the growth of bacteria. (2)

Antibodies Proteins produced by plasma cells. They react specifically with the antigen that stimulated their formation. (30)

Anticodon Triplet of nucleotides in tRNA that recognizes and base pairs with a particular codon in mRNA. (14)

Antidiuretic Hormone (ADH) One of the two hormones released by the posterior pituitary. ADH increases water reabsorption in the kidney, which then produces a more concentrated urine. (25)

Antigen Specific foreign agent that triggers an immune response. (30)

Aorta Largest blood vessel in the body through which blood leaves the heart and enters the systemic circulation. (28)

Apical Dominance The growth pattern in plants in which axillary bud growth is inhibited by the hormone auxin, present in high concentrations in terminal buds. (21)

Apical Meristems Centers of growth located at the tips of shoots, axillary buds, and roots. Their cells divide by mitosis to produce new cells for primary growth in plants. (18)

Aposematic Coloring Warning coloration which makes an organism stand out from its surroundings. (42)

Appendicular Skeleton The bones of the appendages and of the pectoral and pelvic girdles. (26)

Aquatic Living in water. (40)

Archaebacteria Members of the kingdom Monera that differ from typical bacteria in the structure of their membrane lipids, their cell walls, and some characteristics that resemble those of eukaryotes. Their lack of a true nucleus, however, accounts for their assignment to the Moneran kingdom. (36)

Archenteron In gastrulation, the hollow core of the gastrula that becomes an animal's digestive tract. (32)

Arteries Large, thick-walled vessels that carry blood away from the heart. (28)

Arterioles The smallest arteries, which carry blood toward capillary beds. (28)

Arthropoda The most diverse phylum on earth, so called from the presence of jointed limbs. Includes insects, crabs, spiders, centipedes. (39)

Ascospores Sexual fungal spore borne in a sac. Produced by the sac fungi, Ascomycota. (37)

Asexual Reproduction Reproduction without the union of male and female gametes. (31)

Association In ecological communities, a major organization characterized by uniformity and two or more dominant species. (41)

Asymmetric Referring to a body form that cannot be divided to produce mirror images. (39)

Atherosclerosis Condition in which the inner walls of arteries contain a buildup of cholesterol-containing plaque that tends to occlude the channel and act as a site for the formation of a blood clot (thrombus). (7)

Atmosphere The layer of air surrounding the Earth. (40)

Atom The fundamental unit of matter that can enter into chemical reactions; the smallest unit of matter that possesses the qualities of an element. (3)

Atomic Mass Combined number of protons and neutrons in the nucleus of an atom. (3)

Atomic Number The number of protons in the nucleus of an atom. (3)

ATP (see **Adenosine Triphosphate**)

ATPase An enzyme that catalyzes a reaction in which ATP is hydrolyzed. These enzymes are typically involved in reactions where energy stored in ATP is used to drive an energy-requiring reaction, such as active transport or muscle contractility. (7, 26)

ATP Synthase A large protein complex present in the plasma membrane of bacteria, the inner membrane of mitochondria, and the thylakoid membrane of chloroplasts. This complex consists of a baseplate in the membrane, a channel across the membrane through which protons can pass, and a spherical head (F_1 particle) which contains the site where ATP is synthesized from ADP and P_i (8, 9)

Atrioventricular (AV) Node A neurological center of the heart, located at the top of the ventricles. (28)

Atrium A contracting chamber of the heart which forces blood into the ventricle. There are two atria in the hearts of all vertebrates, except fish which have one atrium. (28)

Atrophy The shrinkage in size of structure, such as a bone or muscle, usually as a result of disuse. (26)

Autoantibodies Antibodies produced against the body's own tissue. (30)

Autoimmune Disease Damage to a body tissue due to an attack by autoantibodies. Examples include thyroiditis, multiple sclerosis, and rheumatic fever. (30)

Autonomic Nervous System The nerves that control the involuntary activities of the internal organs. It is composed of the parasympathetic system, which functions during normal activity, and the sympathetic system, which operates in times of emergency or prolonged exertion. (23)

Autosome Any chromosome that is not a sex chromosome. (13)

Autotrophs Organisms that satisfy their own nutritional needs by building organic molecules photosynthetically or chemosynthetically from inorganic substances. (8)

Auxins Plant growth hormones that promote cell elongation by softening cell walls. (21)

Axial Skeleton The bones aligned along the long axis of the body, including the skull, vertebral column, and ribcage. (26)

Axillary Bud A bud that is directly above each leaf on the stem. It can develop into a new stem or a flower. (18)

Axon The long, sometimes branched extension of a neuron which conducts impulses from the cell body to the synaptic knobs. (23)

◄ B ►

Bacteriophage A virus attacking specific bacteria that multiplies in the bacterial host cell and usually destroys the bacterium as it reproduces. (36)

Balanced Polymorphism The maintenance of two or more alleles for a single trait at fairly high frequencies. (33)

Bark Common term for the periderm. A collective term for all plant tissues outside the secondary xylem. (18)

Base Substance that removes hydrogen ions (H^+) from solutions. (3)

Basidiospores Sexual spores produced by basidiomycete fungi. Often found by the millions on gills in mushrooms. (37)

Basophil A phagocytic leukocyte which also releases substances, such as histamine, that trigger an inflammatory response. (28)

Batesian Mimicry The resemblance of a good-tasting or harmless species to a species with unpleasant traits. (42)

Bathypelagic Zone The ocean zone beneath the mesopelagic zone, characterized by no light; inhabited by heterotrophic bacteria and benthic scavengers. (40)

B Cell A lymphocyte that becomes a plasma cell and produces antibodies when stimulated by an antigen. (30)

Benthic Zone The deepest ocean zone; the ocean floor, inhabited by bottom dwelling organisms. (40)

Bicarbonate Ion HCO_{3^-}. (3, 29)

Biennials Plants that live for two years. (18)

Bilateral Symmetry The quality possessed by organisms whose body can be divided into mirror images by only one median plane. (39)

Bile Salts Detergentlike molecules produced by the liver and stored by the gallbladder that function in lipid emulsification in the small intestine. (27)

Binomial A term meaning "two names" or "two words". Applied to the system of nomenclature for categorizing living things with a genus and species name that is unique for each type of organism. (1)

Biochemicals Organic molecules produced by living cells. (4)

Bioconcentration The ability of an organism to accumulate substances within its' body or specific cells. (41)

Biodiversity Biological diversity of species, including species diversity, genetic diversity, and ecological diversity. (43)

Biogeochemical Cycles The exchanging of chemical elements between organisms and the abiotic environment. (41)

Biological Control Pest control through the use of naturally occurring organisms such as predators, parasites, bacteria, and viruses. (41)

Biological Magnification An increase in concentration of slowly degradable chemicals in organisms at successively higher trophic levels; for example, DDT or PCB's. (41)

Bioluminescence The capability of certain organisms to utilize chemical energy to produce light in a reaction catalyzed by the enzyme luciferase. (9)

Biomass The weight of organic material present in an ecosystem at any one time. (41)

Biome Broad geographic region with a characteristic array of organisms. (40)

Biosphere Zone of the earth's soil, water, and air in which living organisms are found. (40)

Biosynthesis Construction of molecular components in the growing cell and the replacement of these compounds as they deteriorate. (6)

Biotechnology A new field of genetic engineering; more generally, any practical application of biological knowledge. (16)

Biotic Environment Living components of the environment. (40)

Biotic Potential The innate capacity of a population to increase tremendously in size were it not for curbs on growth; maximum population growth rate. (43)

Blade Large, flattened area of a leaf; effective in collecting sunlight for photosynthesis. (18)

Blastocoel The hollow fluid-filled space in a blastula. (32)

Blastocyst Early stage of a mammalian embryo, consisting of a mass of cells enclosed in a hollow ball of cells called the trophoblast. (32)

Blastodisk In bird and reptile development, the stage equivalent to a blastula. Because of the large amount of yolk, cleavage produces two flattened layers of cells with a blastocoel between them. (32)

Blastomeres The cells produced during embryonic cleavage. (32)

Blastopore The opening of the archenteron that is the embryonic predecessor of the anus in vertebrates and some other animals. (32)

Blastula An early developmental stage in many animals. It is a ball of cells that encloses a cavity, the blastocoel. (32)

Blood A type of connective tissue consisting of red blood cells, white blood cells, platelets, and plasma. (28)

Blood Pressure Positive pressure within the cardiovascular system that propels blood through the vessels. (28)

Blooms are massive growths of algae that occur when conditions are optimal for algae proliferation. (37)

Body Plan The general layout of a plant's or animal's major body parts. (39)

Bohr Effect Increased release of O_2 from hemoglobin molecules at lower pH. (29)

Bone A tissue composed of collagen fibers, calcium, and phosphate that serves as a means of support, a reserve of calcium and phosphate, and an attachment site for muscles. (26)

Botany Branch of biology that studies the life cycles, structure, growth, and classification of plants. (18)

Bottleneck A situation in which the size of a species' population drops to a very small number of individuals, which has a major impact on the likelihood of the population recovering its earlier genetic diversity. As occurred in the cheetah population. (33)

Bowman's Capsule A double-layered container that is an invagination of the proximal end of the renal tubule that collects molecules and wastes from the blood. (28)

Brain Mass of nerve tissue composing the main part of the central nervous system. (23)

Brainstem The central core of the brain, which coordinates the automatic, involuntary body processes. (23)

Bronchi The two divisions of the trachea through which air enters each of the two lungs. (29)

Bronchioles The smallest tubules of the respiratory tract that lead into the alveoli of the lungs where gas exchange occurs. (29)

Bryophyta Division of non-vascular terrestrial plants that include liverworts, mosses, and hornworts. (38)

Budding Asexual process by which offspring develop as an outgrowth of a parent. (39)

Buffers Chemicals that couple with free hydrogen and hydroxide ions thereby resisting changes in pH. (3)

Bundle Sheath Parenchyma cells that surround a leaf vein which regulate the uptake and release of materials between the vascular tissue and the mesophyll cells. (18)

◄ C ►

C₃ Synthesis The most common pathway for fixing CO_2 in the synthesis reactions of photosynthesis. It is so named because the first detectable organic molecule into which CO_2 is incorporated is a 3-carbon molecule, phosphoglycerate (PGA). (8)

C_4 Synthesis Pathway for fixing CO_2 during the light-independent reactions of photosynthesis. It is so named because the first detectable organic molecule into which CO_2 is incorporated is a 4-carbon molecule. (8)

Calcitonin A thyroid hormone which regulates blood calcium levels by inhibiting its release from bone. (25)

Calorie Energy (heat) necessary to elevate the temperature of one gram of water by one degree Centigrade (1° C). (6)

Calvin Cycle The cyclical pathway in which CO_2 is incorporated into carbohydrate. See C_3 synthesis. (8)

Calyx The outermost whorl of a flower, formed by the sepals. (20)

CAM Crassulacean acid metabolism. A variation of the photosynthetic reactions in plants, biochemically identical to C_4 synthesis except that all reactions occur in the same cell and are separated by time. Because CAM plants open their stomates at night, they have a competitive advantage in hot, dry climates. (8)

Cambium A ring or cluster of meristematic cells that increase the width of stems and roots when they divide to produce secondary tissues. (18)

Camouflage Adaptations of color, shape and behavior that make an organism more difficult to detect. (42)

Cancer A disease resulting from uncontrolled cell divisions. (10,13)

Capillaries The tiniest blood vessels consisting of a single layer of flattened cells. (28)

Capillary Action Tendency of water to be pulled into a small-diameter tube. (3)

Carbohydrates A group of compounds that includes simple sugars and all larger molecules constructed of sugar subunits, e.g. polysaccharides. (4)

Carbon Cycle The cycling of carbon in different chemical forms, from the environment to organisms and back to the environment. (41)

Carbon Dioxide Fixation In photosynthesis, the combination of CO_2 with carbon-accepting molecules to form organic compounds. (8)

Carcinogen A cancer-causing agent. (13)

Cardiac Muscle One of the three types of muscle tissue; it forms the muscle of the heart. (26)

Cardiovascular System The organ system consisting of the heart and the vessels through which blood flows. (28)

Carnivore An animal that feeds exclusively on other animals. (42)

Carotenoid A red, yellow, or orange plant pigment that absorbs light in 400-500 nm wavelengths. (8)

Carpels Central whorl of a flower containing the female reproductive organs. Each separate carpel, or each unit of fused carpels, is called a pistil. (20)

Carrier Proteins Proteins within the plasma membrane that bind specific substances and facilitate their movement across the membrane. (7)

Carrying Capacity The size of a population that can be supported indefinitely in a given environment. (43)

Cartilage A firm but flexible connective tissue. In the human, most cartilage originally present in the embryo is transformed into bones. (26)

Casparian Strip The band of waxy suberin that surrounds each endodermal cell of a plant's root tissue. (18)

Catabolism Metabolic pathways that degrade complex compounds into simpler molecules, usually with the release of the chemical energy that held the atoms of the larger molecule together. (6)

Catalyst A chemical substance that accelerates a reaction or causes a reaction to occur but remains unchanged by the reaction. Enzymes are biological catalysts. (6)

Cation A positively charged ion. (3)

Cecum A closed-ended sac extending from the intestine in grazing animals lacking a rumen (e.g., horses) that enables them to digest cellulose. (27)

Cell The basic structural unit of all organisms. (5)

Cell Body Region of a neuron that contains most of the cytoplasm, the nucleus, and other organelles. It relays impulses from the dendrites to the axon. (23)

Cell Cycle Complete sequence of stages from one cell division to the next. The stages are denoted G_1, S, G_2, and M phase. (10)

Cell Differentiation The process by which the internal contents of a cell become assembled into a structure that allows the cell to carry out a specific set of activities, such as secretion of enzymes or contraction. (32)

Cell Division The process by which one cell divides into two. (10)

Cell Fusion Technique whereby cells are caused to fuse with one another producing a large cell with a common cytoplasm and plasma membrane. (5, 10)

Cell Plate In plants, the cell wall material deposited midway between the daughter cells during cytokinesis. Plate material is deposited by small Golgi vesicles. (5, 10)

Cell Sap Solution that fills a plant vacuole. In addition to water, it may contain pigments, salts, and even toxic chemicals. (5)

Cell Theory The fundamental theory of biology that states: 1) all organisms are composed of one or more cells, 2) the cell is the basic organizational unit of life, 3) all cells arise from pre-existing cells. (5)

Cellular Respiration (See **Aerobic respiration**)

Cellulose The structural polysaccharide comprising the bulk of the plant cell wall. It is the most abundant polysaccharide in nature. (4, 5)

Cell Wall Rigid outer-casing of cells in plants and other organisms which gives support, slows dehydration, and prevents a cell from bursting when internal pressure builds due to an influx of water. (5)

Central Nervous System In vertebrates, the brain and spinal cord. (23)

Centriole A pinwheel-shaped structure at each pole of a dividing animal cell. (10)

Centromere Indented region of a mitotic chromosome containing the kinetochore. (10)

Cephalization The clustering of neural tissues and sense organs at the anterior (leading) end of the animal. (39)

Cerebellum A bulbous portion of the vertebrate brain involved in motor coordination. Its prominence varies greatly among different vertebrates . (23)

Cerebral Cortex The outer, highly convoluted layer of the cerebrum. In the human, this is the center of higher brain functions, such as speech and reasoning. (23)

Cerebrospinal Fluid Fluid present within the ventricles of the brain, central canal of the spinal cord, and which surrounds and cushions the central nervous system. (23)

Cerebrum The most dominant part of the human forebrain, composed of two cerebral hemispheres, generally associated with higher brain functions. (23)

Cervix The lower tip of the uterus. (31)

Chapparal A type of shrubland in California, characterized by drought-tolerant and fire-adapted plants. (40)

Character Displacement Divergence of a physical trait in closely related species in response to competition. (42)

Chemical Bonds Linkage between atoms as a result of electrons being shared or donated. (3)

Chemical Evolution Spontaneous synthesis of increasingly complex organic compounds from simpler molecules. (35)

Chemical Reaction Interaction between chemical reactants. (6)

Chemiosmosis The process by which a pH gradient drives the formation of ATP. (8, 9)

Chemoreceptors Sensory receptors that respond to the presence of specific chemicals. (24)

Chemosynthesis An energy conversion process in which inorganic substances (H, N, Fe, or S) provide energized electrons and hydrogen for carbohydrate formation (9, 36)

Chiasmata Cross-shaped regions within a tetrad, occurring at points of crossing over or genetic exchange. (11)

Chitin Structural polysaccharide that forms the hard, strong external skeleton of many arthropods and the cell walls of fungi. (4)

Chlamydia Obligate intracellular parasitic bacteria that lack a functional ATP-generating system. (36)

Chlorophyll Pigments Major light-absorbing pigments of photosynthesis. (8)

Chlorophyta Green algae, the largest group of algae; members of this group were very likely the ancestors of the modern plant kingdom. (38)

Chloroplasts An organelle containing chlorophyll found in plant cells in which photosynthesis occurs. (5, 8)

Cholecystokinin (CCK) Hormone secreted by endocrine cells in the wall of the small intestine that stimulates the release of digestive products by the pancreas. (27)

Chondrocytes Living cartilage cells embedded within the protein-polysaccharide matrix they manufacture. (26)

Chordamesoderm In vertebrates, the block of mesoderm that underlies the dorsal ectoderm of the gastrula, induces the formation of the nervous system, and gives rise to the notochord. (32)

Chordate A member of the phylum Chordata possessing a skeletal rod of tissue called a notochord, a dorsal hollow nerve cord, gill slits, and a post-anal tail at some stage of its development. (39)

Chorion The outermost of the four extraembryonic membranes. In placental mammals, it forms the embryonic portion of the placenta. (32)

Chorionic Villus Sampling (CVS) A procedure for obtaining fetal cells by removing a small sample of tissue from the developing placenta of a pregnant woman. (17)

Chromatid Each of the two identical subunits of a replicated chromosome. (10)

Chromatin DNA-protein fibers which, during prophase, condense to form the visible chromosomes. (5, 10)

Chromatography A technique for separating different molecules on the basis of their solubility in a particular solvent. The mixture of substances is spotted on a piece of paper or other material, one end of which is then placed in the solvent. As the solvent moves up the paper by capillary action, each substance in the mixture is carried a particular distance depending on its solubility in the moving solvent. (8)

Chromosomes Dark-staining structures in which the organism's genetic material (DNA) is organized. Each species has a characteristic number of chromosomes. (5, 10)

Chromosome Aberrations Alteration in the structure of a chromosome from the normal state. Includes chromosome deletions, duplications, inversions, and translocations. (13)

Chromosome Puff A site on an insect polytene chromosome where the DNA has unraveled and is being transcribed. (15)

Cilia Short, hairlike structures projecting from the surfaces of some cells. They beat in coordinated ways, are usually found in large numbers, and are densely packed. (5)

Ciliated Mucosa Layer of ciliated epithelial cells lining the respiratory tract. The beating of cilia propels an associated mucous layer and trapped foreign particles. (29)

Circadian Rhythm Behavioral patterns that cycle during approximately 24 hour intervals.

Circulatory System The system that circulates internal fluids throughout an organism to deliver oxygen and nutrients to cells and to remove metabolic wastes. (28)

Class (Taxonomic) A level of the taxonomic hierarchy that groups together members of related orders. (1)

Classical Conditioning A form of learning in which an animal develops a response to a new stimulus by repeatedly associating the new stimulus with a stimulus that normally elicits the response. (44)

Cleavage Successive mitotic divisions in the early embryo. There is no cell growth between divisions. (32)

Cleavage Furow Constriction around the middle of a dividing cell caused by constriction of microfilaments. (10)

Climate The general pattern of average weather conditions over a long period of time in a specific region, including precipitation, temperature, solar radiation, and humidity. (40)

Climax Final or stable community of successional stages, that is more or less in equilibrium with existing environmental conditions for a long period of time. (41)

Climax Community Community that remains essentially the same over long periods of time; final stage of ecological succession. (41)

Clitoris A protrusion at the point where the labia minora merge; rich in sensory neurons and erectile tissue. (31)

Clonal Selection Mechanism The mechanism by which the body can synthesize antibodies specific for the foreign substance (antigen) that stimulated their production. (30)

Clones Offspring identical to the parent, produced by asexual processes. (15)

Closed Circulatory System Circulatory system in which blood travels throughout the body in a continuous network of closed tubes. (Compare with open circulatory system). (28)

Clumped Pattern Distribution of individuals of a population into groups, such as flocks or herds. (43)

Cnidaria A phylum that consists of radial symmetrical animals that have two cell layers. There are three classes: 1) Hydrozoa (hydra), 2) Scy-

phozoa (jellyfish), 3) Anthozoa (sea anemones, corals). Most are marine forms that live in warm, shallow water. (39)

Cnidocytes Specialized stinging cells found in the members of the phylum Cnidaria. (39)

Coastal Waters Relatively warm, nutrient-rich shallow water extending from the high-tide mark on land to the sloping continental shelf. The greatest concentration of marine life are found in coastal waters. (40)

Coated Pits Indentations at the surfaces of cells that contain a layer of bristly protein (called clathrin) on the inner surface of the plasma membrane. Coated pits are sites where cell receptors become clustered. (7)

Cochlea Organ within the inner ear of mammals involved in sound reception. (24)

Codominance The simultaneous expression of both alleles at a genetic locus in a heterozygous individual. (12)

Codon Linear array of three nucleotides in mRNA. Each triplet specifies a particular amino acid during the process of translation. (14)

Coelomates Animals in which the body cavity is completely lined by mesodermally-derived tissues. (39)

Coenzyme An organic cofactor, typically a vitamin or a substance derived from a vitamin. (6)

Coevolution Evolutionary changes that result from reciprocal interactions between two species, e.g., flowering plants and their insect pollinators. (33)

Cofactor A non-protein component that is linked covalently or noncovalently to an enzyme and is required by the enzyme to catalyze the reaction. Cofactors may be organic molecules (coenzymes) or metals. (6)

Cohesion The tendency of different parts of a substance to hold together because of forces acting between its molecules. (3)

Coitus Sexual union in mammals. (31)

Coleoptile Sheath surrounding the tip of the monocot seedling, protecting the young stem and leaves as they emerge from the soil. (21)

Collagen The most abundant protein in the human body. It is present primarily in the extracellular space of connective tissues such as bone, cartilage, and tendons. (26)

Collenchyma Living plant cells with irregularly thickened primary cell walls. A supportive cell type often found inside the epidermis of stems with primary growth. Angular, lacunar and laminar are different types of collenchyma cells. (18)

Commensalism A form of symbiosis in which one organism benefits from the union while the other member neither gains nor loses. (42)

Community The populations of all species living in a given area. (41)

Compact Bone The solid, hard outer regions of a bone surrounding the honey-combed mass of spongy bone. (26)

Companion Cell Specialized parenchyma cell associated with a sieve-tube member in phloem. (18)

Competition Interaction among organisms that require the same resource. It is of two types: 1) intraspecific (between members of the same species); 2) interspecific (between members of different species). (42)

Competitive Exclusion Principle (Gause's Principle) Competition in which a winner species captures a greater share of resources, increasing its survival and reproductive capacity. The other species is gradually displaced. (42)

Competitive Inhibition Prevention of normal binding of a substrate to its enzyme by the presence of an inhibitory compound that competes with the substrate for the active site on the enzyme. (6)

Complement Blood proteins with which some antibodies combine following attachment to antigen (the surface of microorganisms). The bound complement punches the tiny holes in the plasma membrane of the foreign cell, causing it to burst. (28)

Complementarity The relationship between the two strands of a DNA molecule determined by the base pairing of nucleotides on the two strands of the helix. A nucleotide with guanine on one strand always pairs with a nucleotide having cytosine on the other strand; similarly with adenine and thymine. (14)

Complete Digestive Systems Systems that have a digestive tract with openings at both ends—a mouth for entry and an anus for exit. (27)

Complete Flower A flower containing all four whorls of modified leaves—sepels, petals, stamen, and carpels. (20)

Compound Chemical substances composed of atoms of more than one element. (3)

Compound Leaf A leaf that is divided into leaflets, with two or more leaflets attached to the petiole. (18)

Concentration Gradient Regions in a system of differing concentration representing potential energy, such as exist in a cell and its environment, that cause molecules to move from areas of higher concentration to lower concentration. (7)

Conditioned Reflex A reflex ("automatic") response to a stimulus that would not normally have elicited the response. Conditioned reflexes develop by repeated association of a new stimulus with an old stimulus that normally elicits the response. (44)

Conformation The three-dimensional shape of a molecule as determined by the spatial arrangement of its atoms. (4)

Conformational Change Change in molecular shape (as occurs, for example, in an enzyme as it catalyzes a reaction, or a myosin molecule during contraction). (6)

Conjugation A method of reproduction in single-celled organisms in which two cells link and exchange nuclear material. (11)

Connective Tissues Tissues that protect, support, and hold together the internal organs and other structures of animals. Includes bone, cartilage, tendons, and other tissues, all of which have large amounts of extracellular material. (22)

Consumers Heterotrophs in a biotic environment that feed on other organisms or organic waste. (41)

Continental Drift The continuous shifting of the earth's land masses explained by the theory of plate tectonics. (35)

Continuous Variation An inheritance pattern in which there is graded change between the two extremes in a phenotype (compare with discontinuous variation). (12)

Contraception The prevention of pregnancy. (31)

Contractile Proteins Actin and myosin, the protein filaments that comprise the bulk of the muscle mass. During contraction of skeletal muscle, these filaments form a temporary association and slide past each other, generating the contractile force. (26)

Control (Experimental) A duplicate of the experiment identical in every way except for the one variable being tested. Use of a control is necessary to demonstrate cause and effect. (2)

Convergent Evolution The evolution of similar structures in distantly related organisms in response to similar environments. (33)

Cork Cambium In stems and roots of perennials, a secondary meristem that produces the outer protective layer of the bark. (18)

Coronary Arteries Large arteries that branch immediately from the aorta, providing oxygen-rich blood to the cardiac muscle. (28)

Corpus Callosum A thick cable composed of hundreds of millions of neurons that connect the right and left cerebral hemispheres of the mammalian brain. (23)

Corpus Luteum In the mammalian ovary, the structure that develops from the follicle after release of the egg. It secretes hormones that prepare the uterine endometrium to receive the developing embryo. (31)

Cortex In the stem or root of plants, the region between the epidermis and the vascular tissues. Composed of ground tissue. In animals, the outermost portion of some organs. (18)

Cotyledon The seed leaf of a dicot embryo containing stored nutrients required for the germinated seed to grow and develop, or a food digesting seed leaf in a monocot embryo. (20)

Countercurrent Flow Mechanism for increasing the exchange of substances or heat from one stream of fluid to another by having the two fluids flow in opposite directions. (29)

Covalent Bonds Linkage between two atoms which share the same electrons in their outermost shells. (3)

Cranial Nerves Paired nerves which emerge from the central stalk of the vertebrate brain and innervate the body. Humans have 12 pairs of cranial nerves. (23)

Cranium The bony casing which surrounds and protects the vertebrate brain. (23)

Cristae The convolutions of the inner membrane of the mitochondrion. Embedded within them are the components of the electron transport system and proton channels for chemiosmosis. (9)

Crossing Over During synapsis, the process by which homologues exchange segments with each other. (11)

Cryptic Coloration A form of camouflage wherein an organism's color or patterning helps it resemble its background. (42)

Cutaneous Respiration The uptake of oxygen across virtually the entire outer body surface. (29)

Cuticle 1) Waxy layer covering the outer cell walls of plant epidermal cells. It retards water vapor loss and helps prevent dehydration. (18) 2) Outer protective, nonliving covering of some animals, such as the exoskeleton of anthropods. (26, 39)

Cyanobacteria A type of prokaryote capable of photosynthesis using water as a source of electrons. Cyanobacteria were responsible for initially creating an O_2-containing atmosphere on earth. (35, 36)

Cyclic AMP (Cyclic adenosine monophosphate) A ring-shaped molecular version of an ATP minus two phosphates. A regulatory molecule formed by the enzyme adenylate cyclase which converts ATP to cAMP. A second messenger. (25)

Cyclic Pathways Metabolic pathways in which the intermediates of the reaction are regenerated while assisting the conversion of the substrate to product. (9)

Cyclic Photophosphorylation A pathway that produces ATP, but not NADPH, in the light reactions of photosynthesis. Energized electrons are shuttled from a reaction center, along a molecular pathway, back to the original reaction center, generating ATP en route. (8)

Cysts Protective, dormant structure formed by some protozoa. (37)

Cytochrome Oxidase A complex of proteins that serves as the final electron carrier in the mitochondrial electron transport system, transferring its electrons to O_2 to form water. (9)

Cytokinesis Final event in eukaryotic cell division in which the cell's cytoplasm and the new nuclei are partitioned into separate daughter cells. (10)

Cytokinins Growth-producing plant hormones which stimulate rapid cell division. (21)

Cytoplasm General term that includes all parts of the cell, except the plasma membrane and the nucleus. (5)

Cytoskeleton Interconnecting network of microfilaments, microtubules, and intermediate filaments that serves as a cell scaffold and provides the machinery for intracellular movements and cell motility. (5)

Cytotoxic (Killer) T Cells A class of T cells capable of recognizing and destroying foreign or infected cells. (30)

◄ D ►

Day Neutral Plants Plants that flower at any time of the year, independent of the relative lengths of daylight and darkness. (21)

Deciduous Trees or shrubs that shed their leaves in a particular season, usually autumn, before entering a period of dormancy. (40)

Deciduous Forest Forests characterized by trees that drop their leaves during unfavorable conditions, and leaf out during warm, wet seasons. Less dense than tropical rain forests. (40)

Decomposers (Saprophytes) Organisms that obtain nutrients by breaking down organic compounds in wastes and dead organisms. Includes fungi, bacteria, and some insects. (41)

Deletion Loss of a portion of a chromosome, following breakage of DNA. (13)

Denaturation Change in the normal folding of a protein as a result of heat, acidity, or alkalinity. Such changes result in a loss of enzyme functioning. (4)

Dendrites Cytoplasmic extensions of the cell body of a neuron. They carry impulses from the area of stimulation to the cell body. (23)

Denitrification The conversion by denitrifying bacteria of nitrites and nitrates into nitrogen gas. (41)

Denitrifying Bacteria Bacteria which take soil nitrogen, usable to plants, and convert it to unusable nitrogen gas. (41)

Density-Dependent Factors Factors that control population growth which are influenced by population size. (43)

Density-Independent Factors Factors that control population growth which are not affected by population size. (43)

Deoxyribonucleic Acid (DNA) Double-stranded polynucleotide comprised of deoxyribose (a sugar), phosphate, and four bases (adenine, guanine, cytosine, and thymine). Encoded in the sequence of nucleotides are the instructions for making proteins. DNA is the genetic material in all organisms except certain viruses. (14)

Depolarization A decrease in the potential difference (voltage) across the plasma membrane of a cell typically due to an increase in the movement of sodium ions into the cell. Acts to excite a target cell. (23)

Dermal Bone Bones of vertebrates that form within the dermal layer of the skin, such as the scales of fishes and certain bones of the skull. (26)

Dermal Tissue System In plants, the epidermis in primary growth, or the periderm in secondary growth. (18)

Dermis In animals, layer of cells below the epidermis in which connective tissue predominates. Embedded within it are vessels, various glands, smooth muscle, nerves, and follicles. (26)

Desert Biome characterized by intense solar radiation, very little rainfall, and high winds. (40)

Detrivore Organism that feeds on detritus, dead organisms or their parts, and living organisms' waste. (41)

Deuterostome One path of development exhibited by coelomate animals (e.g., echinoderms and chordates). (39)

Diabetes Mellitus A disease caused by a deficiency of insulin or its receptor, preventing glucose from being absorbed by the cells. (25)

Diaphragm A sheet of muscle that separates the thoracic cavity from the abdominal wall. (29)

Diastolic Pressure The second number of a blood pressure reading; the lowest pressure in the arteries just prior to the next heart contraction. (28)

Diatoms are golden-brown algae that are distinguished most dramatically by their intricate silica shells. (37)

Dicotyledonae (Dicots) One of the two classes of flowering plants, characterized by having seeds with two cotyledons, flower parts in 4s or 5s, net-veined leaves, one main root, and vascular bundles in a circular array within the stem. (Compare with Monocotylenodonae). (18)

Diffusion Tendency of molecules to move from a region of higher concentration to a region of lower concentration, until they are uniformly dispersed. (7)

Digestion The process by which food particles are disassembled into molecules small enough to be absorbed into the organism's cells and tissues. (27)

Digestive System System of specialized organs that ingests food, converts nutrients to a form that can be distributed throughout the animal's body, and eliminates undigested residues. (27)

Digestive Tract Tubelike channel through which food matter passes from its point of ingestion at the mouth to the elimination of indigestible residues from the anus. (27)

Dihybrid Cross A mating between two individuals that differ in two genetically-determined traits. (12)

Dimorphism Presence of two forms of a trait within a population, resulting from diversifying selection. (33)

Dinoflagellates Single-celled photosynthesizers that have two flagella. They are members of the pyrophyta, phosphorescent algae that sometimes cause red tide, often synthesizing a neurotoxin that accumulates in plankton eaters, causing paralytic shellfish poisoning in people who eat the shellfish. (37)

Dioecious Plants that produce either male or female reproductive structures but never both. (38)

Diploid Having two sets of homologous chromosomes. Often written 2N. (10, 13)

Directional Selection The steady shift of phenotypes toward one extreme. (33)

Discontinuous Variation An inheritance pattern in which the phenomenon of all possible phenotypes fall into distinct categories. (Compare with continuous variation). (12)

Displays The signals that form the language by which animals communicate. These signals are species specific and stereotyped and may be visual, auditory, chemical, or tactile. (44)

Disruptive Coloration Coloration that disguises the shape of an organism by breaking up its outline. (42)

Disruptive Selection The steady shift toward more than one extreme phenotype due to the elimination of intermediate phenotypes as has occurred among African swallowtail butterflies whose members resemble more than one species of distasteful butterfly. (33)

Divergent Evolution The emergence of new species as branches from a single ancestral lineage. (33)

Diversifying Selection The increasing frequency of extreme phenotypes because individuals with average phenotypes die off. (33)

Diving Reflex Physiological response that alters the flow of blood in the body of diving mammals that allows the animal to maintain high levels of activity without having to breathe. (29)

Division (or Phylum) A level of the taxonomic hierarchy that groups together members or related classes. (1)

DNA (see **Deoxyribonucleic Acid**)

DNA Cloning The amplification of a particular DNA by use of a growing population of bacteria. The DNA is initially taken up by a bacterial cell—usually as a plasmid—and then replicated along with the bacteria's own DNA. (16)

DNA Fingerprint The pattern of DNA fragments produced after treating a sample of DNA with a particular restriction enzyme and separating the fragments by gel electrophoresis. Since different members of a population have DNA with a different nucleotide sequence, the pattern of DNA fragments produced by this method can be used to identify a particular individual. (16)

DNA Ligase The enzyme that covalently joins DNA fragments into a continuous DNA strand. The enzyme is used in a cell during replication to seal newly-synthesized fragments and by biotechnologists to form recombinant DNA molecules from separate fragments. (14, 16)

DNA Polymerase Enzyme responsible for replication of DNA. It assembles free nucleotides, aligning them with the complementary ones in the unpaired region of a single strand of DNA template. (14)

Dominant The form of an allele that masks the presence of other alleles for the same trait. (12)

Dormancy A resting period, such as seed dormancy in plants or hibernation in animals, in which organisms maintain reduced metabolic rates. (21)

Dorsal In anatomy, the back of an animal. (39)

Double Blind Test A clinical trial of a drug in which neither the human subjects or the researchers know who is receiving the drug or placebo. (2)

Down Syndrome Genetic disorder in humans characterized by distinct facial appearance and mental retardation, resulting from an extra copy of chromosome number 21 (trisomy 21) in each cell. (11, 17)

Duodenum First part of the human small intestine in which most digestion of food occurs. (27)

Duplication The repetition of a segment of a chromosome. (13)

◄ E ►

Ecdysis Molting process by which an arthropod periodically discards its exoskeleton and replaces it with a larger version. The process is controlled by the hormone ecydysone. (39)

Ecdysone An insect steroid hormone that triggers molting and metamorphosis. (15)

Echinodermata A phylum composed of animals having an internal skeleton made of many small calcium carbonate plates which have jutting spines. Includes sea stars, sea urchins, etc. (39)

Echolocation The use of reflected sound waves to help guide an animal through its environment and/or locate objects. (24)

Ecological Equivalent Organisms that occupy similar ecological niches in different regions or ecosystems of the world. (41)

Ecological Niche The habitat, functional role(s), requirements for environmental resources and tolerance ranges for each abiotic condition in relation to an organism. (41)

Ecological Pyramid Illustration showing the energy content, numbers of organisms, or biomass at each trophic level. (41)

Ecology The branch of biology that studies interactions among organisms as well as the interactions of organisms and their physical environment. (40)

Ecosystem Unit comprised of organisms interacting among themselves and with their physical environment. (41)

Ecotypes Populations of a single species with different, genetically fixed tolerance ranges. (41)

Ectoderm In animals, the outer germ cell layer of the gastrula. It gives rise to the nervous system and integument. (32)

Ectotherms Animals that lack an internal mechanism for regulating body temperature. "Cold-blooded" animals. (28)

Edema Swelling of a tissue as the result of an accumulation of fluid that has moved out of the blood vessels. (28)

Effectors Muscle fibers and glands that are activated by neural stimulation. (23)

Efferent (Motor) Nerves The nerves that carry messages from the central nervous system to the effectors, the muscles, and glands. They are divided into two systems: somatic and autonomic. (23)

Egg Female gamete, also called an ovum. A fertilized egg is the product of the union of female and male gametes (egg and sperm cells). (32)

Electrocardiogram (EKG) Recording of the electrical activity of the heart, which is used to diagnose various types of heart problems. (28)

Electron Acceptor Substances that are capable of accepting electrons transferred from an electron donor. For example, molecular oxygen (O_2) is the terminal electron acceptor during respiration. Electron acceptors also receive electrons from chlorophyll during photosynthesis. Electron acceptors may act as part of an electron transport system by transferring the electrons they receive to another substance. (8, 9)

Electron Carrier Substances (such as NAD^+ and FAD) that transport electrons from one step of a metabolic pathway to the next or from metabolic reactions to biosynthetic reactions. (8, 9)

Electrons Negatively charged particles that orbit the atomic nucleus. (3)

Electron Transport System Highly organized assembly of cytochromes and other proteins which transfer electrons. During transport, which occurs within the inner membranes of mitochondria and chloroplasts, the energy extracted from the electrons is used to make ATP. (8, 9)

Electrophoresis A technique for separating different molecules on the basis of their size and/or electric charge. There are various ways the technique is used. In gel electrophoresis, proteins or DNA fragments are driven through a porous gel by their charge, but become separated according to size; the larger the molecule, the slower it can work its way through the pores in the gel, and the less distance it travels along the gel. (16)

Element Substance composed of only one type of atom. (3)

Embryo An organism in the early stages of development, beginning with the first division of the zygote. (32)

Embryo Sac The fully developed female gametophyte within the ovule of the flower. (20)

Emigration Individuals permanently leaving an area or population. (43)

Endergonic Reactions Chemical reactions that require energy input from another source in order to occur. (6)

Endocrine Glands Ductless glands, which secrete hormones directly into surrounding tissue fluids and blood vessels for distribution to the rest of the body by the circulatory system. (25)

Endocytosis A type of active transport that imports particles or small cells into a cell. There are two types of endocytic processes: phagocytosis, where large particles are ingested by the cell, and pinocytosis, where small droplets are taken in. (7)

Endoderm In animals, the inner germ cell layer of the gastrula. It gives rise to the digestive tract and associated organs and to the lungs. (32)

Endodermis The innermost cylindrical layer of cortex surrounding the vascular tissues of the root. The closely pressed cells of the endodermis have a waxy band, forming a waterproof layer, the Casparian strip. (18)

Endogenous Plant responses that are controlled internally, such as biological clocks controlling flower opening. (21)

Endometrium The inner epithelial layer of the uterus that changes markedly with the uterine (menstrual) cycle in preparation for implantation of an embryo. (31)

Endoplasmic Reticulum (ER) An elaborate system of folded, stacked and tubular membranes contained in the cytoplasm of eukaryotic cells. (5)

Endorphins (Endogenous Morphinelike Substances) A class of peptides released from nerve cells of the limbic system of the brain that can block perceptions of pain and produce a feeling of euphoria. (23)

Endoskeleton The internal support structure found in all vertebrates and a few invertebrates (sponges and sea stars). (26)

Endosperm Nutritive tissue in plant embryos and seeds. (20)

Endosperm Mother Cell A binucleate cell in the embryo sac of the female gametophyte, occurring in the ovule of the ovary in angiosperms. Each nucleus is haploid; after fertilization, nutritive endosperm develops. (20)

Endosymbiosis Theory A theory to explain the development of complex eukaryotic cells by proposing that some organelles once were free-living prokaryotic cells that then moved into another larger such cell, forming a beneficial union with it. (5)

Endotherms Animals that utilize metabolically produced heat to maintain a constant, elevated body temperature. "Warm-blooded" animals. (28)

End Product The last product in a metabolic pathway. Typically a substance, such as an amino acid or a nucleotide, that will be used as a monomer in the formation of macromolecules. (6)

Energy The ability to do work. (6)

Entropy Energy that is not available for doing work; measure of disorganization or randomness. (6)

Environmental Resistance The factors that eventually limit the size of a population. (43)

Enzyme Biological catalyst; a protein molecule that accelerates the rate of a chemical reaction. (6)

Eosiniphil A type of phagocytic white blood cell. (28)

Epicotyl The portion of the embryo of a dicot plant above the cotyledons. The epicotyl gives rise to the shoot. (20)

Epidermis In vertebrates, the outer layer of the skin, containing superficial layers of dead cells produced by the underlying living epithelial cells. In plants, the outer layer of cells covering leaves, primary stem, and primary root. (26, 18)

Epididymis Mass of convoluted tubules attached to each testis in mammals. After leaving the testis, sperm enter the tubules where they finish maturing and acquire motility. (31)

Epiglottis A flap of tissue that covers the glottis during swallowing to prevent food and liquids from entering the lower respiratory tract. (29)

Epinephrine (Adrenalin) Substance that serves both as an excitatory neurotransmitter released by certain neurons of the CNS and as a hormone released by the adrenal medulla that increases the body's ability to combat a stressful situation. (25)

Epipelagic Zone The lighted upper ocean zone, where photosynthesis occurs; large populations of phytoplankton occur in this zone. (40)

Epiphyseal Plates The action centers for ossification (bone formation). (26)

Epistasis A type of gene interaction in which a particular gene blocks the expression of another gene at another locus. (12)

Epithelial Tissue Continuous sheets of tightly packed cells that cover the body and line its tracts and chambers. Epithelium is a fundamental tissue type in animals. (22)

Erythrocytes Red blood cells. (28)

Erythropoietin A hormone secreted by the kidney which stimulates the formation of erythrocytes by the bone marrow. (28)

Essential Amino Acids Eight amino acids that must be acquired from dietary protein. If even one is missing from the human diet, the synthesis of proteins is prevented. (27)

Essential Fatty Acids Linolenic and linoleic acids, which are required for phospholipid construction and must be acquired from a dietary source. (27)

Essential Nutrients The 16 minerals essential for plant growth, divided into two groups: macronutrients, which are required in large quantities, and micronutrients, which are needed in small amounts. (19)

Estrogen A female sex hormone secreted by the ovaries when stimulated by pituitary gonadotrophins. (31)

Estuaries Areas found where rivers and streams empty into oceans, mixing fresh water with salt water. (40)

Ethology The study of animal behavior. (44)

Ethylene Gas A plant hormone that stimulates fruit ripening. (21)

Etiolation The condition of rapid shoot elongation, small underdeveloped leaves, bent shoot-hook, and lack of chlorophyll, all due to lack of light. (21)

Eubacteria Typical procaryotic bacteria with peptidoglycan in their cell walls. The majority of monerans are eubacteria. (36)

Eukaryotic Referring to organisms whose cellular anatomy includes a true nucleus with a nuclear envelope, as well as other membrane-bound organelles. (5)

Eusocial Species Social species that have sterile workers, cooperative care of the young, and an overlap of generations so that the colony labor is a family affair. (44)

Eutrophication The natural aging process of lakes and ponds, whereby they become marshes and, eventually, terrestrial environments.

Evolution A process whereby the characteristics of a species change over time, eventually leading to the formation of new species that go about life in new ways. (33)

Evolutionarily Stable Strategy (ESS) A behavioral strategy or course of action that depends on what other members of the population are doing. By definition, an ESS cannot be replaced by any other strategy when most of the members of the population have adopted it. (44)

Excitatory Neurons Neurons that stimulate their target cells into activity. (23)

Excretion Removal of metabolic wastes from an organism. (28)

Excretory System The organ system that eliminates metabolic wastes from the body. (28)

Exergonic Reactions Chemical reactions that occur spontaneously with the release of energy. (6)

Exocrine Glands Glands which secrete their products through ducts directly to their sites of action, e.g., tear glands. (26)

Exocytosis A form of active transport used by cells to move molecules, particles, or other cells contained in vesicles across the plasma membrane to the cell's environment. (5)

Exogenous Plant responses that are controlled externally, or by environmental conditions. (21)

Exons Structural gene segments that are transcribed and whose genetic information is subsequently translated into protein. (15)

Exoskeletons Hard external coverings found in some animals (e.g., lobsters, insects) for protection, support, or both. Such organisms grow by the process of molting. (26)

Exploitative Competition A competition in which one species manages to get more of a resource, thereby reducing supplies for a competitor. (42)

Exponential Growth An increase by a fixed percentage in a given time period; such as population growth per year. (43)

Extensor Muscle A muscle which, when contracted, causes a part of the body to straighten at a joint. (26)

External Fertilization Fertilization of an egg outside the body of the female parent. (31)

Extinction The loss of a species. (33)

Extracellular Digestion Digestion occurring outside the cell; occurs in bacteria, fungi, and multicellular animals. (27)

Extracellular Matrix Layer of extracellular material residing just outside a cell. (5)

◄ **F** ►

F_1 First filial generation. The first generation of offspring in a genetic cross. (12)

F_2 Second filial generation. The offspring of an F_1 cross. (12)

Facilitated Diffusion The transport of molecules into cells with the aid of "carrier" proteins embedded in the plasma membrane. This carrier-assisted transport does not require the expenditure of energy by the cell. (7)

FAD Flavin adenine dinucleotide. A coenzyme that functions as an electron carrier in metabolic reactions. When it is reduced to $FADH_2$, this molecule becomes a cellular energy source. (9)

Family A level of the taxonomic hierarchy that groups together members of related genera. (1)

Fast-Twitch Fibers Skeletal muscle fibers that depend on anaerobic metabolism to produce ATP rapidly, but only for short periods of time before the onset of fatigue. Fast-twitch fibers generate greater forces for shorter periods than slow-twitch fibers. (9)

Fat A triglyceride consisting of three fatty acids joined to a glycerol. (4)

Fatty Acid A long unbranched hydrocarbon chain with a carboxyl group at one end. Fatty acids lacking a double bond are said to be saturated. (4)

Fauna   The animals in a particular region.

Feedback Inhibition (Negative Feedback)
A mechanism for regulating enzyme activity by temporarily inactivating a key enzyme in a biosynthetic pathway when the concentration of the end product is elevated. (6)

Fermentation   The direct donation of the electrons of NADH to an organic compound without their passing through an electron transport system. (9)

Fertility Rate   In humans, the average number of children born to each woman between 15 and 44 years of age. (43)

Fertilization   The process in which two haploid nuclei fuse to form a zygote. (32)

Fetus   The term used for the human embryo during the last seven months in the uterus. During the fetal stage, organ refinement accompanies overall growth. (32)

Fibrinogen   A rod-shaped plasma protein that, converted to fibrin, generates a tangled net of fibers that binds a wound and stops blood loss until new cells replace the damaged tissue. (28)

Fibroblasts   Cells found in connective tissues that secrete the extracellular materials of the connective tissue matrix. These cells are easily isolated from connective tissues and are widely used in cell culture. (22)

Fibrous Root System   Many approximately equal-sized roots; monocots are characterized by a fibrous root system. Also called diffuse root system. (18)

Filament   The stalk of a stamen of angiosperms, with the anther at its tip. Also, the threadlike chain of cells in some algae and fungi. (20)

Filamentous Fungus   Multicellular members of the fungus kingdom comprised mostly of living threads (hyphae) that grow by division of cells at their tips (see molds). (37)

Filter Feeders   Aquatic animals that feed by straining small food particles from the surrounding water. (27, 39)

Fitness   The relative degree to which an individual in a population is likely to survive to reproductive age and to reproduce. (33)

Fixed Action Patterns   Motor responses that may be triggered by some environmental stimulus, but once started can continue to completion without external stimuli. (44)

Flagella   Cellular extensions that are longer than cilia but fewer in number. Their undulations propel cells like sperm and many protozoans, through their aqueous environment. (5)

Flexor Muscle   A muscle which, when contracted, causes a part of the body to bend at a joint. (26)

Flora   The plants in a particular region. (21)

Florigen Proposed   A chemical hormone that is produced in the leaves and stimulates flowering. (21)

Fluid Mosaic Model   The model proposes that the phospholipid bilayer has a viscosity similar to that of light household oil and that globular proteins float like icebergs within this bilayer. The now favored explanation for the architecture of the plasma membrane. (5)

Follicle (Ovarian)   A chamber of cells housing the developing oocytes. (31)

Food Chain   Transfers of food energy from organism to organism, in a linear fashion. (41)

Food Web   The map of all interconnections between food chains for an ecosystem. (41)

Forest Biomes   Broad geographic regions, each with characteristic tree vegetation: 1) tropical rain forests (lush forests in a broad band around the equator), 2) deciduous forests (trees and shrubs drop their leaves during unfavorable seasons), 3) coniferous forest (evergreen conifers). (40)

Fossil Record   An entire collection of remains from which paleontologists attempt to reconstruct the phylogeny, anatomy, and ecology of the preserved organisms. (34)

Fossils   The preserved remains of organisms from a former geologic age. (34)

Fossorial   Living underground.

Founder Effect   The potentially dramatic difference in allele frequency of a small founding population as compared to the original population. (33)

Founder Population   The individuals, usually few, that colonize a new habitat. (33)

Frameshift Mutation   The insertion or deletion of nucleotides in a gene that throws off the reading frame. (14)

Free Radical   Atom or molecule containing an unpaired electron, which makes it highly reactive. (3)

Freeze-Fracture   Technique in which cells are frozen into a block which is then struck with a knife blade that fractures the block in two. Fracture planes tend to expose the center of membranes for EM examination. (5)

Fronds   The large leaf-like structures of ferns. Unlike true leaves, fronds have an apical meristem and clusters of sporangia called sori. (38)

Fruit   A mature plant ovary (flower) containing seeds with plant embryos. Fruits protect seeds and aid in their dispersal. (20)

Fruiting Body   A spore-producing structure that extends upward in an elevated position from the main mass of a mold or slime mold. (37)

FSH   Follicle stimulating hormone. A hormone secreted by the anterior pituitary that prepares a female for ovulation by stimulating the primary follicle to ripen or stimulates spermatogenesis in males. (31)

Functional Groups   Accessory chemical entities (e.g., —OH, —NH$_2$, —CH$_3$), which help determine the identity and chemical properties of a compound. (4)

Fundamental Niche   The potential ecological niche of a species, including all factors affecting that species. The fundamental niche is usually never fully utilized. (41)

Fungus   Yeast, mold, or large filamentous mass forming macroscopic fruiting bodies, such as mushrooms. All fungi are eukaryotic nonphotosynthetic heterotrophics with cell walls. (37)

◄ **G** ►

G$_1$ Stage   The first of three consecutive stages of interphase. During G$_1$, cell growth and normal functions occur. The duration of this stage is most variable. (10)

G$_2$ Stage   The final stage of interphase in which the final preparations for mitosis occur. (10)

Gallbladder   A small saclike structure that stores bile salts produced by the liver. (27)

Gamete   A haploid reproductive cell--either a sperm or an egg. (10)

Gas Exchange Surface   Surface through which gases must pass in order to enter or leave the body of an animal. It may be the plasma membrane of a protistan or the complex tissues of the gills or the lungs in multicellular animals. (29)

Gastrovascular Cavity   In cnidarians and flatworms, the branched cavity with only one opening. It functions in both digestion and transport of nutrients. (39)

Gastrula   The embryonic stage formed by the inward migration of cells in the blastula. (32)

Gastrulation   The process by which the blastula is converted into a gastrula having three germ layers (ectoderm, mesoderm, and endoderm). (32)

Gated Ion Channels   Most passageways through a plasma membrane that allow ions to pass contain "gates" that can occur in either an open or a closed conformation. (7, 23)

Gel Electrophoresis   (See **Electrophoresis**)

Gene Pool   All the genes in all the individuals of a population. (33)

Gene Regulatory Proteins   Proteins that bind to specific sites in the DNA and control the transcription of nearby genes. (15)

Genes   Discrete units of inheritance which determine hereditary traits. (12, 14)

Gene Therapy   Treatment of a disease by alteration of the person's genotype, or the genotype of particular affected cells. (17)

Genetic Carrier   A heterozygous individual who shows no evidence of a genetic disorder but, because they possess a recessive allele for a disorder, can pass the mutant gene on to their offspring. (17)

Genetic Code   The correspondence between the various mRNA triplets (codons, e.g., UGC) and the amino acid that the triplet specifies (e.g., cysteine). The genetic code includes 64 possible three-letter words that constitute the genetic language for protein synthesis. (14)

Genetic Drift Random changes in allele frequency that occur by chance alone. Occurs primarily in small populations. (33)

Genetic Engineering The modification of a cell or organism's genetic composition according to human design. (16)

Genetic Equilibrium A state in which allele frequencies in a population remain constant from generation to generation. (33)

Genetic Mapping Determining the locations of specific genes or genetic markers along particular chromosomes. This is typically accomplished using crossover frequencies; the more often alleles of two genes are separated during crossing over, the greater the distance separating the genes. (13)

Genetic Recombination The reshuffling of genes on a chromosome caused by breakage of DNA and its reunion with the DNA of a homologue. (11)

Genome The information stored in all the DNA of a single set of chromosomes. (17)

Genotype An individual's genetic makeup. (12)

Genus Taxonomic group containing related species. (1)

Geologic Time Scale The division of the earth's 4.5 billion-year history into eras, periods, and epochs based on memorable geologic and biological events. (35)

Germ Cells Cells that are in the process of or have the potential to undergo meiosis and form gametes. (11, 31)

Germination The sprouting of a seed, beginning with the radicle of the embryo breaking through the seed coat. (21)

Germ Layers Collective name for the endoderm, ectoderm, and mesoderm, from which all the structures of the mature animal develop. (32)

Gibberellins More than 50 compounds that promote growth by stimulating both cell elongation and cell division. (21)

Gills Respiratory organs of aquatic animals. (29)

Globin The type of polypeptide chains that make up a hemoglobin molecule.

Glomerular Filtration The process by which fluid is filtered out of the capillaries of the glomerulus into the proximal end of the nephron. Proteins and blood cells remain behind in the bloodstream. (28)

Glomerulus A capillary bundle embedded in the double-membraned Bowman's capsule, through which blood for the kidney first passes. (28)

Glottis Opening leading to the larynx and lower respiratory tract. (29)

Glucagon A hormone secreted by the Islets of Langerhans that promotes glycogen breakdown to glucose. (25)

Glucocorticoids Steroid hormones which regulate sugar and protein metabolism. They are secreted by the adrenal cortex. (25)

Glycogen A highly branched polysaccharide consisting of glucose monomers that serves as a storage of chemical energy in animals. (4)

Glycolysis Cleavage, releasing energy, of the six-carbon glucose molecule into two molecules of pyruvic acid, each containing three carbons. (9)

Glycoproteins Proteins with covalently-attached chains of sugars. (5)

Glycosidic Bond The covalent bond between individual molecules in carbohydrates. (4)

Golgi Complex A system of flattened membranous sacs, which package substances for secretion from the cell. (5)

Gonadotropin-Releasing Hormone (GnRH) Hypothalmic hormone that controls the secretion of the gonadotropins FSH and LH. (31)

Gonadotropins Two anterior pituitary hormones which act on the gonads. Both FSH (follicle-stimulating hormone) and LH (luteinizing hormone) promote gamete development and stimulate the gonads to produce sex hormones. (25)

Gonads Gamete-producing structures in animals: ovaries in females, testes in males. (31)

Grasslands Areas of densely packed grasses and herbaceous plants. (40)

Gravitropisms (Geotropisms) Changes in plant growth caused by gravity. Growth away from gravitational force is called negative gravitropism; growth toward it is positive. (21)

Gray Matter Gray-colored neural tissue in the cerebral cortex of the brain and in the butterfly-shaped interior of the spinal cord. Composed of nonmyelinated cell bodies and dendrites of neurons. (23)

Greenhouse Effect The trapping of heat in the Earth's troposphere, caused by increased levels of carbon dioxide near the Earth's surface; the carbon dioxide is believed to act like glass in a greenhouse, allowing light to reach the Earth, but not allowing heat to escape. (41)

Ground Tissue System All plant tissues except those in the dermal and vascular tissues. (18)

Growth An increase in size, resulting from cell division and/or an increase in the volume of individual cells. (10)

Growth Hormone (GH) Hormone produced by the anterior pituitary; stimulates protein synthesis and bone elongation. (25)

Growth Ring In plants with secondary growth, a ring formed by tracheids and/or vessels with small lumens (late wood) during periods of unfavorable conditions; apparent in cross section. (18)

Guard Cells Specialized epidermal plant cells that flank each stomated pore of a leaf. They regulate the rate of gas diffusion and transpiration. (18)

Guild Group of species with similar ecological niches. (41)

Guttation The forcing of water and mineral completely out to the tips of leaves as a result of positive root pressure. (19)

Gymnosperms The earliest seed plants, bearing naked seeds. Includes the pines, hemlocks, and firs. (38)

◀ **H** ▶

Habitat The place or region where an organism lives. (41)

Habituation The phenomenon in which an animal ceases to respond to a repetitive stimulus. (23, 44)

Hair Cells Sensory receptors of the inner ear that respond to sound vibration and bodily movement. (24)

Half-Life The time required for half the mass of a radioactive element to decay into its stable, non-radioactive form. (3)

Haplodiploidy A genetic pattern of sex determination in which fertilized eggs develop into females and non-fertilized eggs develop into males (as occurs among bees and wasps). (44)

Haploid Having one set of chromosomes per cell. Often written as 1N. (10)

Hardy-Weinberg Law The maintenance of constant allele frequencies in a population from one generation to the next when certain conditions are met. These conditions are the absence of mutation and migration, random mating, a large population, and an equal chance of survival for all individuals. (33)

Haversian Canals A system of microscopic canals in compact bone that transport nutrients to and remove wastes from osteocytes. (26)

Heart An organ that pumps blood (or hemolymph in arthropods) through the vessels of the circulatory system. (28)

Helper T Cells A class of T cells that regulate immune responses by recognizing and activating B cells and other T cells. (30)

Hemocoel In arthropods, the unlined spaces into which fluid (hemolymph) flows when it leaves the blood vessels and bathes the internal organs. (28)

Hemoglobin The iron-containing blood protein that temporarily binds O_2 and releases it into the tissues. (4, 29)

Hemophilia A genetic disorder determined by a gene on the X chromosome (an X-linked trait) that results from the failure of the blood to form clots. (13)

Herbaceous Plants having only primary growth and thus composed entirely of primary tissue. (18)

Herbivore An organism, usually an animal, that eats primary producers (plants). (42)

Herbivory The term for the relationship of a secondary consumer, usually an animal, eating primary producers (plants). (42)

Heredity The passage of genetic traits to off-spring which consequently are similar or identical to the parent(s). (12)

Hermaphrodites Animals that possess gonads of both the male and the female. (31)

Heterosporous Higher vascular plants producing two types of spores, a megaspore which grows into a female gametophyte and a microspore which grows into a male gametophyte. (38)

Heterozygous A term applied to organisms that possess two different alleles for a trait. Often, one allele (A) is dominant, masking the presence of the other (a), the recessive. (12)

High Intertidal Zone In the intertidal zone, the region from mean high tide to around just below sea level. Organisms are submerged about 10% of the time. (40)

Histones Small basic proteins that are complexed with DNA to form nucleosomes, the basic structural components of the chromatin fiber. (14)

Homeobox That part of the DNA sequence of homeotic genes that is similar (homologous) among diverse animal species. (32)

Homeostasis Maintenance of fairly constant internal conditions (e.g., blood glucose level, pH, body temperature, etc.) (22)

Homeotic Genes Genes whose products act during embryonic development to affect the spatial arrangement of the body parts. (32)

Hominids Humans and the various groups of extinct, erect-walking primates that were either our direct ancestors or their relatives. Includes the various species of *Homo* and *Australopithecus*. (34)

Homo the genus that contains modern and extinct species of humans. (34)

Homologous Structures Anatomical structures that may have different functions but develop from the same embryonic tissues, suggesting a common evolutionary origin. (34)

Homologues Members of a chromosome pair, which have a similar shape and the same sequence of genes along their length. (10)

Homoplasy (see **Analogous Structures**)

Homosporous Plants that manufacture only one type of spore, which develops into a gametophyte containing both male and female reproductive structures. (38)

Homozygous A term applied to an organism that has two identical alles for a particular trait. (12)

Hormones Chemical messengers secreted by ductless glands into the blood that direct tissues to change their activities and correct imbalances in body chemistry. (25)

Host The organism that a parasite lives on and uses for food. (42)

Human Chorionic Gonadotropin (HCG) A hormone that prevents the corpus luteum from degenerating, thereby maintaining an adequate level of progesterone during pregnancy. It is produced by cells of the early embryo. (25)

Human Immunodeficiency Virus (HIV) The infectious agent that causes AIDS, a disease in which the immune system is seriously disabled. (30, 36)

Hybrid An individual whose parents possess different genetic traits in a breeding experiment or are members of different species. (12)

Hybridization Occurs when two distinct species mate and produce hybrid offspring. (33)

Hybridoma A cell formed by the fusion of a malignant cell (a myeloma) and an antibody-producing lymphocyte. These cells proliferate indefinitely and produce monoclonal antibodies. (30)

Hydrogen Bonds Relatively weak chemical bonds formed when two molecules share an atom of hydrogen. (3)

Hydrologic Cycle The cycling of water, in various forms, through the environment, from Earth to atmosphere and back to Earth again. (41)

Hydrolysis Splitting of a covalent bond by donating the H^+ or OH^- of a water molecule to the two components. (4)

Hydrophilic Molecules Polar molecules that are attracted to water molecules and readily dissolve in water. (3)

Hydrophobic Interaction When nonpolar molecules are "forced" together in the presence of a polar solvent, such as water. (3)

Hydrophobic Molecules Nonpolar substances, insoluble in water, which form aggregates to minimize exposure to their polar surroundings. (3)

Hydroponics The science of growing plants in liquid nutrient solutions, without a solid medium such as soil. (19)

Hydrosphere That portion of the Earth composed of water. (40)

Hydrostatic Skeletons Body support systems found usually in underwater animals (e.g., marine worms). Body shape is protected against gravity and other physical forces by internal hydrostatic pressure produced by contracting muscles encircling their closed, fluid-filled chambers. (26)

Hydrothermal Vents Fissures in the ocean floor where sea water becomes superheated. Chemosynthetic bacteria that live in these vents serve as the autotrophs that support a diverse community of ocean-dwelling organisms. (8)

Hyperpolarization An increase in the potential difference (voltage) across the plasma membrane of a cell typically due to an increase in the movement of potassium ions out of the cell. Acts to inhibit a target cell. (23)

Hypertension High blood pressure (above about 130/90). (28)

Hypertonic Solutions Solutions with higher solute concentrations than found inside the cell. These cause a cell to lose water and shrink. (7)

Hypervolume In ecology, a multidimensional area which includes all factors in an organism's ecological niche, or its' potential niche. (41)

Hypocotyl Portion of the plant embryo below the cotyledons. The hypocotyl gives rise to the root and, very often, to the lower part of the stem. (20)

Hypothalamus The area of the brain below the thalamus that regulates body temperature, blood pressure, etc. (25)

Hypothesis A tentative explanation for an observation or a phenomenon, phrased so that it can be tested by experimentation. (2)

Hypotonic Solutions Solutions with lower solute concentrations than found inside the cell. These cause a cell to accumulate water and swell. (7)

◄ I ►

Immigration Individuals permanently moving into a new area or population. (43)

Immune System A system in vertebrates for the surveillance and destruction of disease-causing microorganisms and cancer cells. Composed of lymphocytes, particularly B cells and T cells, and triggered by the introduction of antigens into the body which makes the body, upon their destruction, resistant to a recurrence of the same disease. (30)

Immunoglobulins (IGs) Antibody molecules. (30)

Imperfect Flowers Flowers that contain either stamens or carpels, making them male or female flowers, respectively. (20)

Imprinting A type of learning in which an animal develops an association with an object after exposure to the object during a critical period early in its life. (44)

Inbreeding When individuals mate with close relatives, such as brothers and sisters. May occur when population sizes drastically shrink and results in a decrease in genetic diversity. (33)

Incomplete Digestive Tract A digestive tract with only one opening through which food is taken in and residues are expelled. (27)

Incomplete (Partial) Dominance A phenomenon in which heterozygous individuals are phenotypically distinguishable from either homozygous type. (12)

Incomplete Flower Flowers lacking one or more whorls of sepals, petals, stamen, or pistils. (20)

Independent Assortment The shuffling of members of homologous chromosome pairs in meiosis I. As a result, there are new chromosome combinations in the daughter cells, which later produce offspring with random mixtures of traits from both parents. (11, 12)

Indoleatic Acid (IAA) An auxin responsible for many plant growth responses including apical dominance, a growth pattern in which shoot tips prevent axillary buds from sprouting. (21)

Induction The process in which one embryonic tissue induces another tissue to differentiate along a pathway that it would not otherwise have taken. (32) Stimulation of transcription of a gene in an operon. Occurs when the repressor protein is unable to bind to the operator. (15)

Inflammation   A body strategy initiated by the release of chemicals following injury or infection which brings additional blood with its protective cells to the injured area. (30)

Inhibitory Neurons   Neurons that oppose a response in the target cells. (23)

Inhibitory Neurotransmitters   Substances released from inhibitory neurons where they synapse with the target cell. (23)

Innate Behavior   Actions that are under fairly precise genetic control, typically species-specific, highly stereotyped, and that occur in a complete form the first time the stimulus is encountered. (44)

Insight Learning   The sudden solution to a problem without obvious trial-and-error procedures. (44)

Insulin   One of the two hormones secreted by endocrine centers called Islets of Langerhans; promotes glucose absorption, utilization, and storage. Insulin is secreted by them when the concentration of glucose in the blood begins to exceed the normal level. (25)

Integumentary System   The body's protective external covering, consisting of skin and subcutaneous tissue. (26)

Integuments   Protective covering of the ovule. (20)

Intercellular Junctions   Specialized regions of cell-cell contact between animal cells. (5)

Intercostal Muscles   Muscles that lie between the ribs in humans whose contraction expands the thoracic cavity during breathing. (29)

Interference Competition   One species' direct interference by another species for the same limited resource; such as aggressive animal behavior. (42)

Internal Fertilization   Fertilization of an egg within the body of the female. (31)

Interneurons   Neurons situated entirely within the central nervous system. (23)

Internodes   The portion of a stem between two nodes. (18)

Interphase   Usually the longest stage of the cell cycle during which the cell grows, carries out normal metabolic functions, and replicates its DNA in preparation for cell division. (10)

Interstitial Cells   Cells in the testes that produce testosterone, the major male sex hormone. (31)

Interstitial Fluid   The fluid between and surrounding the cells of an animal; the extracellular fluid. (28)

Intertidal Zone   The region of beach exposed to air between low and high tides. (40)

Intracellular Digestion   Digestion occurring inside cells within food vacuoles. The mode of

digestion found in protists and some filter-feeding animals (such as sponges and clams). (27)

Intraspecific Competition   Individual organisms of one species competing for the same limited resources in the same habitat, or with overlapping niches. (42)

Intrinsic Rate of Increase (r_m)   the maximum growth rate of a population under conditions of maximum birth rate and minimum death rate. (43)

Introns   Intervening sequences of DNA in the middle of structural genes, separating exons. (15)

Invertebrates   Animals that lack a vertebral column, or backbone. (39)

Ion   An electrically charged atom created by the gain or loss of electrons. (3)

Ionic Bond   The noncovalent linkage formed by the attraction of oppositely charged groups. (3)

Islets of Langerhans   Clusters of endocrine cells in the pancreas that produce insulin and glucagon. (25)

Isolating Mechanisms   Barriers that prevent gene flow between populations or among segments of a single population. (33)

Isotopes   Atoms of the same element having a different number of neutrons in their nucleus. (3)

Isotonic Solutions   Solutions in which the solute concentration outside the cell is the same as that inside the cell. (7)

◀ **J** ▶

Joints   Structures where two pieces of a skeleton are joined. Joints may be flexible, such as the knee joint of the human leg or the joints between segments of the exoskeleton of the leg of an insect, or inflexible, such as the joints (sutures) between the bones of the skull. (26)

J-Shaped Curve   A curve resulting from exponential growth of a population. (43)

◀ **K** ▶

Karyotype   A visual display of an individual's chromosomes. (10)

Kidneys   Paired excretory organs which, in humans, are fist-sized and attached to the lower spine. In vertebrates, the kidneys remove nitrogenous wastes from the blood and regulate ion and water levels in the body. (28)

Killer T Cells   A type of lymphocyte that functions in the destruction of virus-infected cells and cancer cells. (30)

Kinases   Enzymes that catalyze reactions in which phosphate groups are transferred from ATP to another molecule. (6)

Kinetic Energy   Energy in motion. (6)

Kinetochore   Part of a mitotic (or meiotic) chromosome that is situated within the centromere and to which the spindle fibers attach. (10)

Kingdom   A level of the taxonomic hierarchy that groups together members of related phyla or divisions. Modern taxonomy divides all organisms into five Kingdoms: Monera, Protista, Fungi, Plantae, and Animalia. (1)

Klinefelter Syndrome   A male whose cells have an extra X chromosome (XXY). The syndrome is characterized by underdeveloped male genitalia and feminine secondary sex characteristics. (17)

Krebs Cycle   A circular pathway in aerobic respiration that completely oxidizes the two pyruvic acids from glycolysis. (9)

K-Selected Species   Species that produce one or a few well-cared for individuals at a time. (43)

◀ **L** ▶

Lacteal   Blind lymphatic vessel in the intestinal villi that receives the absorbed products of lipid digestion. (27)

Lactic Acid Fermentation   The process in which electrons removed during glycolysis are transferred from NADH to pyruvic acid to form lactic acid. Used by various prokaryotic cells under oxygen-deficient conditions and by muscle cells during strenuous activity. (9)

Lake   Large body of standing fresh water, formed in natural depressions in the Earth. Lakes are larger than ponds. (40)

Lamella   In bone, concentric cylinders of calcified collagen deposited by the osteocytes. The laminated layers produce a greatly strengthened structure. (26)

Large Intestine   Portion of the intestine in which water and salts are reabsorbed. It is so named because of its large diameter. The large instestine, except for the rectum, is called the colon. (27)

Larva   A self-feeding, sexually, and developmentally immature form of an animal. (32)

Larynx   The short passageway connecting the pharynx with the lower airways. (29)

Latent (hidden) Infection   Infection by a microorganism that causes no symptoms but the microbe is well-established in the body. (36)

Lateral Roots   Roots that arise from the pericycle of older roots; also called branch roots or secondary roots. (18)

Law of Independent Assortment   Alleles on nonhomologous chromosomes segregate independently of one another. (12)

Law of Segregation   During gamete formation, pairs of alleles separate so that each sperm or egg cell has only one gene for a trait. (12)

Law of the Minimum   The ecological principle that a species' distribution will be limited by whichever abiotic factor is most deficient in the environment. (41)

Laws of Thermodynamics   Physical laws that describe the relationship of heat and mechanical energy. The first law states that energy cannot be created or destroyed, but one form can change into another. The second law states that the total energy the universe decreases as energy conversions occur and some energy is lost as heat. (6)

Leak Channels Passageways through a plasma membrane that do not contain gates and, therefore, are always open for the limited diffusion of a specific substance (ion) through the membrane. (7, 23)

Learning A process in which an animal benefits from experience so that its behavior is better suited to environmental conditions. (44)

Lenticels Loosely packed cells in the periderm of the stem that create air channels for transferring CO_2, H_2O, and O_2. (18)

Leukocytes White blood cells. (28)

LH Luteinizing hormone. A hormone secreted by the anterior pituitary that stimulates testosterone production in males and triggers ovulation and the transformation of the follicle into the corpus luteum in females. (31)

Lichen Symbiotic associations between certain fungi and algae. (37)

Life Cycle The sequence of events during the lifetime of an organism from zygote to reproduction. (39)

Ligaments Strong straps of connective tissue that hold together the bones in articulating joints or support an organ in place. (26)

Light-Dependent Reactions First stage of photosynthesis in which light energy is converted to chemical energy in the form of energy-rich ATP and NADPH. (8)

Light-Independent Reactions Second stage of photosynthesis in which the energy stored in ATP and NADPH formed in the light reactions is used to drive the reactions in which carbon dioxide is converted to carbohydrate. (8)

Limb Bud A portion of an embryo that will develop into either a forelimb or hindlimb. (32)

Limbic System A series of an interconnected group of brain structures, including the thalamus and hypothalamus, controlling memory and emotions. (23)

Limiting Factors The critical factors which impose restraints of the distribution, health, or activities of an organism. (41)

Limnetic Zone Open water of lakes, through which sunlight penetrates and photosynthesis occurs. (40)

Linkage The tendency of genes of the same chromosome to stay together rather than to assort independently. (13)

Linkage Groups Groups of genes located on the same chromosome. The genes of each linkage group assort independently of the genes of other linkage groups. In all eukaryotic organisms, the number of linkage groups is equal to the haploid number of chromosomes. (13)

Lipids A diverse group of biomolecules that are insoluble in water. (4)

Lithosphere The solid outer zone of the Earth; composed of the crust and outermost portion of the mantle. (40)

Littoral Zone Shallow, nutrient-rich waters of a lake, where sunlight reaches the bottom; also the lakeshore. Rooted vegetation occurs in this zone. (40)

Locomotion The movement of an organism from one place to another. (26)

Locus The chromosomal location of a gene. (13)

Logistic Growth Population growth producing a sigmoid, or S-shaped, growth curve. (43)

Long-Day Plants Plants that flower when the length of daylight exceeds some critical period. (21)

Longitudinal Fission The division pattern in flagellated protozoans, where division is along the length of the cell.

Loop of Henle An elongated section of the renal tubule that dips down into the kidney's medulla and then ascends back out to the cortex. It separates the proximal and distal convoluted tubules and is responsible for forming the salt gradient on which water reabsorption in the kidney depends. (28)

Low Density Lipoprotein (LDL) Particles that transport cholesterol in the blood. Each particle consists of about 1,500 cholesterol molecules surrounded by a film of phospholipids and protein. LDLs are taken into cells following their binding to cell surface LDL receptors. (7)

Low Intertidal Zone In the intertidal zone, the region which is uncovered by "minus" tides only. Organisms are submerged about 90% of the time. (40)

Lumen A space within an hollow organ or tube. (28)

Luminescence (see **Bioluminescence**)

Lungs The organs of terrestrial animals where gas exchange occurs. (29)

Lymph The colorless fluid in lymphatic vessels. (28)

Lymphatic System Network of fluid-carrying vessels and associated organs that participate in immunity and in the return of tissue fluid to the main circulation. (28)

Lymphocytes A group of non-phagocytic white blood cells which combat microbial invasion, fight cancer, and neutralize toxic chemicals. The two classes of lymphocytes, B cells and T cells, are the heart of the immune system. (28, 30)

Lymphoid Organs Organs associated with production of blood cells and the lymphatic system, including the thymus, spleen, appendix, bone marrow, and lymph nodes. (30)

Lysis (1) To split or dissolve. (2) Cell bursting.

Lysomes A type of storage vesicle produced by the Golgi complex, containing hydrolytic (digestive) enzymes capable of digesting many kinds of macromolecules in the cell. The membrane around them keeps them sequestered. (5)

◀ **M** ▶

M Phase That portion of the cell cycle during which mitosis (nuclear division) and cytokinesis (cytoplasmic division) takes place. (10)

Macroevolution Evolutionary changes that lead to the appearance of new species. (33)

Macrofungus Filamentous fungus so named for the large size of its fleshy sexual structures; a mushroom, for example. (37)

Macromolecules Large polymers, such as proteins, nucleic acids, and polysaccharides. (4)

Macronutrients Nutrients required by plants in large amounts: carbon, oxygen, hydrogen, nitrogen, potassium, calcium, phosphorus, magnesium, and sulfur. (19)

Macrophages Phagocytic cells that develop from monocytes and present antigen to lymphocytes. (30)

Macroscopic Referring to biological observations made with the naked eye or a hand lens.

Mammals A class of vertebrates that possesses skin covered with hair and that nourishes their young with milk from mammary glands. (39)

Mammary Glands Glands contained in the breasts of mammalian mothers that produce breast milk. (39)

Marsupials Mammals with a cloaca whose young are born immature and complete their development in an external pouch in the mother's skin. (39)

Mass Extinction The simultaneous extinction of a multitude of species as the result of a drastic change in the environment. (33, 35)

Maternal Chromosomes The set of chromosomes in an individual that were inherited from the mother. (11)

Mechanoreceptors Sensory receptors that respond to mechanical pressure and detect motion, touch, pressure, and sound. (24)

Medulla The center-most portion of some organs. (23)

Medusa The motile, umbrella-shaped body form of some members of the phylum Cnidaria, with mouth and tentacles on the lower, concave service. (Compare with polyp.) (39)

Megaspores Spores that divide by mitosis to produce female gametophytes that produce the egg gamete. (20)

Meiosis The division process that produces cells with one-half the number of chromosomes in each somatic cell. Each resulting daughter cell is haploid (1N) (11)

Meiosis I A process of reductional division in which homologous chromosomes pair and then segregate. Homologues are partitioned into separate daughter cells. (11)

Meiosis II Second meiotic division. A division process resembling mitosis, except that the haploid number of chromosomes is present. After the chromosomes line up at the meta-phase plate, the two sister chromatids separate. (11)

Melanin A brown pigment that gives skin and hair its color (12)

Melanoma A deadly form of skin cancer that develops from pigment cells in the skin and is promoted by exposure to the sun. (14)

Memory Cells Lymphocytes responsible for active immunity. They recall a previous exposure to an antigen and, on subsequent exposure to the same antigen, proliferate rapidly into plasma cells and produce large quantities of antibodies in a short time. This protection typically lasts for many years. (30)

Mendelian Inheritance Transmission of genetic traits in a manner consistent with the principles discovered by Gregor Mendel. Includes traits controlled by simple dominant or recessive alleles; more complex patterns of transmission are referred to as Nonmendelian inheritance. (12)

Meninges The thick connective tissue sheath which surrounds and protects the vertebrate brain and spinal cord. (23)

Menstrual Cycle The repetitive monthly changes in the uterus that prepare the endometrium for receiving and supporting an embryo. (31)

Meristematic Region New cells arise from this undifferentiated plant tissue; found at root or shoot apical meristems, or lateral meristems. (18)

Meristems In plants, clusters of cells that retain their ability to divide, thereby producing new cells. One of the four basic tissues in plants. (18)

Mesoderm In animals, the middle germ cell layer of the gastrula. It gives rise to muscle, bone, connective tissue, gonads, and kidney. (32)

Mesopelagic Zone The dimly lit ocean zone beneath the epipelagic zone; large fishes, whales and squid occupy this zone; no phytoplankton occur in this zone. (40)

Mesophyll Layers of cells in a leaf between the upper and lower epidermis; produced by the ground meristem. (18)

Messenger RNA (mRNA) The RNA that carries genetic information from the DNA in the nucleus to the ribosomes in the cytoplasm, where the sequence of bases in the mRNA is translated into a sequence of amino acids. (14)

Metabolic Intermediates Compounds produced as a substrate are converted to end product in a series of enzymatic reactions. (6)

Metabolic Pathways Set of enzymatic reactions involved in either building or dismantling complex molecules. (6)

Metabolic Rate A measure of the level of activity of an organism usually determined by measuring the amount of oxygen consumed by an individual per gram body weight per hour. (22)

Metabolic Water Water produced as a product of metabolic reactions. (28)

Metabolism The sum of all the chemical reactions in an organism; includes all anabolic and catabolic reactions. (6)

Metamorphosis Transformation from one form into another form during development. (32)

Metaphase The stage of mitosis when the chromosomes line-up along the metaphase plate, a plate that usually lies midway between the spindle poles. (10)

Metaphase Plate Imaginary plane within a dividing cell in which the duplicated chromosomes become aligned during metaphase. (10)

Microbes Microscopic organisms. (36)

Microbiology The branch of biology that studies microorganisms. (36)

Microevolution Changes in allele frequency of a species' gene pool which has not generated new species. Exemplified by changes in the pigmentation of the peppered moth and by the acquisition of pesticide resistance in insects. (33)

Microfibrils Bundles formed from the intertwining of cellulose molecules, i.e., long chains of glucose molecules in the cell walls of plants. (5)

Microfilaments Thin actin-containing protein fibers that are responsible for maintenance of cell shape, muscle contraction and cyclosis. (5)

Micrometer One millionth (1/1,000,000) of a meter.

Micronutrients Nutrients required by plants in small amounts: iron, chlorine, copper, manganese, zinc, molybdenum, and boron. (19)

Micropyle A small opening in the integuments of the ovule through which the pollen tube grows to deliver sperm. (21)

Microspores Spores within anthers of flowers. They divide by mitosis to form pollen grains, the male gametophytes that produce the plant's sperm. (20)

Microtubules Thin, hollow tubes in cells; built from repeating protein units of tubulin. Microtubules are components of cilia, flagella, and the cytoskeleton. (5)

Microvilli The small projections on the cells that comprise each villus of the intestinal wall, further increasing the absorption surface area of the small intestine. (27)

Middle Intertidal Zone In the intertidal zone, the region which is covered and uncovered twice a day, the zero of tide tables. Organisms are submerged about 50% of the time. (40)

Migration Movements of a population into or out of an area. (44)

Mimicry A defense mechanism where one species resembles another in color, shape, behavior, or sound. (42)

Mineralocorticoids Steroid hormones which regulate the level of sodium and potassium in the blood. (25)

Mitochondria Organelles that contain the biochemical machinery for the Krebs cycle and the electron transport system of aerobic respiration. They are composed of two membranes, the inner one forming folds, or cristae. (9)

Mitosis The process of nuclear division producing daughter cells with exactly the same number of chromosomes as in the mother cell. (10)

Mitosis Promoting Factor (MPF) A protein that appears to be a universal trigger of cell division in eukaryotic cells. (10)

Mitotic Chromosomes Chromosomes whose DNA-protein threads have become coiled into microscopically visible chromosomes, each containing duplicated chromatids ready to be separated during mitosis. (10)

Molds Filamentous fungi that exist as colonies of threadlike cells but produce no macroscopic fruiting bodies. (37)

Molecule Chemical substance formed when two or more atoms bond together; the smallest unit of matter that possesses the qualities of a compound. (3)

Mollusca A phylum, second only to Arthropoda in diversity. Composed of three main classes: 1) Gastropoda (spiral-shelled), 2) Bivalvia (hinged shells), 3) Cephalopoda (with tentacles or arms and no, or very reduced shells). (39)

Molting (Ecdysis) Shedding process by which certain arthropods lose their exoskeletons as their bodies grow larger. (39)

Monera The taxonomic kingdom comprised of single-celled prokaryotes such as bacteria, cyanobacteria, and archebacteria. (36)

Monoclonal Antibodies Antibodies produced by a clone of hybridoma cells, all of which descended from one cell. (30)

Monocotyledae (Monocots) One of the two divisions of flowering plants, characterized by seeds with a single cotyledon, flower parts in 3s, parallel veins in leaves, many roots of approximately equal size, scattered vascular bundles in its stem anatomy, pith in its root anatomy, and no secondary growth capacity. (18)

Monocytes A type of leukocyte that gives rise to macrophages. (28)

Monoecious Both male and female reproductive structures are produced on the same sporophyte individual. (20, 38)

Monohybrid Cross A mating between two individuals that differ only in one genetically-determined trait. (12)

Monomers Small molecular subunits which are the building blocks of macromolecules. The macromolecules in living systems are constructed of some 40 different monomers. (4)

Monotremes A group of mammals that lay eggs from which the young are hatched. (39)

Morphogenesis The formation of form and internal architecture within the embryo brought about by such processes as programmed cell death, cell adhesion, and cell movement. (32)

Morphology The branch of biology that studies form and structure of organisms.

Mortality Death rate in a population or area. (43)

Motile Capable of independent movement.

Motor Neurons Nerve cells which carry outgoing impulses to their effectors, either glands or muscles. (23)

Mucosa The cell layer that lines the digestive tract and secretes a lubricating layer of mucus. (27)

Mullerian Mimicry Resemblance of different species, each of which is equally obnoxious to predators. (42)

Multicellular Consisting of many cells. (35)

Multichannel Food Chain Where the same primary producer supplies the energy for more than one food chain. (41)

Multiple Allele System Three or more possible alleles for a given trait, such as ABO blood groups in humans. (12)

Multiple Fission Division of the cell's nucleus without a corresponding division of cytoplasm.

Multiple Fruits Fruits that develop from pistils of separate flowers. (20)

Muscle Fiber A muiltinucleated skeletal muscle cell that results from the fusion of several pre-muscle cells during embryonic development. (26)

Muscle Tissue Bundles and sheets of contractile cells that shorten when stimulated, providing force for controlled movement. (26)

Mutagens Chemical or physical agents that induce genetic change. (14)

Mutation Random heritable changes in DNA that introduce new alleles into the gene pool. (14)

Mutualism The symbiotic interaction in which both participants benefit. (42)

Mycology The branch of biology that studies fungi. (37)

Mycorrhizae An association between soil fungi and the roots of vascular plants, increasing the plant's ability to extract water and minerals from the soil. (19)

Myelin Sheath In vertebrates, a jacket which covers the axons of high-velocity neurons, thereby increasing the speed of a neurological impulse. (23)

Myofibrils In striated muscle, the banded fibrils that lie parallel to each other, constituting the bulk of the muscle fiber's interior and powering contraction. (26)

Myosin A contractile protein that makes up the major component of the thick filaments of a muscle cell and is also present in nonmuscle cells. (26)

◀ **N** ▶

NADPH Nicotinamide adenine dinucleotide phosphate. NADPH is formed by reduction of NADP$^+$, and serves as a store of electrons for use in metabolism (see Reducing Power). (9)

NAD$^+$ Nicotinamide adenine dinucleotide. A coenzyme that functions as an electron carrier in metabolic reactions. When reduced to NADH, the molecule becomes a cellular energy source. (9)

Natality Birthrate in a population or area. (43)

Natural Killer (NK) Cells Nonspecific, lymphocytelike cells which destroy foreign cells and cancer cells. (30)

Natural Selection Differential survival and reproduction of organisms with a resultant increase in the frequency of those best adapted to the environment. (33)

Neanderthals A subspecies of Homo sapiens different from that of modern humans that were characterized by heavy bony skeletons and thick bony ridges over the eyes. They disappeared about 35,000 years ago. (34)

Nectary Secretory gland in flowering plants containing sugary fluid that attracts pollinators as a food source. Usually located at the base of the flower. (20)

Negative Feedback Any regulatory mechanism in which the increased level of a substance inhibits further production of that substance, thereby preventing harmful accumulation. A type of homeostatic mechanism. (22, 25)

Negative Gravitropism In plants, growth against gravitational forces, or shoot growth upward. (21)

Nematocyst Within the stinging cell (cnidocyte) of cnidarians, a capsule that contains a coiled thread which, when triggered, harpoons prey and injects powerful toxins. (39)

Nematoda The widespread and abundant animal phylum containing the roundworms. (39)

Nephridium A tube surrounded by capillaries found in an organism's excretory organs that removes nitrogenous wastes and regulates the water and chemical balance of body fluids. (28)

Nephron The functional unit of the vertebrate kidney, consisting of the glomerulus, Bowman's capsule, proximal and distal convoluted tubules, and loop of Henle. (28)

Nerve Parallel bundles of neurons and their supporting cells. (23)

Nerve Impulse A propagated action potential. (23)

Nervous Tissue Excitable cells that receive stimuli and, in response, transmit an impulse to another part of the animal. (23)

Neural Plate In vertebrates, the flattened plate of dorsal ectoderm of the late gastrula that gives rise to the nervous system. (32)

Neuroglial Cells Those cells of a vertebrate nervous system that are not neurons. Includes a variety of cell types including Schwann cells. (23)

Neuron A nerve cell. (23)

Neurosecretory Cells Nervelike cells that secrete hormones rather than neurotransmitter substances when a nerve impulse reaches the distal end of the cell. In vertebrates, these cells arise from the hypothalamus. (25)

Neurotoxins Substances, such as curare and tetanus toxin, that interfere with the transmission of neural impulses. (23)

Neurotransmitters Chemicals released by neurons into the synaptic cleft, stimulating or inhibiting the post-synaptic target cell. (23)

Neurulation Formation by embryonic induction of the neural tube in a developing vertebrate embryo. (32)

Neutrons Electrically neutral (uncharged) particles contained within the nucleus of the atom. (3)

Neutrophil Phagocytic leukocyte, most numerous in the human body. (28)

Niche An organism's habitat, role, resource requirements, and tolerance ranges for each abiotic condition. (42)

Niche Breadth Relative size and dimension of ecological niches; for example, broad or narrow niches. (41)

Niche Overlap Organisms that have the same habitat, role, environmental requirements, or needs. (41)

Nitrogen Fixation The conversion of atmospheric nitrogen gas N_2 into ammonia (NH_3) by certain bacteria and cyanobacteria. (19)

Nitrogenous Wastes Nitrogen-containing metabolic waste products, such as ammonia or urea, that are produced by the breakdown of proteins and nucleic acids. (28)

Nodes The attachment points of leaves to a stem. (18)

Nodes of Ranvier Uninsulated (nonmyelinated) gaps along the axon of a neuron. (23)

Noncovalent Bonds Linkages between two atoms that depend on an attraction between positive and negative charges between molecules or ions. Includes ionic and hydrogen bonds. (3)

Non-Cyclic Photophosphorylation The pathway in the light reactions of photosynthesis in which electrons pass from water, through two photosystems, and then ultimately to NADP+. During the process, both ATP and NADPH are produced. It is so named because the electrons do not return to their reaction center. (8)

Nondisjunction Failure of chromosomes to separate properly at meiosis I or II. The result is that one daughter will receive an extra chromosome and the other gets one less. (11, 13)

Nonpolar Molecules Molecules which have an equal charge distribution throughout their structure and thus lack regions with a localized positive or negative charge. (3)

Notochord A flexible rod that is below the dorsal surface of the chordate embryo, beneath the nerve cord. In most chordates, it is replaced by the vertebral column. (32)

Nuclear Envelope A double membrane pierced by pores that separates the contents of the nucleus from the rest of the eukaryotic cell. (5)

Nucleic Acids DNA and RNA; linear polymers of nucleotides, responsible for the storage and expression of genetic information. (4, 14)

Nucleoid A region in the prokaryotic cell that contains the genetic material (DNA). It is unbounded by a nuclear membrane. (36)

Nucleoplasm The semifluid substance of the nucleus in which the particulate structures are suspended. (5)

Nucleosomes Nuclear protein complex consisting of a length of DNA wrapped around a central cluster of 8 histones. (14)

Nucleotides Monomers of which DNA and RNA are built. Each consists of a 5-carbon sugar, phosphate, and a nitrogenous base. (4)

Nucleous (pl. nucleoli) One or more darker regions of a nucleus where each ribosomal subunit is assembled from RNA and protein. (5)

Nucleus The large membrane-enclosed organelle that contains the DNA of eukaryotic cells. (5)

Nucleus, Atomic The center of an atom containing protons and neutrons. (3)

◄ O ►

Obligate Symbiosis A symbiotic relationship between two organisms that is necessary for the survival or both organisms. (42)

Olfaction The sense of smell. (24)

Oligotrophic Little nourished, as a young lake that has few nutrients and supports little life. (40)

Omnivore An animal that obtains its nutritional needs by consuming plants and other animals. (42)

Oncogene A gene that causes cancer, perhaps activated by mutation or a change in its chromosomal location. (10)

Oocyte A female germ cell during any of the stages of meiosis. (31)

Oogenesis The process of egg production. (31)

Oogonia Female germ cells that have not yet begun meiosis. (31)

Open Circulatory System Circulatory system in which blood travels from vessels to tissue spaces, through which it percolates prior to returning to the main vessel (compare with closed circulatory system). (28)

Operator A regulatory gene in the operon of bacteria. It is the short DNA segment to which the repressor binds, thus preventing RNA polymerase from attaching to the promoter. (15)

Operon A regulatory unit in prokaryotic cells that controls the expression of structural genes. The operon consists of structural genes that produce enzymes for a particular metabolic pathway, a regulator region composed of a promoter and an operator, and R (regulator) gene that produces a repressor. (15)

Order A level of the taxonomic hierarchy that groups together members of related families. (1)

Organ Body part composed of several tissues that performs specialized functions. (22)

Organelle A specialized part of a cell having some particular function. (5)

Organic Compounds Chemical compounds that contain carbon. (4)

Organism A living entity able to maintain its organization, obtain and use energy, reproduce, grow, respond to stimuli, and display homeostatis. (1)

Organogenesis Organ formation in which two or more specialized tissue types develop in a precise temporal and spatial relationship to each other. (32)

Organ System Group of functionally related organs. (22)

Osmoregulation The maintenance of the proper salt and water balance in the body's fluids. (28)

Osmosis The diffusion of water through a differentially permeable membrane into a hypertonic compartment. (7)

Ossification Synthesis of a new bone. (26)

Osteoclast A type of bone cell which breaks down the bone, thereby releasing calcium into the bloodstream for use by the body. Osteoclasts are activated by hormones released by the parathyroid glands. (26)

Osteocytes Living bone cells embedded within the calcified matrix they manufacture. (26)

Osteoporosis A condition present predominantly in postmenopausal women where the bones are weakened due to an increased rate of bone resorption compared to bone formation. (26)

Ovarian Cycle The cycle of egg production within the mammalian ovary. (31)

Ovarian Follicle In a mammalian ovary, a chamber of cells in which the oocyte develops. (31)

Ovary In animals, the egg-producing gonad of the female. In flowering plants, the enlarged base of the pistil, in which seeds develop. (20)

Oviduct (Fallopian Tube) The tube in the female reproductive organ that connects the ovaries and uterus and where fertilization takes place. (31)

Ovulation The release of an egg (ovum) from the ovarian follicle. (31)

Ovule In seed plants, the structure containing the female gametophyte, nucellus, and integuments. After fertilization, the ovule develops into a seed. (20, 38)

Ovum An unfertilized egg cell; a female gamete. (31)

Oxidation The removal of electrons from a compound during a chemical reaction. For a carbon atom, the fewer hydrogens bonded to a carbon, the greater the oxidation state of the atom. (6)

Oxidative Phosphorylation The formation of ATP from ADP and inorganic phosphate that occurs in the electron-transport chain of cellular respiration. (8, 9)

Oxyhemoglobin A complex of oxygen and hemoglobin, formed when blood passes through the lungs and is dissociated in body tissues, where oxygen is released. (29)

Oxytocin A female hormone released by the posterior pituitary which triggers uterine contractions during childbirth and the release of milk during nursing. (25)

◄ P ►

P680 Reaction Center (P = Pigment) Special chlorophyll molecule in Photosystem II that traps the energy absorbed by the other pigment molecules. It absorbs light energy maximally at 680 nm. (8)

Palisade Parenchyma In dicot leaves, densely packed, columnar shaped cells functioning in photosynthesis. Found just beneath the upper epidermis. (18)

Pancreas In vertebrates, a large gland that produces digestive enzymes and hormones. (27)

Parallel Evolution When two species that have descended from the same ancestor independently acquire the same evolutionary adaptations. (33)

Parapatric Speciation The splitting of a population into two species' populations under conditions where the members of each population reside in adjacent areas. (33)

Parasite An organism that lives on or inside another called a host, on which it feeds. (39, 42)

Parasitism A relationship between two organisms where one organism benefits, and the other is harmed. (42)

Parasitoid Parasitic organisms, such as some insect larvae, which kill their host. (42)

Parasympathetic Nervous System Part of the autonomic nervous system active during relaxed activity. (23)

Parathyroid Glands Four glands attached to the thyroid gland which secrete parathyroid hormone (PTH). When blood calcium levels are low, PTH is secreted, causing calcium to be released from bone. (25)

Parenchyma The most prevalent cell type in herbaceous plants. These thin-walled, polygonal-shaped cells function in photosynthesis and storage. (18)

Parthenogenesis Process by which offspring are produced without egg fertilization. (31)

Passive Immunity Immunity achieved by receiving antibodies from another source, as occurs with a newborn infant during nursing. (30)

Paternal Chromosomes The set of chromosomes in an individual that were inherited from the father. (11)

Pathogen A disease-causing microorganism. (36)

Pectoral Girdle In humans, the two scapulae (shoulder blades) and two clavicles (collarbones) which support and articulate with the bones of the upper arm. (26)

Pedicel A shortened stem carrying a flower. (20)

Pedigree A diagram showing the inheritance of a particular trait among the members of the family. (13)

Pelagic Zone The open oceans, divided into three layers: 1) photo- or epipelagic (sunlit), 2) mesopelagic (dim light), 3) aphotic or bathypelagic (always dark). (40)

Pelvic Girdle The complex of bones that connect a vertebrate's legs with its backbone. (26)

Penis An intrusive structure in the male animal which releases male gametes into the female's sex receptacle. (31)

Peptide Bond The covalent bond between the amino group of one amino acid and the carboxyl group of another. (4)

Peptidoglycan A chemical component of the prokaryotic cell wall. (36)

Percent Annual Increase A measure of population increase; the number of individuals (people) added to the population per 100 individuals. (43)

Perennials Plants that live longer than two years. (18)

Perfect Flower Flowers that contain both stamens and pistils. (20)

Perforation Plate In plants, that portion of the wall of vessel members that is perforated, and contains an area with neither primary nor secondary cell wall; a "hole" in the cell wall. (18)

Pericycle One or more layers of cells found in roots, with phloem or xylem to its' inside, and the endodermis to its' outside. Functions in producing lateral roots and formation of the vascular cambium in roots with secondary growth. (18)

Periderm Secondary tissue that replaces the epidermis of stems and roots. Consists of cork, cork cambium, and an internal layer of parenchyma cells. (18)

Peripheral Nervous System Neurons, excluding those of the brain and spinal cord, that permeate the rest of the body. (23)

Peristalsis Sequential waves of muscle contractions that propel a substance through a tube. (27)

Peritoneum The connective tissue that lines the coelomic cavities. (39)

Permeability The ability to be penetrable, such as a membrane allowing molecules to pass freely across it. (7)

Petal The second whorl of a flower, often brightly colored to attract pollinators; collectively called the corolla. (20)

Petiole The stalk leading to the blade of a leaf. (18)

pH A scale that measures the concentration of hydrogen ions in a solution. The pH scale extends from 0 to 14. Acidic solutions have a pH of less than 7; alkaline solutions have a pH above 7; neutral solutions have a pH equal to 7. (3)

Phagocytosis Engulfing of food particles and foreign cells by amoebae and white blood cells. A type of endocytosis. (5)

Pharyngeal Pouches In the vertebrate embryo, outgrowths from the walls of the pharynx that break through the body surface to form gill slits. (32)

Pharynx The throat; a portion of both the digestive and respiratory system just behind the oral cavity. (29)

Phenotype An individual's observable characteristics that are the expression of its genotype. (12)

Pheromones Chemicals that, when released by an animal, elicit a particular behavior in other animals of the same species. (44)

Phloem The vascular tissue that transports sugars and other organic molecules from sites of photosynthesis and storage to the rest of the plant. (18)

Phloem Loading The transfer of assimilates to phloem conducting cells, from photosynthesizing source cells. (19)

Phloem Unloading The transfer of assimilates to storage (sink) cells, from phloem conducting cells. (19)

Phospholipids Lipids that contain a phosphate and a variable organic group that form polar, hydrophilic regions on an otherwise nonpolar, hydrophobic molecule. They are the major structural components of membranes. (4)

Phosphorylation A chemical reaction in which a phosphate group is added to a molecule or atom. (6)

Photoexcitation Absorption of light energy by pigments, causing their electrons to be raised to a higher energy level. (8)

Photolysis The splitting of water during photosynthesis. The electrons from water pass to Photosystem II, the protons enter the lumen of the thylakoid and contribute to the proton gradient across the thylakoid membrane, and the oxygen is released into the atmosphere. (8)

Photon A particle of light energy. (8)

Photoperiod Specific lengths of day and night which control certain plant growth responses to light, such as flowering or germination. (21)

Photoperiodism Changes in the behavior and physiology of an organism in response to the relative lengths of daylight and darkness, i.e., the photoperiod. (21)

Photoreceptors Sensory receptors that respond to light. (24)

Photorespiration The phenomenon in which oxygen binds to the active site of a CO_2-fixing enzyme, thereby competing with CO_2 fixation, and lowering the rate of photosynthesis. (8)

Photosynthesis The conversion by plants of light energy into chemical energy stored in carbohydrate. (8)

Photosystems Highly organized clusters of photosynthetic pigments and electron/hydrogen carriers embedded in the thylakoid membranes of chloroplasts. There are two photosystems, which together carry out the light reactions of photosynthesis. (8)

Photosystem I Photosystem with a P700 reaction center; participates in cyclic photophosphorylation as well as in noncyclic photophosphorylation. (8)

Photosystem II Photosystem activated by a P680 reaction center; participates only in noncyclic photophosphorylation and is associated with photolysis of water. (8)

Phototropism The growth responses of a plant to light. (21)

Phyletic Evolution The gradual evolution of one species into another. (33)

Phylogeny Evolutionary history of a species. (35)

Phylum The major taxonomic divisions in the Animal kingdom. Members of a phylum share common, basic features. The Animal kingdom is divided into approximately 35 phyla. (39)

Physiology The branch of biology that studies how living things function. (22)

Phytochrome A light-absorbing pigment in plants which controls many plant responses, including photoperiodism. (21)

Phytoplankton Microscopic photosynthesizers that live near the surface of seas and bodies of fresh water. (37)

Pineal Gland An endocrine gland embedded within the brain that secretes the hormone melatonin. Hormone secretion is dependent on levels of environmental light. In amphibians and reptiles, melatonin controls skin coloration. In humans, pineal secretions control sexual maturation and daily rhythms. (25)

Pinocytosis Uptake of small droplets and dissolved solutes by cells. A type of endocytosis. (5)

Pistil The female reproductive part and central portion of a flower, consisting of the ovary, style and stigma. May contain one carpel, or one or more fused carpels. (20)

Pith A plant tissue composed of parenchyma cells, found in the central portion of primary growth stems of dicots, and monocot roots. (18)

Pith Ray Region between vascular bundles in vascular plants. (18)

Pituitary Gland (see **Posterior and Anterior Pituitary**).

Placenta In mammals (exclusive of marsupials and monotremes), the structure through which nutrients and wastes are exchanged between the mother and embryo/fetus. Develops from both embryonic and uterine tissues. (32)

Plant Multicellular, autotrophic organism able to manufacture food through photosynthesis. (38)

Plasma In vertebrates, the liquid portion of the blood, containing water, proteins (including fibrinogen), salts, and nutrients. (28)

Plasma Cells Differentiated antibody-secreting cells derived from B lymphocytes. (30)

Plasma Membrane The selectively permeable, molecular boundary that separates the cytoplasm of a cell from the external environment. (5)

Plasmid A small circle of DNA in bacteria in addition to its own chromosome. (16)

Plasmodesmata Openings between plant cell walls, through which adjacent cells are connected via cytoplasmic threads. (19)

Plasmodium Genus of protozoa that causes malaria. (37)

Plasmodium A huge multinucleated "cell" stage of a plasmodial slime mold that feeds on dead organic matter. (37)

Plasmolysis The shrinking of a plant cell away from its cell wall when the cell is placed in a hypertonic solution. (7)

Platelets Small, cell-like fragments derived from special white blood cells. They function in clotting. (28)

Plate Tectonics The theory that the earth's crust consists of a number of rigid plates that rest on an underlying layer of semimolten rock. The movement of the earth's plates results from the upward movement of molten rock into the solidified crust along ridges within the ocean floor. (35)

Platyhelminthes The phylum containing simple, bilaterally symmetrical animals, the flatworms. (39)

Pleiotropy Where a single mutant gene produces two or more phenotypic effects. (12)

Pleura The double-layered sac which surrounds the lungs of a mammal. (29, 39)

Pneumocytis Pneumonia (PCP) A disease of the respiratory tract caused by a protozoan that strikes persons with immunodeficiency diseases, such as AIDS. (30)

Point Mutations Changes that occur at one point within a gene, often involving one nucleotide in the DNA. (14)

Polar Body A haploid product of meiosis of a female germ cell that has very little cytoplasm and disintegrates without further function. (31)

Polar Molecule A molecule with an unequal charge distribution that creates distinct positive and negative regions or poles. (3)

Pollen The male gametophyte of seed plants, comprised of a generative nucleus and a tube nucleus surrounded by a tough wall. (20)

Pollen Grain The male gametophyte of conifers and angiosperms, containing male gametes. In angiosperms, pollen grains are contained in the pollen sacs of the anther of a flower. (20)

Pollination The transfer of pollen grains from the anther of one flower to the stigma of another. The transfer is mediated by wind, water, insects, and other animals. (20)

Polygenic Inheritance An inheritance pattern in which a phenotype is determined by two or more genes at different loci. In humans, examples include height and pigmentation. (12)

Polymer A macromolecule formed of monomers joined by covalent bonds.. Includes proteins, polysaccharides, and nucleic acids. (4)

Polymerase Chain Reaction (PCR) Technique to amplify a specific DNA molecule using a temperature-sensitive DNA polymerase obtained from a heat-resistant bacterium. Large numbers of copies of the initial DNA molecule can be obtained in a short period of time, even when the starting material is present in vanishingly small amounts, as for example from a blood stain left at the scene of a crime. (16)

Polymorphic Property of some protozoa to produce more than one stage of organism as they complete their life cycles. (37)

Polymorphic Genes Genes for which several different alleles are known, such as those that code for human blood type. (17)

Polyp Stationary body form of some members of the phylum Cnidaria, with mouth and tentacles facing upward. (Compare with medusa.) (39)

Polypeptide An unbranched chain of amino acids covalently linked together and assembled on a ribosome during translation. (4)

Polyploidy An organism or cell containing three or more complete sets of chromosomes. Polyploidy is rare in animals but common in plants. (33)

Polysaccharide A carbohydrate molecule consisting of monosaccharide units. (4)

Polysome A complex of ribosomes found in chains, linked by mRNA. Polysomes contain the ribosomes that are actively assembling proteins. (14)

Polytene Chromosomes Giant banded chromosomes found in certain insects that form by the repeated duplication of DNA. Because of the multiple copies of each gene in a cell, polytene chromosomes can generate large amounts of a gene product in a short time. Transcription occurs at sites of chromosome puffs. (13)

Pond Body of standing fresh water, formed in natural depressions in the Earth. Ponds are smaller than lakes. (40)

Population Individuals of the same species inhabiting the same area. (43)

Population Density The number of individual species living in a given area. (43)

Positive Gravitropism In plants, growth with gravitational forces, or root growth downward. (21)

Posterior Pituitary A gland which manufactures no hormones but receives and later releases hormones produced by the cell bodies of neurons in the hyopthalamus. (25)

Potential Energy Stored energy, such as occurs in chemical bonds. (6)

Preadaptation A characteristic (adaptation) that evolved to meet the needs of an organism in one type of habitat, but fortuitously allows the organism to exploit a new habitat. For example, lobed fins and lungs evolved in ancient fishes to help them live in shallow, stagnant ponds, but also facilitated the evolution of terrestrial amphibians. (33, 39)

Precells Simple forerunners of cells that, presumably, were able to concentrate organic molecules, allowing for more frequent molecular reactions. (35)

Predation Ingestion of prey by a predator for energy and nutrients. (42)

Predator An organism that captures and feeds on another organism (prey). (42)

Pressure Flow In the process of phloem loading and unloading, pressure differences resulting from solute increases in phloem conducting cells and neighboring xylem cells cause the flow of water to phloem. A concentration gradient is created between xylem and phloem cells. (19)

Prey An organism that is captured and eaten by another organism (predator). (42)

Primary Consumer Organism that feeds exclusively on producers (plants). Herbivores are primary consumers. (41)

Primary Follicle In the mammalian ovary, a structure composed of an oocyte and its surrounding layer of follicle cells. (31)

Primary Growth Growth from apical meristems, resulting in an increase in the lengths of shoots and roots in plants. (18)

Primary Immune Response Process of antibody production following the first exposure to an antigen. There is a lag time from exposure until the appearance in the blood of protective levels of antibodies. (30)

Primary Oocyte Female germ cell that is either in the process of or has completed the first meiotic division. In humans, germ cells may remain in this stage in the ovary for decades. (31)

Primary Producers All autotrophs in a biotic environment that use sunlight or chemical energy to manufacture food from inorganic substances. (41)

Primary Sexual Characteristics Gonads, reproductive tracts, and external genitals. (31)

Primary Spermatocyte Male germ cell that is either in the process of or has completed the first meiotic division. (31)

Primary Succession The development of a community in an area previously unoccupied by any community; for example, a "bare" area such as rock, volcanic material, or dunes. (41)

Primary Tissues Tissues produced by primary meristems of a plant, which arise from the shoot and root apical meristems. In general, primary tissues are a result of an increase in plant length. (18)

Primary Transcript An RNA molecule that has been transcribed but not yet subjected to any type of processing. The primary transcript corresponds to the entire stretch of DNA that was transcribed. (15)

Primates Order of mammals that includes humans, apes, monkeys, and lemurs. (39)

Primitive An evolutionary early condition. Primitive features are those that were also present in an early ancestor, such as five digits on the feet of terrestrial vertebrates. (34)

Prions An infectious particle that contains protein but no nucleic acid. It causes slow diseases of animals, including neurological disease of humans. (36)

Processing-Level Control Control of gene expression by regulating the pathway by which a primary RNA transcript is processed into an mRNA. (15)

Products In a chemical reaction, the compounds into which the reactants are transformed. (6)

Profundal Zone Deep, open water of lakes, where it is too dark for photosynthesis to occur. (40)

Progesterone A hormone produced by the corpus luteum within the ovary. It prepares and maintains the uterus for pregnancy, participates in milk production, and prevents the ovary from releasing additional eggs late in the cycle or during pregnancy. (25)

Prokaryotic Referring to single-celled organisms that have no membrane separating the DNA from the cytoplasm and lack membrane-enclosed organelles. Prokaryotes are confined to the kingdom Monera; they are all bacteria. (36)

Prokaryotic Fission The most common type of cell division in bacteria (prokaryotes). Duplicated DNA strands are attached to the plasma membrane and become separated into two cells following membrane growth and cell wall formation. (10, 36)

Prolactin A hormone produced by the anterior pituitary, stimulating milk production by mammary glands. (25)

Promoter A short segment of DNA to which RNA polymerase attaches at the start of transcription. (15)

Prophase Longest phase of mitosis, involving the formation of a spindle, coiling of chromatin fibers into condensed chromosomes, and movement of the chromosomes to the center of the cell. (10)

Prostaglandins Hormones secreted by endocrine cells scattered throughout the body responsible for such diverse functions as contraction of uterine muscles, triggering the inflammatory response, and blood clotting. (25)

Prostate Gland A muscular gland which produces and releases fluids that make up a substantial portion of the semen. (31)

Proteins Long chains of amino acids, linked together by peptide bonds. They are folded into specific shapes essential to their functions. (4)

Prothallus The small, heart-shaped gametophyte of a fern. (38)

Protists A member of the kingdom Protista; simple eukaryotic organisms that share broad taxonomic similarities. (36, 37)

Protocooperation Non-compulsory interactions that benefit two organisms, e.g., lichens. (42)

Proton Gradient A difference in hydrogen ion (proton) concentration on opposite sides of a membrane. Proton gradients are formed during photosynthesis and respiration and serve as a store of energy used to drive ATP formation. (8, 9)

Protons Positively charged particles within the nucleus of an atom. (3)

Protostomes One path of development exhibited by coelomate animals (e.g., mollusks, annelids, and arthropods). (39)

Protozoa Member of protist kingdom that is unicellular and eukaryotic; vary greatly in size, motility, nutrition and life cycle. (37)

Provirus DNA copy of a virus' nucleic acid that becomes integrated into the host cell's chromosome. (36)

Pseudocoelamates Animals in which the body cavity is not lined by cells derived from mesoderm. (39)

Pseudopodia (psuedo = false, pod = foot). Pseudopodia are fingerlike extensions of cytoplasm that flow forward from the "body" of an amoeba; the rest of the cell then follows. (37)

Puberty Development of reproductive capacity, often accompanied by the appearance of secondary sexual characteristics. (31)

Pulmonary Circulation The loop of the circulatory system that channels blood to the lungs for oxygenation. (28)

Punctuated Equilibrium Theory A theory to explain the phenomenon of the relatively sudden appearance of new species, followed by long periods of little or no change. (33)

Punnett Square Method A visual method for predicting the possible genotypes and their expected ratios from a cross. (12)

Pupa In insects, the stage in metamorphosis between the larva and the adult. Within the pupal case, there is dramatic transformation in body form as some larval tissues die and others differentiate into those of the adult. (32)

Purine A nitrogenous base found in DNA and RNA having a double ring structure. Adenine and guanine are purines. (14)

Pyloric Sphincter Muscular valve between the human stomach and small intestine. (27)

Pyrimidine A nitrogenous base found in DNA and RNA having a single ring structure. Cytosine, thymine, and uracil are pyrimidines. (14)

Pyramid of Biomass Diagrammatic representation of the total dry weight of organisms at each trophic level in a food chain or food web. (41)

Pyramid of Energy Diagrammatic representation of the flow of energy through trophic levels in a food chain or food web. (41)

Pyramid of Numbers Similar to a pyramid of energy, but with numbers of producers and consumers given at each trophic level in a food chain or food web. (41)

 Q ▶

Quiescent Center The region in the apical meristem of a root containing relatively inactive cells. (18)

◄ R ►

R-Group The variable portion of a molecule. (4)

r-Selected Species Species that possess adaptive strategies to produce numerous off-spring at once. (43)

Radial Symmetry The quality possessed by animals whose bodies can be divided into mirror images by more than one median plane. (39)

Radicle In the plant embryo, the tip of the hypocotyl that eventually develops into the root system. (20)

Radioactivity A property of atoms whose nucleus contains an unstable combination of particles. Breakdown of the nucleus causes the emission of particles and a resulting change in structure of the atom. Biologists use this property to track labeled molecules and to determine the age of fossils. (3)

Radiodating The use of known rates of radioactive decay to date a fossil or other ancient object. (3, 34)

Radioisotope An isotope of an element that is radioactive. (3)

Radiolarian A prozoan member of the protistan group Sarcodina that secretes silicon shells through which it captures food.

Rainshadow The arid, leeward (downwind) side of a mountain range. (40)

Random Distribution Distribution of individuals of a population in a random manner; environmental conditions must be similar and individuals do not affect each other's location in the population. (43)

Reactants Molecules or atoms that are changed to products during a chemical reaction. (6)

Reaction A chemical change in which starting molecules (reactants) are transformed into new molecules (products). (6)

Reaction Center A special chlorophyll molecule in a photosystem (P_{700} in Photosystem I, P_{680} in Photosystem II). (8)

Realized Niche Part of the fundamental niche of an organism that is actually utilized. (41)

Receptacle The base of a flower where the flower parts are attached; usually a widened area of the pedicel. (20)

Receptor-Mediated Endocytosis The uptake of materials within a cytoplasmic vesicle (endocytosis) following their binding to a cell surface receptor. (7)

Receptor Site A site on a cell's plasma membrane to which a chemical such as a hormone binds. Each surface site permits the attachment of only one kind of hormone. (5)

Recessive An allele whose expression is masked by the dominant allele for the same trait. (12)

Recombinant DNA A DNA molecule that contains DNA sequences derived from different biological sources that have been joined together in the laboratory. (16)

Recombination The rejoining of DNA pieces with those of a different strand or with the same strand at a point different from where the break occurred. (11, 13)

Red Marrow The soft tissue in the interior of bones that produces red blood cells. (26)

Red Tide Growth of one of several species of reddish brown dinoflagellate algae so extensive that it tints the coastal waters and inland lakes a distinctive red color. Often associated with paralytic shellfish poisoning (see dinoflagellates). (37)

Reducing Power A measure of the cell's ability to transfer electrons to substrates to create molecules of higher energy content. Usually determined by the available store of NADPH, the molecule from which electrons are transferred in anabolic (synthetic) pathways. (6)

Reduction The addition of electrons to a compound during a chemical reaction. For a carbon atom, the more hydrogens that are bonded to the carbon, the more reduced the atom. (6)

Reduction Division The first meiotic division during which a cell's chromosome number is reduced in half. (11)

Reflex An involuntary response to a stimulus. (23)

Reflex Arc The simplest example of central nervous system control, involving a sensory neuron, motor neuron, and usually an interneuron. (23)

Regeneration Ability of certain animals to replace injured or lost limbs parts by growth and differentiation of undifferentiated stem cells. (15)

Region of Elongation In root tips, the region just above the region of cell division, where cells elongate and the root length increases. (18)

Region of Maturation In root tips, the region above the region of elongation; cells differentiate and root hairs occur in this region. (18)

Regulatory Genes Genes whose sole function is to control the expression of structural genes. (15)

Releaser A sign stimulus that is given by an individual to another member of the same species, eliciting a specific innate behavior. (44)

Releasing Factors Hormones secreted by the tips of hypothalmic neurosecretory cells that stimulate the anterior pituitary to release its hormones. GnRH, for example, stimulates the release of gonadotropins. (25)

Renal Referring to the kidney. (28)

Replication Duplication of DNA, usually prior to cell division. (14)

Replication Fork The site where the two strands of a DNA helix are unwinding during replication. (14)

Repression Inhibition of transcription of a gene which, in an operon, occurs when repressor protein binds to the operator. (15)

Repressor Protein encoded by a bacterial regulatory gene that binds to an operator site of an operon and inhibits transcription. (15)

Reproduction The process by which an organism produces offspring. (31)

Reproductive Isolation Phenomenon in which members of a single population become split into two populations that no longer interbreed. (33)

Reproductive System System of specialized organs that are utilized for the production of gametes and, in some cases, the fertilization and/or development of an egg. (31)

Reptiles Members of class Reptilia, scaly, air-breathing, egg-laying vertebrates such as lizards, snakes, turtles, and crocodiles. (39)

Resolving Power The ability of an optical instrument (eye, microscopes) to discern whether two very close objects are separate from each other. (APP.)

Resource Partitioning Temporal or spatial sharing of a resource by different species. (42)

Respiration Process used by organisms to exchange gases with the environment; the source of oxygen required for metabolism. The process organisms use to oxidize glucose to CO_2 and H_2O using an electron transport system to extract energy from electrons and store it in the high-energy bonds of ATP. (29)

Respiratory System The specialized set of organs that function in the uptake of oxygen from the environment. (29)

Resting Potential The electrical potential (voltage) across the plasma membrane of a neuron when the cell is not carrying an impulse. Results from a difference in charge across the membrane. (23)

Restriction Enzyme A DNA-cutting enzyme found in bacteria. (16)

Restriction Fragment Length Polymorphism (RFLP) Certain sites in the DNA tend to have a highly variable sequence from one individual to another. Because of these differences, restriction enzymes cut the DNA from different individuals into fragments of different length. Variations in the length of particular fragments (RFLPs) can be used as genetic signposts for the identification of nearby genes of interest. (17)

Restriction Fragments The DNA fragments generated when purified DNA is treated with a particular restriction enzyme. (16)

Reticular Formation A series of interconnected sites in the core of the brain (brainstem) that selectively arouse conscious activity. (23)

Retroviruses RNA viruses that reverse the typical flow of genetic information; within the infected cell, the viral DNA serves as a template for synthesis of a DNA copy. Examples include HIV, which causes AIDS, and certain cancer viruses. (36)

Reverse Genetics Determining the amino acid sequence and function of a polypeptide from the nucleotide sequence of the gene that codes for that polypeptide. (17)

Reverse Transcriptase An enzyme present in retroviruses that transcriibes a strand of DNA, using viral RNA as the template. (36)

Rhizoids Slender cells that resemble roots but do not absorb water or minerals. (36)

Rhodophyta Red algae; seaweeds that can absorb deeper penetrating light rays than most aquatic photosynthesizers. (36)

Rhyniophytes Ancient plants having vascular tissue which thrived in marshy areas during the Silurian period.

Ribonucleic Acid (RNA) Single-stranded chain of nucleotides each comprised of ribose (a sugar), phosphate, and one of four bases (adenine, guanine, cytosine, and uracil). The sequence of nucleotides in RNA is dictated by DNA, from which it is transcribed. There are three classes of RNA: mRNA, tRNA, and rRNA, all required for protein synthesis. (4, 14)

Ribosomal RNA (rRNA) RNA molecules that form part of the ribosome. Included among the rRNAs is one that is thought to catalyze peptide bond formation. (14)

Ribosomes Organelles involved in protein synthesis in the cytoplasm of the cell. (14)

Ribozymes RNAs capable of catalyzing a chemical reaction, such as peptide bond formation or RNA cutting and splicing. (15)

Rickettsias A group of obligate intracellular parasites, smaller than the typical prokaryote. They cause serious diseases such as typhus. (36)

River Flowing body of surface fresh water; rivers are formed from the convergence of streams. (40)

RNA Polymerase The enzyme that directs transcription and assembling RNA nucleotides in the growing chain. (14)

RNA Processing The process by which the intervening (noncoding) portions of a primary RNA transcript are removed and the remaining (coding) portions are spliced together to form an mRNA. (15)

Root Cap A protective cellular helmet at the tip of a root that surrounds delicate meristematic cells and shields them from abrasion and directs the growth downward. (18)

Root Hairs Elongated surface cells near the tip of each root for the absorption of water and minerals. (18)

Root Nodules Knobby structures on the roots of certain plants. They house nitrogen-fixing bacteria which supply nitrogen in a form that can be used by the plant. (19)

Root Pressure A positive pressure as a result of continuous water supply to plant roots that assists (along with transpirational pull) the pushing of water and nutrients up through the xylem. (19)

Root System The below-ground portion of a plant, consisting of main roots, lateral roots, root hairs, and associated structures and systems such as root nodules or mycorrhizae. (18)

Rough ER (RER) Endoplasmatic reticulum with many ribosomes attached. As a result, they appear rough in electron micrographs. (5)

Ruminant Grazing mammals that possess an additional stomach chamber called rumen which is heavily fortified with cellulose-digesting microorganisms. (27)

◄ S ►

S Phase The second stage of interphase in which the materials needed for cell division are synthesized and an exact copy of cell's DNA is made by DNA replication. (10)

Sac Body The body plan of simple animals, like cnidarians, where there is a single opening leading to and from a digestion chamber.

Saltatory Conduction The "hopping" movement of an impulse along a myelinated neuron from one Node of Ranvier to the next one. (23)

Sap Fluid found in xylem or sieve of phloem. (20)

Saprophyte Organisms, mainly fungi and bacteria, that get their nutrition by breaking down organic wastes and dead organisms, also called decomposers. (42)

Saprobe Organism that obtains its nutrients by decomposing dead organisms. (37)

Sarcolemma The plasma membrane of a muscle fiber. (26)

Sarcomere The contractile unit of a myofibril in skeletal muscle. (26)

Sarcoplasmic Reticulum (SR) In skeletal muscle, modified version of the endoplasmic reticulum that stores calcium ions. (26)

Savanna A grassland biome with alternating dry and rainy seasons. The grasses and scattered trees support large numbers of grazing animals. (40)

Scaling Effect A property that changes disproportionally as the size of organisms increase. (22)

Scanning Electron Microscope (SEM) A microscope which operates by showering electrons back and forth across the surface of a specimen prepared with a thin metal coating. The resultant image shows three-dimensional features of the specimen's surface. (APP.)

Schwann Cells Cells which wrap themselves around the axons of neurons forming an insulating myelin sheath composed of many layers of plasma membrane. (23)

Sclereids Irregularly-shaped sclerenchyma cells, all having thick cell walls; a component of seed coats and nuts. (18)

Sclerenchyma Component of the ground tissue system of plants. They are thick walled cells of various shapes and sizes, providing support or protection. They continue to function after the cell dies. (18)

Sclerenchyma Fibers Non-living elongated plant cells with tapering ends and thick secondary walls. A supportive cell type found in various plant tissues. (18)

Sebaceous Glands Exocrine glands of the skin that produce a mixture of lipids (sebum) that oil the hair and skin. (26)

Secondary Cell Wall An additional cell wall that improves the strength and resiliency of specialized plant cells, particularly those cells found in stems that support leaves, flowers, and fruit. (5)

Secondary Consumer Organism that feeds exclusively on primary consumers; mostly animals, but some plants. (41)

Secondary Growth Growth from cambia in perennials; results in an increase in the diameter of stems and roots. (18)

Secondary Meristems (vascular cambium, cork cambrium) Rings or clusters of meristematic cells that increase the width of stems and roots when the divide. (18)

Secondary Sex Characteristics Those characteristics other than the gonads and reproductive tract that develop in response to sex hormones. For example, breasts and pubic hair in women and a deep voice and pubic hair in men. (31)

Secondary Succession The development of a community in an area previously occupied by a community, but which was disturbed in some manner; for example, fire, development, or clear-cutting forests. (41)

Secondary Tissues Tissues produced to accommodate new cell production in plants with woody growth. Secondary tissues are produced from cambia, which produce vascular and cork tissues, leading to an increase in plant girth. (18)

Second Messenger Many hormones, such as glucagon and thyroid hormone, evoke a response by binding to the outer surface of a target cell and causing the release of another substance (which is the second messenger). The best-studied second messenger is cyclic AMP which is formed by an enzyme on the inner surface of the plasma membrane following the binding of a hormone to the outer surface of the membrane. The cyclic AMP diffuses into the cell and activates a protein kinase. (25)

Secretion The process of exporting materials produced by the cell. (5)

Seed A mature ovule consisting of the embryo, endosperm, and seed coat. (20)

Seed Dormancy Metabolic inactivity of seeds until favorable conditions promote seed germination. (20)

Secretin Hormone secreted by endocrine cells in the wall of the intestine that stimulates the release of digestive products from the pancreas. (25)

Segmentation A condition in which the body is constructed, at least in part, from a series of repeating parts. Segmentation occurs in annelids, arthropods, and vertebrates (as revealed during embryonic development). (39)

Selectively Permeable A term applied to the plasma membrane because membrane proteins control which molecules are transported. Enables a cell to import and accumulate the molecules essential for normal metabolism. (7)

Semen The fluid discharged during a male orgasm. (31)

Semiconsevative Replication The manner in which DNA replicates; half of the original DNA strand is conserved in each new double helix. (14)

Seminal Vesicles The organs which produce most of the ejaculatory fluid. (31)

Seminiferous Tubules Within the testes, highly coiled and compacted tubules, lined with a self-perpetuating layer of spermatogonia, which develop into sperm. (31)

Senescence Aging and eventual death of an organism, organ or tissue. (3, 18)

Sense Strand The one strand of a DNA double helix that contains the information that encodes the amino sequence of a polypeptide. This is the strand that is selectively transcribed by RNA polymerase forming an mRNA that can be properly translated. (14)

Sensory Neurons Neurons which relay impulses to the central nervous system. (23)

Sensory Receptors Structures that detect changes in the external and internal environment and transmit the information to the nervous system. (24)

Sepal The outermost whorl of a flower, enclosing the other flower parts as a flower bud; collectively called the calyx. (20)

Sessile Sedentary, incapable of independent movement. (39)

Sex Chromosomes The one chromosomal pair that is not identical in the karyotypes of males and females of the same animal species. (10, 13)

Sex Hormones Steroid hormones which influence the production of gametes and the development of male or female sex characteristics. (25)

Sexual Dimorphism Differences in the appearance of males and females in the same species. (33)

Sexual Reproduction The process by which haploid gametes are formed and fuse during fertilization to form a zygote. (31)

Sexual Selection The natural selection of adaptations that improve the chances for mating and reproducing. (33)

Shivering Involuntary muscular contraction for generating metabolic heat that raises body temperature. (28)

Shoot In angiosperms, the system consisting of stems, leaves, flowers and fruits. (18)

Shoot System The above-ground portion of an angiosperm plant consisting of stems with nodes, including branches, leaves, flowers and fruits. (18)

Short-Day Plants Plants that flower in late summer or fall when the length of daylight becomes shorter than some critical period. (21)

Shrubland A biome characterized by densely growing woody shrubs in mediterranean type climate; growth is so dense that understory plants are not usually present. (40)

Sickle Cell Anemia A genetic (recessive autosomal) disorder in which the beta globin genes of adult hemoglobin molecules contain an amino acid substitution which alters the ability of hemoglobin to transport oxygen. During times of oxygen stress, the red blood cells of these individuals may become sickle shaped, which interferes with the flow of the cells through small blood vessels. (4, 17)

Sieve Plate Found in phloem tissue in plants, the wall between sieve-tube members, containing perforated areas for passage of materials. (18)

Sieve-Tube Member A living, food-conducting cell found in phloem tissue of plants; associated with a companion cell. (18)

Sigmoid Growth Curve An S-shaped curve illustrating the lag phase, exponential growth, and eventual approach of a population to its carrying capacity. (43)

Sign Stimulus An object or action in the environment that triggers an innate behavior. (44)

Simple Fruits Fruits that develop from the ovary of one pistil. (20)

Simple Leaf A leaf that is undivided; only one blade attached to the petiole. (18)

Sinoatrial (SA) Node A collection of cells that generates an action potential regulating heart beat; the heart's pacemaker. (28)

Skeletal Muscles Separate bundles of parallel, striated muscle fibers anchored to the bone, which they can move in a coordinated fashion. They are under voluntary control. (26)

Skeleton A rigid form of support found in most animals either surrounding the body with a protective encasement or providing a living girder system within the animal. (26)

Skull The bones of the head, including the cranium. (26)

Slow-Twitch Fibers Skeletal muscle fibers that depend on aerobic metabolism for ATP production. These fibers are capable of undergoing contraction for extended periods of time without fatigue, but generate lesser forces than fast-twitch fibers. (9)

Small Intestine Portion of the intestine in which most of the digestion and absorption of nutrients takes place. It is so named because of its narrow diameter. There are three sections: duodenum, jejunum, and ilium. (27)

Smell Sense of the chemical composition of the environment. (24)

Smooth ER (SER) Membranes of the endoplasmic reticulum that have no ribosomes on their surface. SER is generally more tubular than the RER. Often acts to store calcium or synthesize steroids. (5)

Smooth Muscle The muscles of the internal organs (digestive tract, glands, etc.). Composed of spindle-shaped cells that interlace to form sheets of visceral muscle. (26)

Social Behavior Behavior among animals that live in groups composed of individuals that are dependent on one another and with whom they have evolved mechanisms of communication. (44)

Social Learning Learning of a behavior from other members of the species. (44)

Social Parasitism Parasites that use behavioral mechanisms of the host organism to the parasite's advantage, thereby harming the host. (42)

Solute A substance dissolved in a solvent. (3)

Solution The resulting mixture of a solvent and a solute. (3)

Solvent A substance in which another material dissolves by separating into individual molecules or ions. (3)

Somatic Cells Cells that do not have the potential to form reproductive cells (gametes). Includes all cells of the body except germ cells. (11)

Somatic Nervous System The nerves that carry messages to the muscles that move the skeleton either voluntarily or by reflex. (23)

Somatic Sensory Receptors Receptors that respond to chemicals, pressure, and temperature that are present in the skin, muscles, tendons, and joints. Provides a sense of the physiological state of the body. (24)

Somites In the vertebrate embryo, blocks of mesoderm on either side of the notochord that give rise to muscles, bones, and dermis. (32)

Speciation The formation of new species. Occurs when one population splits into separate populations that diverge genetically to the point where they become separate species. (33)

Species Taxonomic subdivisions of a genus. Each species has recognizable features that distinguish it from every other species. Members of one species generally will not interbreed with members of other species. (33)

Specific Epithet In taxonomy, the second term in an organism's scientific name identifying its species within a particular genus. (1)

Spermatid Male germ cell that has completed meiosis but has not yet differentiated into a sperm. (31)

Spermatogenesis The production of sperm. (31)

Spermatogonia Male germ cells that have not yet begun meiosis. (31)

Spermatozoa (Sperm) Male gametes. (31)

Sphinctors Circularly arranged muscles that close off the various tubes in the body.

Spinal Cord A centralized mass of neurons for processing neurological messages and linking the brain with that part of peripheral nervous system not reached by the cranial nerves. (23)

Spinal Nerves Paired nerves which emerge from the spinal cord and innervate the body. Humans have 31 pairs of spinal nerves. (23)

Spindle Apparatus In dividing eukaryotic cells, the complex rigging, made of microtubules, that aligns and separates duplicated chromosomes. (10)

Splash Zone In the intertidal zone, the uppermost region receiving splashes and sprays of water to the mean of high tides. (40)

Spleen One of the organs of the lymphatic system that produces lymphocytes and filters blood; also produces red blood cells in the human fetus. (28)

Splicing The step during RNA processing in which the coding segments of the primary transcript are covalently linked together to form the mRNA. (15)

Spongy Parenchyma In monocot and dicot leaves, loosely arranged cells functioning in photosynthesis. Found above the lower epidermis and beneath the palisade parenchyma in dicots, and between the upper and lower epidermis in monocots. (18)

Spontaneous Generation Disproven concept that living organisms can arise directly from inanimate materials. (2)

Sporangiospores Black, asexual spores of the zygomycete fungi. (37)

Sporangium A hollow structure in which spores are formed. (37)

Spores In plants, haploid cells that develop into the gametophyte generation. In fungi, an asexual or sexual reproductive cell that gives rise to a new mycelium. Spores are often lightweight for their dispersal and adapted for survival in adverse conditions. (37)

Sporophyte The diploid spore producing generation in plants. (38)

Stabilizing Selection Natural selection favoring an intermediate phenotype over the extremes. (33)

Starch Polysaccharides used by plants to store energy. (4)

Stamen The flower's male reproductive organ, consisting of the pollen-producing anther supported by a slender stalk, the filament. (20)

Stem In plants, the organ that supports the leaves, flowers, and fruits. (18)

Stem Cells Cells which are undifferentiated and capable of giving rise to a variety of different types of differentiated cells. For example, hematopoietic stem cells are capable of giving rise to both red and white blood cells. (17)

Steroids Compounds classified as lipids which have the basic four-ringed molecular skeleton as represented by cholesterol. Two examples of steroid hormones are the sex hormones; testosterone in males and estrogen in females. (4, 25)

Stigma The sticky area at the top of each pistil to which pollen adheres. (20)

Stimulus Any change in the internal or external environment to which an organism can respond. (24)

Stomach A muscular sac that is part of the digestive system where food received from the esophagus is stored and mixed, some breakdown of food occurs, and the chemical degradation of nutrients begins. (27)

Stomates (Pl. Stomata) Microscopic pores in the epidermis of the leaves and stems which allow gases to be exchanged between the plant and the external environment. (18)

Stratified Epithelia Multicellular layered epithelium. (22)

Stream Flowing body of surface fresh water; streams merge together into larger streams and rivers. (40)

Stretch Receptors Sensory receptors embedded in muscle tissue enabling muscles to respond reflexively when stretched. (23, 24)

Striated Referring to the striped appearance of skeletal and cardiac muscle fibers. (26)

Strobilus In lycopids, terminal, cone-like clusters of specialized leaves that produce sporangia.

Stroma The fluid interior of chloroplasts. (8)

Stromatolites Rocks formed from masses of dense bacteria and mineral deposits. Some of these rocky masses contain cells that date back over three billion years revealing the nature of early prokaryotic life forms. (35)

Structural Genes DNA segments in bacteria that direct the formation of enzymes or structural proteins. (15)

Style The portion of a pistil which joins the stigma to the ovary. (20)

Substrate-Level Phosphorylation The formation of ATP by direct transfer of a phosphate group from a substrate, such as a sugar phosphate, to ADP. ATP is formed without the involvement of an electron transport system. (9)

Substrates The reactants which bind to enzymes and are subsequently converted to products. (6)

Succession The orderly progression of communities leading to a climax community. It is one of two types: primary, which occurs in areas where no community existed before; and secondary, which occurs in disturbed habitats where some soil and perhaps some organisms remain after the disturbance. (41)

Succulents Plants having fleshy, water-storing stems or leaves. (40)

Suppressor T Cells A class of T cells that regulate immune responses by inhibiting the activation of other lymphocytes. (30)

Surface Area-to-Volume Ratio The ratio of the surface area of an organism to its volume, which determines the rate of exchange of materials between the organism and its environment. (22)

Surface Tension The resistance of a liquid's surface to being disrupted. In aqueous solutions, it is caused by the attraction between water molecules. (3)

Survivorship Curve Graph of life expectancy, plotted as the number of survivors versus age. (43)

Sweat Glands Exocrine glands of the skin that produce a dilute salt solution, whose evaporation cools in the body. (26)

Symbiosis A close, long-term relationship between two individuals of different species. (42)

Symmetry Referring to a body form that can be divided into mirror image halves by at least one plane through its body. (39)

Sympathetic Nervous System Part of the autonomic nervous system that tends to stimulate bodily activities, particularly those involved with coping with stressful situations. (23)

Sympatric Speciation Speciation that occurs in populations with overlapping distributions. It is common in plants when polyploidy arises within a population. (33)

Synapse Juncture of a neuron and its target cell (another neuron, muscle fiber, gland cell). (23)

Synapsis The pairing of homologous chromosomes during prophase of meiosis I. (11)

Synaptic Cleft Small space between the synaptic knobs of a neuron and its target cell. (23)

Synaptic Knobs The swellings that branch from the end of the axon. They deliver the neurological impulse to the target cell. (23)

Synaptonemal Complex Ladderlike structure that holds homologous chromosomes together as a tetrad during crossing over in prophase I of meiosis. (11)

Synovial Cavities Fluid-filled sacs around joints, the function of which is to lubricate and separate articulating bone surfaces. (26)

Systemic Circulation Part of the circulatory system that delivers oxygenated blood to the tissues and routes deoxygenated blood back to the heart. (28)

Systolic Pressure The first number of a blood pressure reading; the highest pressure attained in the arteries as blood is propelled out of the heart. (28)

◄ **T** ►

Taiga A biome found south of tundra biomes; characterized by coniferous forests, abundant precipitation, and soils that thaw only in the summer. (40)

Tap Root System Root system of plants having one main root and many smaller lateral roots. Typical of conifers and dicots. (18)

Taste Sense of the chemical composition of food. (24)

Taxonomy The science of classifying and grouping organisms based on their morphology and evolution. (1)

T Cell Lymphocytes that carry out cell-mediated immunity. They respond to antigen stimulation by becoming helper cells, killer cells, and memory cells. (30)

Telophase The final stage of mitosis which begins when the chromosomes reach their spindle poles and ends when cytokinesis is completed and two daughter cells are produced. (10)

Tendon A dense connective tissue cord that connects a skeletal muscle to a bone. (26)

Teratogenic Embryo deforming. Chemicals such as thalidomide or alcohol are teratogenic because they disturb embryonic development and lead to the formation of an abnormal embryo and fetus. (32)

Terminal Electron Acceptor In aerobic respiration, the molecule of O_2 which removes the electron pair from the final cytochrome of the respiratory chain. (9)

Terrestrial Living on land. (40)

Territory (Territoriality) An area that an animal defends against intruders, generally in the protection of some resource. (42, 44)

Tertiary Consumer Animals that feed on secondary consumers (plant or animal) or animals only. (41)

Test Cross An experimental procedure in which an individual exhibiting a dominant trait is crossed to a homozygous recessive to determine whether the first individual is homozygous or heterozygous. (12)

Testis In animals, the sperm-producing gonad of the male. (23)

Testosterone The male sex hormone secreted by the testes when stimulated by pituitary gonadotropins. (31)

Tetrad A unit of four chromatids formed by a synapsed pair of homologous chromosomes, each of which has two chromatids. (11)

Thallus In liverworts, the flat, ground-hugging plant body that lacks roots, stems, leaves, and vascular tissues. (38)

Theory of Tolerance Distribution, abundance and existence of species in an ecosystem are determined by the species' range of tolerance of chemical and physical factors. (41)

Thermoreceptors Sensory receptors that respond to changes in temperature. (24)

Thermoregulation The process of maintaining a constant internal body temperature in spite of fluctuations in external temperatures. (28)

Thigmotropism Changes in plant growth stimulated by contact with another object, e.g., vines climbing on cement walls. (21)

Thoracic Cavity The anterior portion of the body cavity in which the lungs are suspended. (39)

Thylakoids Flattened membrane sacs within the chloroplast. Embedded in these membranes are the light-capturing pigments and other components that carry out the light-dependent reactions of photosynthesis. (8)

Thymus Endocrine gland in the chest where T cells mature. (30)

Thyroid Gland A butterfly-shaped gland that lies just in front of the human windpipe, producing two metabolism-regulating hormones, thyroxin and triodothyronine. (25)

Thyroid Hormone A mixture of two iodinated amino acid hormones (thyroxin and triiodothyronine) secreted by the thyroid gland. (25)

Thyroid Stimulating Hormone (TSH) An anterior pituitary hormone which stimulates secretion by the thyroid gland. (25)

Tissue An organized group of cells with a similar structure and a common function. (22)

Tissue System Continuous tissues organized to perform a specific function in plants. The three plant tissue systems are: dermal, vascular, and ground (fundamental). (18)

Tolerance Range The range between the maximum and minimum limits for an environmental factor that is necessary for an organism's survival. (41)

Totipotent The genetic potential for one type of cell from a multicellular organism to give rise to any of the organism's cell types, even to generate a whole new organism. (15)

Trachea The windpipe; a portion of the respiratory tract between the larynx and bronchii. (29)

Tracheal Respiratory System A network of tubes (tracheae) and tubules (tracheoles) that carry air from the outside environment directly to the cells of the body without involving the circulatory system. (29, 39)

Tracheid A type of conducting cell found in xylem functioning when a cell is dead to transport water and dissolved minerals through its hollow interior. (18)

Tracheophytes Vascular plants that contain fluid-conducting vessels. (38)

Transcription The process by which a strand of RNA assembles along one of the DNA strands. (14)

Transcriptional-Level Control Control of gene expression by regulating whether or not a specific gene is transcribed and how often. (15)

Transduction A type of genetic recombination resulting from transfer of genes from one organism to another by a virus.

Transfer RNA (tRNA) A type of RNA that decodes mRNA's codon message and translates it into amino acids. (14)

Transgenic Organism An organism that possesses genes derived from a different species. For example, a sheep that carries a human gene and secretes the human protein in its milk is a transgenic animal. (16)

Translation The cell process that converts a sequence of nucleotides in mRNA into a sequence of amino acids in a polypeptide. (14)

Translational-Level Control Control of gene expression by regulating whether or not a specific mRNA is translated into a polypeptide. (15)

Translocation The joining of segments of two nonhomologous chromosomes (13)

Transmission Electron Microscope (TEM) A microscope that works by shooting electrons through very thinly sliced specimens. The result is an enormously magnified image, two-dimensional, of remarkable detail. (App.)

Transpiration Water vapor loss from plant surfaces. (19)

Transpiration Pull The principle means of water and mineral transport in plants, initiated by transpiration. (19)

Transposition The phenomenon in which certain DNA segments (mobile genetic elements, or jumping genes) tend to move from one part of the genome to another part. (15)

Transverse Fission The division pattern in ciliated protozoans where the plane of division is perpendicular to the cell's length.

Trimester Each of the three stages comprising the 266-day period between conception and birth in humans. (32)

Triploid Having three sets of chromosomes, abbreviated 3N. (11)

Trisomy Three copies of a particular chromosome per cell. (17)

Trophic Level Each step along a feeding pathway. (41)

Trophozoite The actively growing stage of polymorphic protozoa. (37)

Tropical Rain Forest Lush forests that occur near the equator; characterized by high annual rainfall and high average temperature. (40)

Tropical Thornwood A type of shrubland occurring in tropical regions with a short rainy season. Plants lose their small leaves during dry seasons, leaving sharp thorns. (40)

Tropic Hormones Hormones that act on endocrine glands to stimulate the production and release of other hormones. (25)

Tropisms Changes in the direction of plant growth in response to environmental stimuli, e.g., light, gravity, touch. (21)

True-Breeder Organisms that, when bred with themselves, always produce offspring identical to the parent for a given trait. (12)

Tubular Reabsorption The process by which substances are selectively returned from the fluid in the nephron to the bloodstream. (28)

Tubular Secretion The process by which substances are actively and selectively transported from the blood into the fluid of the nephron. (28)

Tumor-Infiltrating Lymphocytes (TILs) Cytotoxic T cells found within a tumor mass that have the capability to specifically destroy the tumor cells. (30)

Tumor-Suppressor Genes Genes whose products act to block the formation of cancers. Cancers form only when both copies of these genes (one on each homologue) are mutated. (13)

Tundra The marshy, unforested biome in the arctic and at high elevations. Frigid temperatures for most of the year prevent the subsoil from thawing, which produces marshes and ponds. Dominant vegetation includes low growing plants, lichens, and mosses. (40)

Turgor Pressure The internal pressure in a plant cell caused by the diffusion of water into the cell. Because of the rigid cell wall, pressure can increase to where it eventually stops the influx of more water. (7)

Turner Syndrome A person whose cells have only one X chromosome and no second sex chromosome (XO). These individuals develop as immature females. (17)

◀ U ▶

Ultimate (Top) Consumer The final carnivore trophic level organism, or organisms that escaped predation; these consumers die and are eventually consumed by decomposers. (41)

Ultracentrifuge An instrument capable of spinning tubes at very high speeds, delivering centrifugal forces over 100,000 times the force of gravity. (9)

Unicellular The description of an organism where the cell is the organism. (35)

Uniform Pattern Distribution of individuals of a population in a uniform arrangement, such as individual plants of one species uniformly spaced across a region. (43)

Urethra In mammals, a tube that extends from the urinary bladder to the outside. (28)

Urinary Tract The structures that form and export urine: kidneys, ureters, urinary bladder, and urethra. (28)

Urine The excretory fluid consisting of urea, other nitrogenous substances, and salts dissolved in water. It is formed by the kidneys. (28)

Uterine (Menstrual) Cycle The repetitive monthly changes in the uterus that prepare the endometrium for receiving and sustaining an embryo. (31)

Uterus An organ in the female reproductive system in which an embryo implants and is maintained during development. (31)

◀ V ▶

Vaccines Modified forms of disease-causing microbes which cannot cause disease but retain the same antigens of it. They permit the immune system to build memory cells without diseases developing during the primary immune response. (30)

Vacoconstriction Reduction in the diameter of blood vessels, particularly arterioles. (28)

Vacuole A large organelle found in mature plant cells, occupying most of the cell's volume, sometimes more than 90% of it. (5)

Vagina The female mammal's copulatory organ and birth canal. (31)

Variable (Experimental) A factor in an experiment that is subject to change, i.e., can occur in more than one state. (2)

Vascular Bundles Groups of vascular tissues (xylem and phloem) in the shoot of a plant. (19)

Vascular Cambium In perennials, a secondary meristem that produces new vascular tissues. (18)

Vascular Cylinder Groups of vascular tissues in the central region of the root. (18)

Vascular Plants Plants having a specialized conducting system of vessels and tubes for transporting water, minerals, food, etc., from one region to another. (18)

Vascular Tissue System All the vascular tissues in a plant, including xylem, phloem, and the vascular cambium or procambium. (18)

Vasodilation Increase in the diameter of blood vessels, particularly arterioles. (28)

Veins In plants, vascular bundles in leaves. In animals, blood vessels that return blood to the heart. (28)

Venation The pattern of vein arrangement in leaf blades. (18)

Ventricle Lower chamber of the heart which pumps blood through arteries. There is one ventricle in the heart of lower vertebrates and two ventricles in the four-chambered heart of birds and mammals. (28)

Venules Small veins that collect blood from the capillaries. They empty into larger veins for return to the heart. (28)

Vertebrae The bones that form the backbone. In the human there are 33 bones arranged in a gracefully curved line along the bone, cushioned from one another by disks of cartilage. (26)

Vertebral Column The backbone, which encases and protects the spinal cord. (26)

Vertebrates Animals with a backbone. (39)

Vesicles Small membrane-enclosed sacs which form from the ER and Golgi complex. Some store chemicals in the cells; others move to the surface and fuse with the plasma membrane to secrete their contents to the outside. (5)

Vessel A tube or connecting duct containing or circulating body fluids. (18)

Vessel Member A type of conducting cell in xylem functioning when the cell is dead to transport water and dissolved minerals through its hollow interior; also called a vessel element. (18)

Vestibular Apparatus A portion of the inner ear of vertebrates that gathers information about the position and movement of the head for use in maintaining balance and equilibrium. (24)

Vestigial Structure Remains of ancestral structures or organs which were, at one time, useful. (34)

Villi Finger-like projections of the intestinal wall that increase the absorption surface of the intestine. (27)

Viroids are associated with certain diseases of plants. Each viroid consists solely of a small single-stranded circle of RNA unprotected by a protein coat. (36)

Virus Minute structures composed of only heredity information (DNA or RNA), surrounded by a protein or protein/lipid coat. After infection, the viral mucleic acid subverts the metabolism of the host cell, which then manufactures new virus particles. (36)

Visible Light The portion of the electromagnetic spectrum producing radiation from 380 nm to 750 nm detectable by the human eye.

Vitamins Any of a group of organic compounds essential in small quantities for normal metabolism. (27)

Vocal Cords Muscular folds located in the larynx that are responsible for sound production in mammals. (29)

Vulva The collective name for the external features of the human female's genitals. (31)

◀ **W** ▶

Water Vascular System A system for locomotion, respiration, etc., unique to echinoderms. (39)

Wavelength The distance separating successive crests of a wave. (8)

Waxes A waterproof material composed of a number of fatty acids linked to a long chain alcohol. (4)

White Matter Regions of the brain and spinal cord containing myelinated axons, which confer the white color. (23)

Wild Type The phenotype of the typical member of a species in the wild. The standard to which mutant phenotypes are compared. (13)

Wilting Drooping of stems or leaves of a plant caused by water loss. (7)

Wood Secondary xylem. (18)

◀ **X** ▶

X Chromosome The sex chromosome present in two doses in cells of a female, and in one dose in the cells of a male. (13)

X-Linked Traits Traits controlled by genes located on the X chromosome. These traits are much more common in males than females. (13)

Xylem The vascular tissue that transports water and minerals from the roots to the rest of the plant. Composed of tracheids and vessel members. (18)

◀ **Y** ▶

Y Chromosome The sex chromosome found in the cells of a male. When Y-carrying sperm unite with an egg, all of which carry a single X chromosome, a male is produced. (13)

Y-Linked Inheritance Genes carried only on Y chromosomes. There are relatively few Y-linked traits; maleness being the most important such trait in mammals. (13)

Yeast Unicellular fungus that forms colonies similar to those of bacteria. (37)

Yolk A deposit of lipids, proteins, and other nutrients that nourishes a developing embryo. (32)

Yolk Sac A sac formed by an extraembryonic membrane. In humans, it manufactures blood cells for the early embryo and later helps to form the umbilical cord. (32)

◀ **Z** ▶

Zero Population Growth In a population, the result when the combined positive growth factors (births and immigration). (43)

Zooplankton Protozoa, small crustaceans and other tiny animals that drift with ocean currents and feed on phytoplankton. (37, 40)

Zygospore The diploid spores of the zygomycete fungi, which include Rhizopus, a common bread mold. After a period of dormancy, the zygospore undergoes meiosis and germinates. (36)

Zygote A fertilized egg. The diploid cell that results from the union of a sperm and egg. (32)

Photo Credits

Part 1 Opener Norbert Wu. **Chapter 1** Fig. 1.1: Jeff Gnass. Fig. 1.2a: Frans Lanting/Minden Pictures, Inc. Fig. 1.2b: David Muench. Fig. 1.3a: Courtesy Dr. Alan Cheetham, National Museum of National History, Smithsonian Institution. Fig. 1.3b: Manfred Kage/Peter Arnold. Fig. 1.3c: Stephen Dalton/NHPA. Fig. 1.3d: Charles Summers, Jr. Fig. 1.3e: Larry West/Photo Researchers. Fig. 1.3f: Bianca Lavies. Fig. 1.3g: Steve Allen/Peter Arnold. Fig. 1.3h: George Grall. Fig. 1.3i: Michio Hoshino/Minden Pictures, Inc. Fig. 1.6a: CNRI/Science Photo Library/Photo Researchers. Fig. 1.6b: M. Abbey/Visuals Unlimited. Fig. 1.6c: Steve Kaufman/Peter Arnold. Fig. 1.6d: Willard Clay. Fig. 1.6e: Jim Bradenburg/Minden Pictures, Inc. Fig. 1.7: (pages 16-17) Charles A. Mauzy; (page 16, top) Anthony Mercieca/Natural Selection; (page 16, center) Wolfgang Bayer/Bruce Coleman, Inc.; (page 16, bottom) Dr. Jeremy Burgess/Science Photo Library/Photo Researchers; (page 17, top) Dr. Eckart Pott/Bruce Coleman, Ltd.; (page 17, bottom) Art Wolfe. Fig. 1..8: Paul Chesley/Photographers Aspen. Fig. 1.9a: Rainbird/Robert Harding Picture Library. Fig. 1.11: Dick Luria/FPG International. **Chapter 2** Fig. 2.2a: Couresy Institut Pasteur. Fig. 2.3a: Topham/The Image Works. Fig. 2.3b: Leonard Lessin/Peter Arnold. Fig. 2.3c: Bettmann Archive. Fig. 2.5a: Laurence Gould/Earth Scenes/Animals Animals. Fig. 2.5b: Ted Horowitz/The Stock Market. **Part 2 Opener:** Nancy Kedersha. **Chapter 3** Fig. 3.1: Franklin Viola. Fig. 3.2: Courtesy Pachyderm Scientific Industries. Bioline: Jacan-Yves Kerban/The Image Bank. Fig. 3.7: Courtesy Stephen Harrison, Harvard Biochemistry Department. Fig. 3.11: Gary Milburn/Tom Stack & Associates. Fig. 3.12: Courtesy R. S. Wilcox, Biology Department, SUNY Binghampton. **Chapter 4** Fig. 4.1: Alastair Black/Tony Stone World Wide. Fig. 4.5a: Don Fawcett/Visuals Unlimited. Fig. 4.5b: Jeremy Burgess/Photo Researchers. Fig. 4.5c: Cabisco/Visuals Unlimited. Fig. 4.6: Tony Stone World Wide. Fig. 4.7: Zig Leszcynski/Animals Animals. Fig. 4.8: Robert & Linda Mitchell. Human Perspective: (a) Photofest (b) Bill Davila/Retna. Fig. 4.12a: Frans Lanting/AllStock, Inc. Fig. 4.12b: Mantis Wildlife Films/Oxford Scientific Films/Animals Animals. Fig. 4.17: Stanley Flegler/Visuals Unlimited. Fig. 4.18a: Courtesy Nelson Max, Lawrence Livermore Laboratory. Fig. 4.18b: Tsuned Hayashida/The Stock Market. **Chapter 5** Fig. 5.1a: The Granger Collection. Fig. 5.1b: Bettmann Archive. Fig. 5.2a: Dr. Jeremey Burgess/Photo Researchers. Fig. 5.2b: CNRI/Photo Researchers. Fig. 5.4: Omikron/Photo Researchers. Fig. 5.6a: Courtesy Richard Chao, California State University at Northridge. Fig. 5.6c: Courtesy Daniel Branton, University of Berkeley. Fig. 5.7: Courtesy G. F. Bohr. Fig. 5.8: Courtesy

Michael Mercer, Zoology Department, Arizona State University. Fig. 5.9: Courtesy D. W. Fawcett, Harvard Medical School. Fig. 5.10a: Courtesy U.S. Department of Agriculture, Fig. 5.11: Courtesy Dr. Birgit H. Satir, Albert Einstein College of Medicine. Fig. 5.13: Courtesy Lennart Nilsson, from *A Child Is Born*. Fig. 5.14a: K. R. Porter/Photo Researchers. Fig. 5.14c: Courtesy Lennart Nilsson, BonnierAlba. Fig. 5.14d: Courtesy Lennart Nilsson, From *A Child Is Born*. Fig. 5.15a: Courtesy J. Elliot Weier. Fig. 5.16a: Courtesy J.V. Small. Fig. 5.17a: Peter Parks/Animals Animals. Fig. 5.17b: Courtesy Dr. Manfred Hauser, RUHR-Universitat Rochum. Fig. 5.18: Courtesy Jean Paul Revel, Division of Biology, California Institute of Technology. Fig. 5.19a: Courtesy E. Vivier, from *Paramecium* by W.J. Wagtendonk, Elsevier North Holland Biomedical Press, 1974. Fig. 5.19b: David Phillips/Photo Researchers. Fig. 5.20: Courtesy C.J. Brokaw and T.F. Simonick, *Journal of Cell Biology*, 75:650 (1977). Reproduced with permission. Fig. 5.21: Courtesy L.G. Tilney and K. Fujiwara. Fig. 5.22a: Courtesy W. Gordon Whaley, University of Texas, Austin. Fig. 5.22c: Courtesy R.D. Preston, University of Leeds, London. **Chapter 6** Fig. 6.1a: Alex Kerstitch. Fig. 6.1b: Peter Parks/Earth Scenes. Fig. 6.2: Marty Stouffer/Animals Animals. Fig. 6.3: Kay Chernush/The Image Bank. Fig. 6.4b: Courtesy Computer Graphics Laboratory, University of California, San Francisco. Fig. 6.8: Courtesy Stan Koszelek, Ph.D., University of California, Riverside. Fig. 6.10: Swarthout/The Stock Market. **Chapter 7** Fig. 7.5: Ed Reschke. Fig. 7.10a: Lennart Nilsson, ©Boehringer Ingelheim, International Gmbh; from *The Incredible Machine*. Human Perspective: (Fig. 1a) Martin Rotker/Phototake; (Fig. 1b) Cabisco/Visuals Unlimited. Fig. 7.11a: Courtesy Dr. Ravi Pathak, Southwestern Medical Center, University of Texas. **Chapter 8** Fig. 8.1a: Carr Clifton. Fig. 8.1b: Shizuo Lijima/Tony Stone World Wide. Fig. 8.5: Joe Englander/Viesti Associates, Inc. Fig. 8.6: Courtesy T. Elliot Weier. Fig. 8.10: Courtesy Lawrence Berkeley Laboratory, University of California. Fig. 8.13: Pete Winkel/Atlanta/Stock South. Bioline: Courtesy Woods Hole Oceanographic Institution. **Chapter 9** Fig. 9.1: Stephen Frink/AllStock, Inc. Bioline: (Fig. 1) Frank Oberle/Bruce Coleman, Inc.; (Fig. 2) Movie Stills Archive. Fig. 9.6: Topham/The Image Works. Fig. 9.9: Courtesy Gerald Karp. Human Perspective: (top and bottom insets) Courtesy MacDougal; (top) Jerry Cooke; (bottom) Richard Kane/Sportchrome East/West. Fig. 9.11: Peter Parks/NHPA. Fig. 9.12a: Grafton M. Smith/The Image Bank. **Chapter 10** Fig. 10.1: Sipa Press. Fig. 10.2a: Institut Pasteur/CNRI/Phototake. Fig. 10.3a: Dr. R. Vernon/Phototake.

Fig. 10.3b: CNRI/Science Photo Library/Photo Researchers. Fig. 10.6: Courtesy Dr. Andrew Bajer, University of Oregon, Fig. 10.7: CNRI/Science Photo Library/Photo Researchers. Fig. 10.8: Dr. G. Shatten/Science Photo Library/Photo Researchers. Fig. 10.9: Courtesy Professor R.G. Kessel. Fig. 10.10: David Phillips/Photo Researchers. **Chapter 11** Fig. 11.3: Courtesy Science Software Systems. Fig. 11.5a: Courtesy Dr. A.J. Solari, from *Chromosoma*, vol. 81, p. 330 (1980), Human Perspective: Donna Zweig/Retna. Fig. 11.7: Cabisco/Visuals Unlimited. **Part 3 Opener:** David Scharf. **Chapter 12** Fig. 12.1: Courtesy Dr. Ing Jaroslav Krizenecky. Fig. 12.6a: Doc Pele/Retna. Fig. 12.6b: FPG International. Fig. 12.9: Sydney Freelance/Gamma Liaison. Fig. 12.11 Hans Reinhard/Bruce Coleman, Inc. **Chapter 13** Fig. 13.2a: Robert Noonan. Fig. 13.6: Biological Photo Service. Bioline: Norbert Wu. Fig. 13.9: Historical Pictures Service. Fig. 13.11a: Courtesy M.L. Barr. Fig. 13.11b: Jean Pragen/Tony Stone World Wide. Fig. 13.12: Courtesy Lawrence Livermore National Laboratory. **Chapter 14** Fig. 14.1b: Lee D. Simon/Science Photo Library/Photo Researchers. Fig. 14.4: David Leah/Science Photo Library/Photo Researchers. Fig. 14.5: Dr. Gopal Murti/Science Photo Library/Photo Researchers. Fig. 14.6b: Fawcett/Olins/Photo Researchers. Fig. 14.7: Courtesy U.K. Laemmli. Fig. 14.10a: From M. Schnos and R.B. Inman, *Journal of Molecular Biology*, 51:61-73 (1970), ©Academic Press. Fig. 14.10b: Courtesy Professor Joel Huberman, Roswell Park Memorial Institute. Human Perspective: (Fig. 2) Courtesy Skin Cancer Foundation; (Fig. 3) Mark Lewis/Gamma Liaison. Fig. 14.17: Courtesy Dr. O.L. Miller, Oak Ridge National Laboratory. **Chapter 15** Fig. 15.1: Courtesy Richard Goss, Brown University. Fig. 15.2a: (top left) Oxford Scientific Films/Animals Animals; (top right) F. Stuart Westmorland/Tom Stack & Associates. Fig. 15.2b: Courtesy Dr. Cecilio Barrera, Department of Biology, New Mexico State University. Fig. 15.3b: Courtesy Michael Pique, Research Institute of Scripps Clinic. Fig. 15.7b: Courtesy Wen Su and Harrison Echols, University of California, Berkeley. Fig. 15.8a: Courtesy Stephen Case, University of Mississippi Medical Center. Fig. 15.9: Roy Morsch/The Stock Market. **Chapter 16** Fig. 16.1a: David M. Dennis/Tom Stack & Associates. Fig. 16.1b: Courtesy Lakshmi Bhatnagor, Ph.D., Michigan Biotechnology Institute. Fig. 16.2: Art Wolfe/All Stock, Inc. Fig. 16.3: Ken Graham. Fig. 16.4a: Courtesy R.L. Brinster, Laboratory for Reproductive Physiology, University of Pennsylvania. Fig. 16.4b: John Marmaras/Woodfin Camp & Associates. Fig. 16.5: Courtesy Robert Hammer, School of Veterinary Medicine, University of

Pennsylvania. Human Perspective: Courtesy Dr. James Asher, Michigan State University. Fig. 16.6b: Professor Stanley N. Cohen/Photo Researchers. Fig. 16.10: Ted Speigel/Black Star. Fig. 16.11b: Philippe Plailly/Science Photo Library/Photo Researchers. Bioline: Photograph by Anita Corbin and John O'Grady, courtesy the British Council. **Chapter 17** ig. 17.2d: Courtesy Howard Hughs Medical Institution. Fig. 17.3: Tom Raymond/Medichrome/The Stock Shop. Fig. 17.7a: Culver Pictures. Fig. 17.7b: Courtesy Roy Gumpel. Fig. 17.9a: Will & Deni McIntyre/Photo Researchers. Human Perspective: Gamma Liaison. **Part 4 Opener:** J. H. Carmichael, Jr./The Image Bank. **Chapter 18** Fig. 18.1a: Ralph Perry/Black Star. Fig. 18.1b: Jeffrey Hutcherson/DRK Photo. Fig. 18.4: Courtesy Professor Ray Evert, University of Wisconsin, Madison. Fig. 18.5: Walter H. Hodge/Peter Arnold; (inset) Courtesy Carolina Biological Supply Company. Bioline: (Fig. 1) Doug Wilson/West Light; (Fig. 2) Courtesy Gil Brum. Fig. 18.7: Luiz C. Marigo/Peter Arnold. Fig. 18.8: Dr. Jeremy Burgess/Science Photo Library/Photo Researchers. Fig. 18.9: Saul Mayer/The Stock Market; (inset) David Scharf/Peter Arnold. Fig. 18.10a: Carr Clifton. Fig. 18.10b: Ed Reschke. Fig. 18.11: Courtesy Thomas A. Kuster, Forest Products Laboratory/USDA. Fig. 18.12: A.J. Belling/Photo Researchers. Fig. 18.13: Courtesy DSIR Library Centre. Fig. 18.15: Biophoto Associates/Photo Researchers; (inset) Courtesy C.Y. Shih and R.G. Kessel, reproduced from *Living Images*, Science Books International, 1982. Fig. 18.16: Courtesy Industrial Research Ltd, New Zealand. Fig. 18.19: Robert & Linda Mitchell; (inset) From *Botany Principles* by Roy H. Saigo and Barbara Woodworth Saigo, ©1983 Prentice-Hall, Inc. Reproduced with permission. Fig. 18.22a: Jerome Wexler/Photo Researchers. Fig. 18.22b: Fritz Polking/Peter Arnold. Fig. 18.22c: Dwight R. Kuhn. Fig. 18.22d: Richard Kolar/Earth Scenes. Fig. 18.24: Biophoto Associates/Science Source/Photo Researchers. **Chapter 19** Fig. 19.1a: Courtesy of C.P. Reid, School of Natural Resources, University of Arizona. Fig. 19.3a: Peter Beck/The Stock Market. Fig. 19.3b: Dr. J. Burgess/Photo Researchers. Fig. 19.4: D. Cavagnaro/DRK Photo. Fig. 19.7a: Biophoto Associates/Photo Researchers. Fig. 19.7b: Alfred Pasieka/Peter Arnold. Fig. 19.7c: Courtesy DSIR Library Centre. Fig. 19.8: Martin Zimmerman. **Chapter 20** Fig. 20.1: G.I. Bernard/Earth Scenes/Animals Animals. Fig. 20.2a: Dwight R. Kuhn/DRK Photo. Bioline (Cases 1 and 3) Biological Photo Service; (Case 2) P.H. and S.L. Ward/Natural Science Photos; (Case 4): G.I. Bernard/Animals Animals. Fig. 20.3: E.R. Degginger. Fig. 20,4a: William E. Ferguson. Fig. 20.4b: Phil Degginger. Fig. 20.4c: E.R. Degginger. Fig. 20.5a: Runk/Shoenberger/Grant Heilman Photography. Fig. 20.5b: William E. Ferguson. Fig. 20.6a: Robert Harding Picture Library. Fig. 20,6b Richard Parker/Photo Researchers. Fig. 20.6c: Jack Wilburn/Earth Scenes/Animals Animals. Fig. 20.9a: Visuals Unlimted. Fig. 20.9b: Manfred Kage/

Peter Arnold. Human Perspective: Harold Sund/The Image Bank. Fig. 20.15: Courtesy Media Resources, California State Polytechnic University. Fig. 20.16a: John Fowler/Valan Photos. Fig. 20.16b: Richard Kolar/Earth Scenes/Animals Animals. Fig. 20.17: Breck Kent. **Chapter 21** Fig. 21.1a: Gary Milburn/Tom Stack & Associates. Fig. 21.1b: Schafer & Hill/Peter Arnold. Human Perspective: (Fig. 1) Enrico Ferorelli; (Fig. 2) Michael Nichols/Magnum Photos, Inc. Fig. 21.5: Runk/Schoenberger/Grant Heilman Photography. Fig. 21.6: Courtesy Dr. Harlan K. Pratt, University of California, Davis. Fig. 21.7: Scott Camazine/Photo Researchers. Fig. 21.9a: E.R. Degginger. Fig. 21.9b: Fritze Prenze/Earth Scenes/Animals Animals. Fig. 21.12a: Pam Hickman/Valan Photos. Fig. 21.12b: R.F. Head/Earth Scenes/Animals Animals. **Part 5 Opener:** Joe Devenney/The Image Bank. **Chapter 22** Fig. 22.2b: Dr. Mary Notter/Phototake. Fig. 22.3b: Professor P. Motta, Department of Anatomy, University of "La Sapienza", Rome/Photo Researchers. Fig. 22.6a: Fred Bavendam/Peter Arnold. Fig. 22.6b: Michael Fogden/DRK Photo. Fig. 22.6c: Peter Lamberti/Tony Stone World Wide. Fig. 22.6d: Gerry Ellis Nature Photography. Fig. 22.7a: Mike Severns/Tom Stack & Associates. Fig. 22.7b: Norbert Wu. **Chapter 23** ig 23.1:Fawcett/Coggeshall/Photo Researchers. Fig. 23.5: (top left) Vu/©T. Reese and D.W. Fawcett/Visuals Unlimited; (bottom left) Courtesy Lennart Nilsson, *Behold Man*, Little Brown & Co., Boston. Fig. 23.9: Courtesy Lennart Nilsson, *Behold Man*, Little, Brown and Co., Boston. Fig. 23.10b: Alan & Sandy Carey. Human Perspective: Science Vu/Visuals Unlimited. Fig. 23.14: Focus on Sports. Fig. 23.15: The Image Works. Fig. 23.20a: Doug Wechsler/Animals Animals. Fig. 23.20b: Robert F. Sisson/National Geographic Society. **Chapter 24** Fig. 24.1a: Anthony Bannister/NHPA. Fig. 24.1b: Giddings/The Image Bank. Fig. 24.1c: Michael Fogden/Bruce Coleman, Inc. Fig. 24.4: (left) Omikron/Photo Researchers. Fig. 24.4: (inset) Don Fawcett/K. Saito/K. Hama/Photo Researchers. Fig. 24.5a: (top) Courtesy Lennart Nilsson, *Behold Man*, Little Brown & Co., Boston; (center) Don Fawcett/K. Saito/K. Hama/Photo Researchers. Fig. 24.5b: Star File. Fig. 24.6: Philippe Petit/Sipa Press. Fig. 24.7b: Jerome Shaw. Fig. 24.10a: Oxford Scientific Films/Animals Animals. Fig. 24.10b: Kjell Sandved/Photo Researchers. Fig. 24.10c: George Shelley; (inset) Raymond A. Mendez/Animals Animals. **Chapter 25** Fig. 25.1: Robert & Linda Mitchell. Fig. 25.7: Courtesy Circus World Museum. Fig. 25.8: Courtesy A.I. Mindelhoff and D.E.. Smith, *American Journal of Medicine*, 20: 133 (1956). Fig. 25.13: Graphics by T. Hynes and A.M. de Vos using University of California at San Francisco MIDAS-plus software; photo courtesy of Genentech, Inc. **Chapter 26** Fig. 26.1: Ed Reschke/Peter Arnold. Fig. 26.2a: Joe Devenney/The Image Bank. Fig. 26.2b: John Cancalosi/DRK Photo. Fig. 26.3: Wallin/Taurus Photos. Fig. 26.4a: E.S. Ross/Phototake. Fig. 26.5a: D. Holden Bailey/Tom

Stack & Associates. Fig. 26.5b: Jany Sauvanet/Photo Researchers. Fig. 26.6a: (top) Courtesy Lennart Nilsson, *Behold Man*, Little Brown & Co., Boston; (center) Michael Abbey/Science Source/Photo Researchers. Fig. 26.6b: F. & A. Michler/Peter Arnold. Human Perspective: Courtesy Professor Philip Osdoby, Department of Biology, Washington University, St. Louis. Fig. 26.8: Duomo Photography, Inc. Fig. 26.12: A. M. Siegelman/FPG International. Fig. 26.19a: Alese & Mort Pechter/The Stock Market. Fig. 26.19b-c: Stephen Dalton/NHPA. **Chapter 27** Bioline, Fig. 27.3, and Fig. 27.5: Courtesy Lennart Nilsson, *Behold Man*, Little Brown & Co., Boston, Fig. 27.6: Micrograph by S.L. Palay, courtesy D.W. Fawcett, from *The Cell*, © W..B. Saunders Co. Fig. 27.8a: Courtesy Gregory Antipa, San Francisco State University. Fig. 27.8b: Gregory Ochocki/Photo Researchers. Human Perspective: Derik Muray Photography, Inc. Fig. 27.9a: Warren Garst/Tom Stack & Associates. Fig. 27.9b: Hervé Chaumeton/Jacana. Fig. 27.9c: D. Wrobel/Biological Photo Service. Fig. 27.10: Sari Levin. **Chapter 28** Fig. 28.2: Biophoto Associates/Photo Researchers. Fig. 28.4: Courtesy Lennart Nilsson, *Behold Man*, LIttle Brown & Co., Boston. Fig. 28.10: CNRI/Science Photo Library/Photo Researchers. Fig. 28.11a-b: Howard Sochurek. Fig. 28.11c: Dan McCoy/Rainbow. Fig. 28.12: Jean-Claude Revy/Phototake. Fig. 28.15: Dr. Tony Brain/Science Photo Library/Photo Researchers. Fig. 28.16: Alan Kearney/FPG International. Fig. 28.17a: Ed Reschke/Peter Arnold. Fig. 28.18: Manfred Kage/Peter Arnold. Fig. 28.19b: Runk/Schoenberger/Grant Heilman Photography. **Chapter 29** Fig. 29.1: Richard Kane/Sports Chrome, Inc. Fig. 29.2: Courtesy Ewald R. Weibel, Anatomisches Institut der Universität Bern, Switzerland, from "Morphological Basis of Alveolar-Capillary Gas Exchange", *Physiological Review*, Vol. 53, No. 2, April 1973, p. 425. Fig. 29.3: (top left and right insets) Courtesy Lennart Nilsson, *Behold Man*, Little Brown & Co., Boston; (center) CNRI/Science Photo Library/Photo Researchers. Fig. 29.7: Four By Five/SUPERSTOCK. Human Perspective: (Fig. 2) Vu/O. Auerbach/Visusals Unlimited; (Fig. 3) Photofest. Fig. 29.8: Kjell Sandved/Bruce Coleman, Inc. Fig. 29.10: Robert F. Sisson/National Geographic Society. Fig. 29.11a: Robert & Linda Mitchell. **Chapter 30** Fig. 30.1 and Fig. 30.5: Courtesy Lennart Nilsson, Boehringer Ingelheim International GmbH. Bioline: Courtesy Steven Rosenberg, National Cancer Institute. Fig. 30.6: G. Robert Bishop/AllStock, Inc. Fig. 30.7b: Courtesy A.J. Olson, TSRI, Scripps Institution of Oceanography. Fig. 30.10: Alan S. Rosenthal, from "Regulation of the Immune Response—Role of the Macrophage', *New England Journal of Medicine*, November 1980, Vol. 303, #20, p. 1154, courtesy Merck Institute for Therapeutic Research. Fig. 30.11: Nancy Kedersha. Human Perspective: (Fig. 1a) David Scharf/Peter Arnold; (Fig. 1b) Courtesy Acarology Laboratory, Museum of Biological Diversity, Ohio State University, Columbus; (Fig. 1c) Scott Camazine/

Photo Researchers; (Fig. 2) Courtesy Lennart Nilsson, Boehringer Ingelheim International GmbH. **Chapter 31** Fig. 31.1a: R. La Salle/Valan Photos. Fig. 31.1b: Richard Campbell/Biological Photo Service. Fig. 31.1c: Robert Harding Picture Library. Fig. 31.2a: Doug Perrine/DRK Photo. Fig. 31.2b: George Grall. Fig. 31.2c: Biological Photo Service. Fig. 31.2d: Densey Clyne/Ocford Scientific Films/Animals Animals. Fig. 31.3a: Chuck Nicklin. Fig. 31.3b: Robert & Linda Mitchell. Bioline: Hans Pfletschinger/Peter Arnold. Fig. 31.5: Courtesy Richard Kessel and Randy Kandon, from *Tissues and Organs*, W.H. Freeman. Fig. 31.9: C. Edelmann/Photo Researchers. Fig. 31.13: From A.P. McCauley and J.S. Geller, "Guide to Norplant Counseling", *Population Reports*, Sept. '92, Series K, No. 4, p. 4; photo courtesy Johns Hopkins University Population Information Program. **Chapter 32** Fig. 32.1a: Courtesy Jonathan Van Blerkom, University of Colorado, Boulder. Fig. 32.1b: Doug Perrine/DRK Photo. Fig. 32.2a: G. Shih and R. Kessel/Visuals Unlimited. Fig. 32.2b: Courtesy A.L. Colwin and L.H. Colwin. Fig. 32.3: Courtesy E.M. Eddy and B.M. Shapiro. Fig. 32.4: Courtesy Richard Kessel and Gene Shih. Fig. 32.10b: Courtesy L. Saxen and S. Toivonen. Fig. 32.13b: Courtesy K. Tosney, from *Tissue Interactions and Development* by Norman Wessels. Fig. 32.14: Dwight Kuhn. Fig 32.17: Courtesy Lennart Nilsson, from *A Child Is Born*. **Part 6 Opener:** David Doubilet/National Geographic Society. **Chapter 33** Fig. 33.1: Ivan Polunin/NHPA. Fig. 33.3: Courtesy of Victor A. McKusick, Medical Genetics Department, Johns Hopkins University. Fig. 33.4a: Sharon Cummings/Dembinsky Photo Associates. Fig. 33.4b: © Ron Kimball Studios. Fig. 33.5b-c: Courtesy Professor Lawrence Cook, University of Manchester. Fig. 33.8a: COMSTOCK, Inc. Fig. 33.8b: Tom McHugh/AllStock, Inc. Fig. 33.8c: Kevin Schafer & Martha Hill/Tom Stack & Associates; (inset) Courtesy K.W. Barthel, Museum beim Solenhofer Aktienverein, Germany. Bioline: (Fig. a) Robert Shallenberger; (Fig. b) Scott Camazine/Photo Researchers; (Fig. c) Jane Burton/Bruce Coleman, Inc.; (Fig. d) R. Konig/Jacana/The Image Bank; (Fig. e) Nancy Sefton/Photo Researchers; (Fig. f) Seaphoto Limited/Planet Earth Pictures. Fig. 33.10a: John Garrett/Tony Stone World Wide. Fig. 33.10b: Heather Angel. Fig. 33.11a: Frans Lanting/Minden Pictures, Inc. Fig. 33.11b: S. Nielsen/DRK Photo. Fig. 33.12: Zeisler/AllStock, Inc. Fig. 33.13a: Zig Leszczynski/Animals Animals. Fig. 33.13b: Art Wolfe/AllStock, Inc. Fig. 33.16a: Gary Milburn/Tom Stack & Associates. Fig. 33.16b: Tom McHugh/Photo Researchers. Fig. 33.16c: Nick Bergkessel/Photo Researchers. Fig. 33.16d: C.S. Pollitt/Australasian Nature Transparencies. **Chapter 34** Fig. 34.1a: Leonard Lee Rue III/Earth Scenes/Animals Animals. Fig. 34.1b: Bradley Smith/Earth Scenes/Animals Animals. Fig. 34.4a: Courtesy Professor George Poinar, University of California, Berkeley. Fig. 34.4b: David Muench Photography. Fig. 34.5a: William E. Ferguson. Fig. 34.7a: Courtesy Merlin Tuttle.

Fig. 34.7b: Stephen Dalton/Photo Researchers. Fig. 34.9: David Brill. Fig. 34.10: Courtesy Institute of Human Origins. Fig. 34.12: John Reader/Science Photo Library/Photo Researchers. **Chapter 35** Fig. 35.2a: Courtesy Dr. S.M. Awramik, University of California, Santa Barbara. Fig. 35.2b: William E. Ferguson. Fig. 35.4: Kim Taylor/Bruce Coleman, Inc. Fig. 35.5: Courtesy Professor Seilacher. Biuoline: Louie Psihoyos/Matrix. Fig. 35.11 and 35.13 Carl Buell. **Part 7 Opener:** Y. Arthus/Peter Arnold. **Chapter 36** Fig. 36.1: Courtesy Zoological Society of San Diego. Fig. 36.2a-b: David M. Phillips/Visuals Unlimited. Fig. 36.2c: Omikron/Science Source/Photo Researchers. Fig. 36.3a: Courtesy Wellcom Institute for the History of Medicine. Fig. 36.3b: Courtesy Searle Corporation. Fig. 36.5a: A. M. Siegelman/Visuals Unlimited. Fig. 36.5b: Science Vu/Visuals Unlimited. Fig. 36.5c: John D. Cunningham/Visuals Unlimited. Fig. 36.6: (top) Courtesy Dr. Edward J. Bottone, Mount Sinai Hospital, New York; (bottom) Courtesy R.S. Wolfe and J.C. Ensign. Fig. 36.7: CNRI/Science Source/Photo Researchers. Fig. 36.8: Courtesy C.C. Remsen, S.W. Watson, J.N. Waterbury and H.S. Tuper, from *J. Bacteriology*, vol 95, p. 2374, 1968. Human Perspective: Rick Rickman/Duomo Photography, Inc. Fig. 36.9: Sinclair Stammers/Science Source/Photo Researchers. Fig. 36.10: Courtesy R.P. Blakemore and Nancy Blakemore, University of New Hampshire. Bioline: (Fig. 1) Courtesy B. Ben Bohlool, NiftAL; (Fig. 2) Courtesy Communication Arts. Fig. 36.11: (left) Courtesy Dr. Russell, Steere, Advanced Biotechnologies, Inc.; (inset) CNRI/Science Source/Photo Researchers. 36.12b: Richard Feldman/Phototake. **Chapter 37** Fig. 37.1a: Jerome Paulin/Visuals Unlimited. Fig. 37.1b: Stanley Flegler/Visuals Unlimited. Fig. 37.1c: Victor Duran/Sharnoff Photos. Fig. 37.1d: Michael Fogden/DRK Photo. Figure 37.2: R. Kessel and G. Shih/Visuals Unlimited. Fig. 37.3: Courtesy Romano Dallai, Department of Biology, Universitá di Siena. Fig. 37.5a: M. Abbey/Visuals Unlimited. Fig. 37.5b: James Dennis/CNRI/Phototake. Fig. 37.8a: Manfred Kage/Peter Arnold. Fig. 37.8b: Eric Grave/Science Source/Photo Researchers. Fig. 37.9: Mark Conlin. Fig. 37.10: John D. Cunningham/Visuals Unlimited. Fig. 37.11a: David M. Phillips/Visuals Unlimited. Fig. 37.11b: Kevin Schafer/Tom Stack & Associates. Fig. 37.12a: E.R. Degginger. Fig. 37.12b: Waaland/Biological Photo Service. Fig. 37.13: Dr. Jeremy Burgess/Science Source/Photo Researchers. Fig. 37.14: Sylvia Duran Sharnoff. Fig. 37.15: Herb Charles Ohlmeyer/Fran Heyl Associates. Fig. 37.16 and 37.17a: Michael Fogden/Earth Scenes/Animals Animals. Fig. 37.17b: Stephen Dalton/NHPA. **Chapter 38** Fig. 38.1a: (left) Courtesy Dr. C.H. Muller, University of California, Santa Barbara; (inset) William E. Ferguson. Fig. 38.1b: Doug Wechsler/Earth Scenes. Fig. 38.2: T. Kitchin/Tom Stack & Associates. Fig. 38.3a: Michael P. Gadomski/Earth Scenes/Animals Animals. Fig. 38.3b: Courtesy Edward S. Ross, California

Academy of Sciences. Fig. 38.3c: Rod Planck/Photo Researchers. Fig. 38.3d: Kurt Coste/Tony Stone World Wide. Bioline: (Fig. a) Michael P. Gadomski/Photo Researchers; (Fig. b-c) E.R. Degginger; (Fig. d) COMSTOCK, Inc. Fig. 38.6a: Kim Taylor/Bruce Coleman, Inc. Fig. 38.6b: Breck Kent. Fig. 38.6c: Flip Nicklin/Minden Pictures, Inc. Fig. 38.6d: D.P. Wilson/Eric & David Hosking/Photo Researchers. Fig. 38.7: J.R. Page/Valan Photos; (inset) Stan E. Elems/Visuals Unlimited. Fig. 38.8a and c: Runk/Schoenberger/Grant Heilman Photography. Fig. 38.8b: John Gerlach/Tom Stack & Associates. Fig. 38.11: John H. Trager/Visuals Unlimited. Fig. 38.13: John D. Cunningham/Visuals Unlimited. Fig. 38.14: Milton Rand/Tom Stack & Associates. Fig. 38.17: Cliff B. Frith/Bruce Coleman, Inc. Fig. 38.18: Leonard Lee Rue/Photo Researchers. Fig. 38.19: Anthony Bannister/Earth Scenes. Fig. 38.21a: J. Carmichael/The Image Bank. Fig. 38.21b: Sebastio Barbosa/The Image Bank. Fig. 38.21c: Holt Studios/Earth Scenes. Fig. 38.21d: Gwen Fidler/COMSTOCK, Inc. **Chapter 39** Fig. 39.1a: Courtesy Karl G. Grell, Universität Tübingen Institut für Biologie. Fig. 39.1b: François Gohier/Photo Researchers. Fig. 39.2a: Larry Ulrich/DRK Photo. Fig. 39.2b: Christopher Newbert/Four by Five/SUPERSTOCK. Fig. 39.6a: Chuck Davis. Fig. 39.7a: Robert & Linda Mitchell. Fig. 39.7b: Runk/Schoenberger/Grant Heilman Photography. Fig. 39.7c: Fred Bavendam/Peter Arnold. Fig. 39.10a: Carl Roessler/Tom Stack & Associates. Fig. 39.10b: Goivaux Communication/Phototake. Fig. 39.10c: Biomedia Associates. Fig. 39.14a: Gary Milburn/Tom Stack & Associates. Fig. 39.14b: E.R. Degginger. Fig. 39.14c: Christopher Newbert/Four by Five/SUPERSTOCK. Fig. 39.15: Courtesy D. Phillips. Fig. 39.16a: Marty Snyderman/Visuals Unlimited. Fig. 39.16b: Robert & Linda Mitchell. Fig. 39.18a: John Cancalosi/Tom Stack & Associates. Fig. 39.18b: John Shaw/Tom Stack & Associates. Fig. 39.18c: Patrick Landman/Gamma Liaison, Fig. 39.20a: Robert & Linda Mitchell. Fig. 39.20b: Edward S. Ross, National Geographic, *The Praying of Predators*, February 1984, p. 277. Fig. 39.20c: Maria Zorn/Animals Animals. Fig. 39.21: Courtesy Philip Callahan. Fig. 39.22a: Biological Photo Service. Fig. 39.22b: Kim Taylor/Bruce Coleman, Inc. Fig. 39.22c: Tom McHugh/Photo Researchers. Fig. 39.23a: E.R. Degginger. Fig. 39.23b: Christopher Newbert/Four by Five/SUPERSTOCK. Fig. 39.26b: Dave Woodward/Tom Stack & Associates. Fig. 39.26c: Terry Ashley/Tom Stack & Associates. Fig. 39.28a: Russ Kinne/COMSTOCK, Inc. Fig. 39.28b: Ken Lucas/Biological Photo Service. Fig. 39.30: Marc Chamberlain/Tony Stone World Wide. Fig. 39.31a: W. Gregory Brown/Animals Animals. Fig. 39.31b: Chris Newbert/Four by Five/SUPERSTOCK. Fig. 39.32a: Zig Leszczynski/Animals Animals. Fig. 39.32b: Chris Mattison/Natural Science Photos. Fig. 39.33a: Michael Fogden/Animals Animals. Fig. 39.33b: Jonathan Blair/Woodfin Camp & Associates. Fig. 39.34a: Bob and Clara Calhoun/Bruce Coleman, Inc. Fig.

39.35a-b: Dave Watts /Tom Stack & Associates. Fig. 39.35c: Boyd Norton. **Part 8 Opener:** Carr Clifton. **Chapter 40** Fig. 40.1: Courtesy Earth Satellite Corporation. Fig. 40.4: Norbert Wu/Tony Stone World Wide. Fig. 40.5: Stephen Frink/AllStock, Inc. Fig. 40.6a-b: Nicholas De Vore III/Bruce Coleman, Inc. Fig. 40.6c: Juergen Schmitt/The Image Bank. Fig. 40.7a: Frans Lanting/Minden Pictures, Inc. Fig. 40.7b: D. Cavagnaro/DRK Photo. Fig. 40.7c: Jeff Gnass Photography. Fig. 40.7d: Larry Ulrich/DRK Photo. Fig. 40.9: Joel Arrington/Visuals Unlimited. Fig. 40.11a: K. Gunnar/Bruce Coleman, Inc. Fig. 40.11b: Claude Rives/Agence C.E.D.R.I. Fig. 40.11c: Maurice Mauritius/Rapho/Gamma Liaison. Human Perspective: Will McIntyre/AllStock, Inc. Fig. 40.13: Jim Zuckerman/West Light. Fig. 40.14: James P. Jackson/Photo Researchers. Fig. 40.15: Steve McCutcheon/Alaska Pictorial Service. Fig. 40.16: Linda Mellman/Bill Ruth/Bruce Coleman, Inc. Fig. 40.17: Bill Bachman/Photo Researchers. Fig. 40.18: Nigel Dennis/NHPA. Fig. 40.19: Bob Daemmrich/The Image Works. Fig. 40.20: B.G. Murray, Jr./Animals Animals. Fig. 40.21a: Breck Kent. Fig. 40.21b: Carr Clifton. **Chapter 41** Fig 41.1a: Gary Braasch/AllStock, Inc. Fig. 41.1b: George Bernard/NHPA. Fig. 41.1c: Stephen Krasemann/NHPA. Fig. 41.2: Robert & Linda stures, Inc. Fig. 41.7a: E.R. Degginger/Photo Researchers. Fig. 41.7b: Anthony Mercieca/Natural Selection. Fig. 41.9: Walter H. Hodge/Peter Arnold. Fig. 41.16: Larry Ulrich/DRK Photo. Fig. 41.18: © Gary Braasch. **Chapter 42** Fig. 42.1a: Andy Callow/NHPA. Fig. 42.1b: Al Grotell. Fig. 42.5a: Cosmos Blank/Photo Researchers. Fig. 42.5b: Karl and Steve Maslowski/Photo Researchers. Fig. 42.5c: Michael Fogden/Bruce Coleman, Inc. Fig. 42.5d: Gregory G. Dimijian, M.D./Photo Researchers. Fig. 42.6: B&C Calhoun/Bruce Coleman , Inc. Fig. 42.7: Kjell Sandved/Photo Researchers. Fig. 42.8a: Wolfgang Bayer/Bruce Coleman, Inc. Fig. 42.8b: J. Carmichael/The Image Bank. Fig. 42.9: Anthony Bannister/NHPA. Fig. 42.10a: Adrian Warren/Ardea London. Fig. 42.10b: Peter Ward/Bruce Coleman, Inc. Fig. 42.10c: Michael Fogden. Fig. 42.11: P.H. & S.L. Ward/Natural Science Photos. Fig 42.12: Tom McHugh/Photo Researchers. Fig. 42.13: E.R. Degginger. Fig. 42.14: William H. Amos. Fig. 42.16: Dr. J.A.L. Cooke/Oxford Scientific Films/Animals Animals. Fig. 42.17: David Cavagnaro/Visuals Unlimited. Fig. 42.19: Stephen G. Maka. Fig. 42.20: Raymond A. Mendez/Animals Animals; (inset): Eric Grave/Phototake. **Chapter 43** Fig. 43.1a: Steve Krasemann/DRK Photo. Fig. 43.1b: Carr Clifton. Fig. 43.1c: Willard Clay. Fig. 43.3: Robert & Linda Mitchell. **Chapter 44** Fig. 44.1: Johnny Johnson/DRK Photo. Fig. 44.2: Dwight R. Kuhn. Fig. 44.3: Mike Severns/Tom Stack & Associates. Fig. 44.4: Bettmann Archive. Fig. 44.5: Robert Maier/Animals Animals. Fig. 44.6: Courtesy Masao Kawai, Primate Research Institute, Kyoto University. Fig. 44.7: Glenn Oliver/Visuals Unlimited. Fig. 44.8: Konrad Wothe/Bruce Coleman, Ltd. Fig. 44.9: Gordon Wiltsie/Bruce Coleman, Inc. Fig. 44.10: Hans Reinhard/Bruce Coleman, Ltd. Fig. 44.11: Oxford Scientific Films/Animals Animals. Fig. 44.12: Patricia Caulfield. Fig. 44.13: Roy P. Fontaine/Photo Researchers. Fig. 44.14: Anup and Manoj Shah/Animals Animals. **Appendix B** Fig. b: Bob Thomason/Tony Stone World Wide. Fig. c: Dr. Gennavo/Science Photo Library/Photo Researchers. Fig. d: F.R. Turner, Indiana University/Biological Photo Service.

Index

Note: A t following a page number denotes a table, f denotes a figure, and n denotes a footnote.

◀ A ▶

Abiotic environment, 934–936, 936–937
ABO blood group antigens, 240
"Abortion" pill, 679
Abscisic acid, 427t, 433
Abscission, 433
Absorption of nutrients, 568, 574
Accidental scientific discovery, 35–37
Accommodation, 503
Acetylcholine, 475t
Acetyl coenzyme A (acetyl CoA), 181
Acid rain, 392, 918
Acids, 61–62
Acoelomate, 870
Acquired immune deficiency syndrome (AIDS), 657, 794, 796–797
Acromegaly, 526, 527f
Acrosome, 670
ACTH (adrenocorticotropic hormone), 522t, 526
Actin, 555
Action potential, 471–473
Activation energy, 123, 124f
Active immunity, 653
Active site of enzyme, 124f, 125
Active transport, 142–144
"Adam's apple," 625
Adaptation, 4–5, 728–729
    and evolution, 4–5, 18
    to food type, 577, 581
    and nitrogenous wastes, 613
    and osmoregulation, 611–613
    and photosynthesis, 164–166
    of respiratory system, 632–635
Adaptive radiation, 734
Addison's disease, 528–529
Adenine, 269, 270f
    base pairing
        in DNA, 271f, 272
        in RNA, 278
Adenosine triphosphate, *see* ATP
Adenylate cyclase, 534, 535f
ADH, *see* Antidiuretic hormone
Adipocytes, 76
Adrenal gland, 522t, 526–529
    cortex, 522t, 526, 528–529
    medulla, 522t, 529
Adrenaline (epinephrine), 522t, 529
Adrenocorticotropic hormone (ACTH), 522t, 526
Adventitious root system, 373, 374f
Aerobes, appearance on earth, 174
Aerobic respiration, 180–184
    versus fermentation, 175, 178
Aestivation, 927
Agar, 841
Age of Mammals, 774
Age of Reptiles, 771
Agent Orange, 431
Age–sex structure of population, 988
    for humans, 999, 1000f
Aggregate fruits, 412, 413f
Agnatha, 886–887
AIDS (acquired immune deficiency syndrome), 657, 794, 796–797
AIDS virus, *see* Human immunodeficiency virus

"Air breathers," 633, 635
Alarm calls, 1023
Albinism, 240
Alcohol, in pregnancy, 704
Alcoholic fermentation, 178, 179
Aldosterone, 522t, 528, 610–611
Algae, 816
    as bacteria, 792
    as lichens, 821
    as plants, 834, 835–842
    as protists, 816–819
Algin, 840
Alkalinity, 61
Allantois, 703
Allele frequency, 717–718
Alleles, 234–235, 240, 241
    linkage groups and, 249, 250
Allelochemicals, 972
Allelopathy, 974–975
Allergies, 656
Alligators, 889
Allopatric speciation, 732
Allosteric site, 130f
Alpha-helix, 67, 80
Alpine tundra, 922
Alternation of generations, 833
Altruism, 1022–1026
Alveolar-capillary gas exchange, 625–626, 627
Alveoli, 625–626
Alzheimer's disease, 101, 483, 484
Ames test, 286–287
Amino acids, 79–80
    assembly into proteins, 78f, 277, 278, 284–285, 286f
    and diet, 578
    and genetic code, 278–279, 282f
    sequencing, 66–67, 84, 750
Amino group, 78f, 79
Ammonia, excretion, 607, 613
Amniocentesis, 347f, 348
Amnion, 703
Amniotic fluid, 703
Amoebas, 813, 816
Amphibians, 888–889
    heart in, 605f
Anabolic pathways, 128
Anabolic reactions, electron transfer in, 128–129
Anaerobes, 174, 791
Anaerobic respiration, 179n
Analogous, versus homologous, 458, 744–746
Analogous features, 744
Anaphase, of mitosis, 203f, 206
Anaphase I, of meiosis, 217f, 218–219
Anaphase II, of meiosis, 217f, 221
Anaphylaxis, 656
Anatomy, and evolution, 749–750
Androgen-insensitivity syndrome, 305
Anecdotal evidence, 31
Angiosperms, 849, 852–853
Angiotensin, 591
Animal behavior, 1006–1030
Animal kingdom, 858–897
    phyla, 863, 864t
    phylogenetic relations, 865f
    phylum Annelida, 864t, 875–876
    phylum Arthropoda, 864t, 876–882
    phylum Chordata, 864t, 884–886
    phylum Cnidaria, 864t, 867–869

    phylum Echinodermata, 864t, 882–884
    phylum Mollusca, 864t, 874–875
    phylum Nematoda, 864t, 871–873
    phylum Platyhelminthes, 864t, 869–871
    phylum Porifera, 864t, 864–867
Animal models, 320
Animals:
    basic characteristics, 860–861
    body plan, 861–863
    cognition in, 1024
    desert adaptations, 926–927
    evolution, 861, 865f, 873
    form and function, 448–464
    growth and development, 686–709
Annelida, 864t, 875–876, 877f
Annual plants, 361
Annulus of fern, 847, 848f
Antarctic ozone hole, 902–903
Anteater, 893
Antennae of insects, 879
Antenna pigments, 158f, 159
Anther, 403
Antheridium, 842
Anthophyta, 839t, 852–853
Anthropoids, 894f
Antibiotics, discovery of, 37
Antibodies, 646, 647, 650–654
    formation of, 650–652
        DNA rearrangement, 651, 652f
    monoclonal, 654
    to oneself (autoantibodies), 654
    production of, 652–653
    specificity, 650
    structure, 650, 651f
Anticodon, 279–280, 283f
Antidiuretic hormone (ADH; vasopressin), 522t, 525, 610, 611
Antigen-binding site, 650, 651f
Antigens, 650
    and antibody production, 652
    defined, 646
    presentation of, 652, 654f
Anti-oncogenes, 260
Antioxidants, and free radicals, 56
Aorta, 595
Apes, 893
    genetic relation to human, 259, 329
Aphids:
    and ants, 962f
    and phloem research, 394
Apical dominance, 430, 431f
Apical meristem, 362, 372f, 373, 375f
Aposematic coloring, 970
Appendicitis, 575
Appendicular skeleton, 548–550
Appendix, 574–575, 576f
Aquatic animals, respiration in, 620–621, 632–633
Aquatic ecosystems, 908–916
Archaebacteria, 791, 796–798
Archaeopteryx, 746–747, 748f
Archegonium, 842
Archenteron, 693f
Arctic tundra, 922
Arteries, 589–590, 595
Arterioles, 590
Arthropoda, 864t, 876–882
Artificial selection, 721
Asbestosis, 101

Ascaris, 873
Ascocarp, 823
Ascomycetes, 822t, 823–824
Ascorbic acid (vitamin C), 540–541, 579t
    and cancer, 31, 39–40
    and colds, 30–31
Ascospores, 823
Ascus, 823
Asexual reproduction, 199, 664–665
    in flowering plants, 417–418
Association, defined, 933
Asthma, 656
Atherosclerosis, 75, 146, 590
Athletes and steroids, 528
Atmosphere, 904
Atolls, 912
Atomic basis of life, 48–64
Atomic mass, 52
Atomic number, 52
Atoms, structure of, 51–55
ATP (adenosine triphosphate), 121–123
    in active transport, 142–143, 144f
    chemical structure, 121f
    formation of
        by chemiosmosis, 161, 184
        by electron transport, 182–184
        by glycolysis, 175–178
        by Krebs cycle, 180f, 181, 182f
        energy yield, 178, 179, 180, 182, 185, 188
        proton gradient and, 161, 184
        site of, 172–173, 184f
    and glucose formation, 163–164
    hydrolysis of, 121f, 122–123
    and muscle, 186–187, 552, 556
    photosynthesis and, 160f, 161–163
ATPase, 142
ATP synthase, 160f, 161, 173, 184
Atrioventricular (AV) node, 599
Atrioventricular (AV) valves, 597
Atrium of heart, 593f, 594
Australopithecines, 752, 754, 755f
Australopithecus species, 743, 752, 754, 755f
Autoantibodies, 654
Autoimmune diseases, 656–657
Autonomic nervous system, 490–491
Autosomal dominant disorders, 343–345
Autosomal recessive disorders, 343
Autosomes, 253
Autotoxicity, 974
Autotrophs, 154–155, 764
Auxins, 425, 427, 430–431
Aves, 889–892
AV (atrioventricular) node, 599
AV (atrioventricular) valves, 597
Axial skeleton, 548, 549f
Axillary bud, 377
Axon, 468, 469f
AZT (azidothymidine), 657

◀ B ▶

Backbone, 886
    origin of, 694
Bacteria, 786–800. See also Prokaryotes
    beneficial, 799
    chemosynthetic, 167–168, 798
    in digestive tract, 575, 581
    disease-causing, 792
    enzyme inhibition in, 126
    in fermentation, 179
    in genetic engineering, 314, 316, 324
    in genetics research, 266–267
    gene transfer between, 804
    growth rate, 788
    metabolic diversity, 791

motility, 792
    and nitrogen fixation, 390, 799, 949–951
    structure of, 793f
    taxonomic criteria, 788–791
    ubiquity of, 799
Bacterial chlorophyll, 791–792
Bacteriophage, 268, 269f, 801
Balance, 506, 508–509
Ball-and-socket joint, 550
Bark, 369
Barnacles, 876
Barr body, 256f
Base pairing:
    in DNA, 271f, 272
    in RNA, 278
Bases, acids and, 61–62
Basidia, 824
Basidiocarp, 824–825
Basidiomycetes, 822t, 824–825
Basidiospores, 824
Basophils, 601–602
Bats, 498–499, 750, 751f
B cells, 647, 652–653
Beagle (HMS), voyage of, 20f, 21
Bean seedling development, 416, 417f
Beards, and gene activation, 305
Bees:
    dances, 1019, 1020f
    haplodiploidy in, 1025–1026
Behavioral adaptation, 729
Behavior of animals, 1006–1030
    development of, 1014–1015
    genes and, 1008–1009, 1010–1011
    mechanisms, 1008–1011
Beri-beri, 566–567
Beta-pleated sheet, 80
Bicarbonate ion, in blood, 629
Biennial plants, 361
Bile salts, 574
Binomial system of nomenclature, 13
Biochemicals, defined, 68
Biochemistry, 66–86, 750
Bioconcentration, 946
Biodiversity Treaty, 997
Bioethics, 25, 40–41. See also Ethical issues
Biogeochemical cycles, 947–952
Biogeography, 750–752
Biological classification, 11–15
Biological clock, of plants, 434, 438
Biological magnification, 946
Biological organization, 7, 11, 12f
Biological species concept, 732, 830
Biology, defined, 6
Biomass, 943
    pyramid of, 944
Biomes, 907f, 916–927, 934
Biosphere, 11, 902–930
    boundaries, 904
Biotechnology:
    applications, 314–321
    DNA "fingerprints," 328, 329
    genetic engineering techniques, 321–329
    opposition to, 321
    patenting genetic information, 315
    recombinant DNA techniques, 321–325. *See also*
        Recombinant DNA
    without recombinant DNA, 325–329
Biotic environment, 934–936, 936–937
Biotic (reproductive) potential, 986, 989–990
Biotin, 579t
Birds, 889–892
    fossil, 746–747, 748f
Birth:
    canal, 673
    control, 677–681

control pills, 678–679
    in human, 705
Bivalves, 874
Blade of leaf, 377
Blastocoel, 690
Blastocyst, 700
Blastodisk, 691f
Blastomere, 690, 692
Blastopore, 693f, 873
Blastula, 690, 691f
Blastulation, 690
"Blind" test, 39
Blood:
    clotting, 450, 602–603
    composition, 600–603
    flow
        through circulatory system, 594f
        through heart, 593f
    gas exchange in, 627–629
Blood-brain barrier, 110
Blood:
    pressure, 589–590. See also Hypertension
    type, 94, 240
    vessels, 588–593
Blooms of algae, 818
Blue-green algae, *see* Cyanobacteria
Body:
    cavities, 861
    defense mechanisms
        nonspecific, 644–645
        specific, 646–650
    fluids:
        compartments, 588
        regulation of, 606–613
    mechanics, 558–559
    organization, 696–698
        genetic control of, 686–687
    plan, 861–863
    size, 460
    surface area, 460–462
        and respiration, 622–623, 626
    symmetry, 861, 862f
    temperature, 449, 450, 451f, 588, 614
    weight, 76, 460
Body contact and social bonds, 1021
Body planes, 862f
Body segmentation, 863, 875–876
Bohr effect, 629
Bonding, 1021–1022
Bones, 546–548
    human skeleton, 548, 549f
    repair and remodeling, 550
Bony fishes, 887, 888f
Bottleneck, genetic, 720
Botulism, 478
Bowman's capsule, 608–609
Brain, 481–487
    evolution, 492
    left versus right, 504
    and sensory stimuli, 510–511
Brainstem, 481f, 486
Breast-feeding, and immunity, 655
Breathing:
    mechanics of, 626–627
    regulation of, 629, 632
Bristleworms, 875–876
Bronchi, 625
Bronchioles, 625
Brown algae, 838–840
Bryophyta, 839t
Bryophytes, 835, 842–843
Budding, in animals, 664
Buffers, 62
Bundle-sheath cells, 165f, 166
Bursitis, 551

◄ C ►

C_3 synthesis, 163–164
C_4 synthesis, 163t, 164–166
Caecilians, 888–889
Calciferol (vitamin D), 579t
Calcitonin, 522t, 531
Calcium:
  blood levels, 520, 530f, 531
  in diet, 580, 580t
  in muscle contraction, 557
Calorie, 60
Calvin–Benson cycle, 164
Calyx of flower, 403
Cambia, 362, 372–373
Camel's hump, 927
Camouflage, 967–968
cAMP (cyclic AMP), 533–534, 535f, 819
CAM synthesis, 163t, 166
Cancer, *see also* Carcinogens
  and cell cycle, 202
  and chromosomal aberrations, 259, 260
  genetic predisposition to, 346
  immunotherapy for, 648
  and lymphatic system, 606
  of skin, 280–281
  and smoking, 630
  viruses causing, 803
Capillaries, 587, 590–593
  lymphatic, 606
Capillary action, 60
Capsid, viral, 801
Capsule, bacterial, 792
Carbohydrases, 573
Carbohydrates, 70–74
  as energy source, 188
  in human diet, 578
  of plasma membrane, 94
  synthesis in plants, 163–166
Carbon:
  biological importance, 68–69
  in oxidation and reduction, 128
  properties, 68–69
Carbon cycle, 948–949
Carbon dating of fossils, 54
Carbon dioxide:
  and bicarbonate ions, 629
  in respiration, 628f, 629, 632
Carbon dioxide fixation, 163–164
  and leaf structure, 165–166
Carbonic acid, 629
Carbonic anhydrase, 629
Carbon monoxide poisoning, 627
Carboxyl group, 78f, 79
Carcinogens, 286–287
  testing for, 37–39, 286–287
Cardiac muscle, 558, 559f
Cardiac sphincter, 569
Cardiovascular system, *see* Circulatory system
Careers in biology, Appendix D
Carnivore, 966
Carotenoids, 157
Carpels, 403
Carrier of genetic trait, 254–255, 343
Carrier proteins of plasma membrane, 141, 143f
Carrying capacity (*K*) of environment, 992, 999–1001
Carson, Rachel, 714–715
Cartilage, 548
Cartilaginous fishes, 887
Casparian strips, 374, 376f, 387, 389f
Catabolic pathways, 128
Catalysts, enzymes as, 123
Cavemen, 742
Cecum, 574–575, 576f, 581
Cell body of neuron, 468, 469f
Cell cycle, 199–202

Cell death, and development, 697
Cell differentiation, 302–303, 698–699
Cell division, 194–210, 212–225. *See also* Meiosis;
      Mitosis
  and cancer, 202
  defined, 196
  in eukaryotes, 197–199
  in prokaryotes, 197
  regulation of, 201–202
  triggering factor, 194–195
  types, 196–197
Cell fate, 690–691
  induction and, 695
Cell fusion, 89
Cell-mediated immunity, 647
Cell plate, in plant cytokinesis, 206
Cells, 88–114. *See also* Eukaryotic cells; Prokaryotic
      cells
  discovery of, 90–91
  exterior structures, 108–110
  as fundamental unit of life, 91
  genetically engineered, 314
  how large particles enter, 136–137, 144–147
  how molecules enter, 138–144
  junctions between, 109–110
  locomotion, 106–107
  membrane, *see* Plasma membrane
  as property of living organism, 7, 11
  size limits, 92–93
  specialization, 302–303
  totipotent, 297
Cell surface receptor, 94, 147–148, 533–534, 535f
Cell theory, 90–91
Cellular defenses
  nonspecific, 645
  specific (T cells), 647, 649–650
Cellular slime molds, 819
Cellulose, 73, 119f
  in diet, 578, 581
Cell wall, 108
  of bacteria, 787
  of plants, 108–109
Cenozoic Era, 774
Centipedes, 876, 881–882
Central nervous system (CNS), 479, 481–489
  brain, *see* Brain
  spinal cord, 488–489
Centromere, 198, 204f, 218f
  versus kinetochore, 205n
Cephalization, 492, 861
Cephalopods, 874–875
Cerebellum, 481f, 486
Cerebral cortex, 481f, 482
Cerebrospinal fluid, 482
Cerebrum, 482–485
Certainty and science, 40–41
Cervix, 673
Cestodes, 869, 870f, 871
Chain (polypeptide) elongation and termination,
      285
Chain of reactions, 1009
Chaparral, 924, 925f
Character displacement, 965
Chargaff's rules, 272
Charging enzymes, 280, 286f
Cheetah, 730
Chelicerates, 876, 878f, 881
Chemical bonds, 55–58
Chemical defenses, 972, 974–975
Chemical energy, from light energy, 156–163
Chemical evolution, 760–761
Chemical reactions:
  electron transfer in, 128–129
  endergonic, 122–123
  energy in, 121–123

exergonic, 122
  favored versus unfavorable, 122
Chemiosmosis, 161
Chemoreception, 509–510, 512
Chemoreceptors, 500, 501t
  and breathing control, 632
Chemosynthesis, 167–168
Chemosynthetic bacteria, 167–168, 798
Chest cavity, 626
Chiasmata (of chromosomes), 218f
Chimpanzee, 894f
Chitin, 73
Chlamydia infections, 795
Chlorophylls, 157
  bacterial, 791–792
Chlorophyta, 835, 837
Chloroplasts, 104
  forerunners of, 792, 793f
  and origin of eukaryote cell, 110, 765, 962
  in photosynthesis, 155, 156, 158f
Cholecystokinin, 522t, 573
Cholera, 573
Cholesterol, 75, 77f, 136–137
  and atherosclerosis, 75, 146
Chondrichthyes, 888f
Chondrocytes, 548
Chordamesoderm, 694–695
Chordata, 864t, 884–886
Chorion, 700, 703
Chorionic villus sampling, 347f, 348
Chromatids, 198, 204f, 216, 218
Chromatin, 97
  and DNA packing, 274f
Chromosomal exchange, 250, 292
Chromosomal puffs, 305
Chromosomes, 97, 246–264, 334–353. *See also* Sex
      chromosomes; Homologous chromosomes;
      Meiosis; Mitosis
  abnormalities, 221, 257–259
    deletions, 258
    duplications, 258
    inversions, 259
    translocations, 259
    wrong number, 221, 336–337
      detection in fetus, 349
      nondisjunction, 236
      polyploidy, 260–261
  banding patterns, 198f, 253f
  condensation, 203–204
  discovery of functions, 246–247
  DNA in, 273, 274f, 275f
  in eukaryotes, 273
  giant (polytene), 252–253, 305
  mapping, 252, 338
    Human Genome Project, 340
  mitotic, 198–199, 204
  number, 199
    in humans, 212–213
  in prokaryotes, 273
  structure, 273, 274f, 275f
Chrysophytes, 817–818
Chyme, 570
Chymotrypsin, 573
Cigarette smoking, 630–631, 704
Cilia:
  of cell, 106–107
  of protozoa, 814–816
Ciliata, 815t
Circadian plant activities, 438
Circulatory (cardiovascular) systems, 455, 588–605
  discovery of, 586–587
  evolution of, 603–605
  in humans, 456f, 588–600
  open versus closed, 605
Circumcision, 669

Citric acid cycle, 181
Clams, 864t, 874
Class, in taxonomic hierarchy, 15
Classical conditioning, 1011–1012
Cleavage, 690
Cleavage furrow, in cell division, 206
Climate, 905–907, 916
Climax community, 952
Clitoris, 673
Clock, biological, 434, 438
Clonal deletion, 654
Clonal selection theory, 651–653
Clones, 296–297
Clotting factors, 602
Clotting of blood, 450, 602–603
Club fungi, 824
Club mosses, 845–846
Clumped distribution patterns, 987
Cnidaria, 864t, 867–869
CNS, *see* Central nervous system; Brain
Coagulation (clotting) of blood, 450, 602–603
Coastal water habitats, 908, 910–912
Cobalamin (vitamin B$_{12}$), 579t
Cocaine, 487
Cochlea, 507f, 508
Codominance, 239–240
Codons, 278
    role in mutations, 286
    role in translation, 279–281, 284–285
Coelacanth fish, 859
Coelom, 861, 863f, 871–872, 875–876
Coenzyme, 127
Coenzyme A (CoA), 181
Coevolution, 734f, 735
Cofactors, 127
Cognition in animals, 1024
Coherence theory of truth, 41
Cohesion of water, 60
Coitus interruptus, 681
Cold (disease), 803
Coleoptile, 417
Coleoptile tip, 424–425
Collagen, 453–454, 541
Collecting duct, 608–609
Collenchyma cells, 364
Colon, 576f
Color adaptations, 968, 970
Colorblindness, 255f, 256f
Combat, 1017, 1018–1019
Commensalism, 976
Communication, 1018–1022
    functions, 1018–1019
    types of signals, 1019–1021
Communities, 11, 932–933, 960–982
    changes in, 952
    climax, 952
    pioneer, 952
Compact bone, 546, 547f
Companion cell of plant, 372
"Compass" bacterium, 798
Competition, 960–961, 963–965
Competitive exclusion, 941, 961, 964
Complement, 645
Complementarity in DNA structure, 272
Complete flower, 407
Composite flower, 406f
Compound (chemical), 55
Compound leaf, 377
Concentration gradient, 138, 142, 144f
    and energy storage, 143–144, 145f
    of protons, 160f
Condensation of molecules, 69, 70f
Conditioned response, 1011–1012
Conditioning, 1011–1012
Condoms, 681
Cones of gymnosperms, 849

Cones of retina, 503, 505f, 506
Confirmation in scientific method, 33
Conformational changes in proteins, 81
Coniferophyta, 839t, 851–852
Coniferous forests, 917, 920, 921f
Conjugation in protists, 222
Connective tissue, 453–454
    and vitamin C, 541
Consumers, 937, 942
Continental drift, 767–770
    and natal homing, 312–313
Continuous variation, 240
Contraception, 677–681
Control group, 33, 37
Controlled experiments, 37–39
Convergent evolution, 734f, 735, 736f
Conversion charts, Appendix A
Cooperation between animals, 1023–1024
Copepods, 881
Coral reefs, 911–912
Corals, 864t, 867–869
Corepressor of operon, 300f
Cork cambium, 362
Corn seedling development, 417, 418f
Corolla of flower, 403
Coronary arteries, 595, 596f
Coronary bypass surgery, 597
Corpolites, 746
Corpus callosum, 481f, 482, 504
Corpus luteum, 676f, 677
Cortex of plants, 372
Cortisol (hydrocortisone), 522t, 526, 528
Cotyledons, 411
Countercurrent flow, 633
Coupling, in chemical reactions, 122
Courtship behavior, 668, 1018
Covalent bonds, 55
    in oxidation and reduction, 128
Covalent modification of enzyme activity, 129
Cowper's gland, 672
Crabs, 864t, 876, 881
Cranial nerves, 489–490
Cranium, skull, 482
Crassulacean acid metabolism, 163t, 166
Cretinism, 531
Crick, Francis, 269–272
Cri du chat syndrome, 258
Crinoids, 864t
Crocodiles, 889
Crops, biotechnology and, 316–319, 418–419
Crosses, genetic
    dihybrid, 238
    monohybrid, 233, 234f
    test cross, 237
Crossing over, 216, 218f, 250, 251f, 292
    and genetic mapping, 252
Crustaceans, 876, 878f, 881
Cryptic coloration, 968
Curare, 478
Cutaneous respiration, 622
Cuticle of arthropods, 545
Cuticle of plants, 366
    and transpiration, 392
Cyanide, 184
Cyanobacteria, 791, 792, 796
    role in evolution, 174, 764
Cycadophyta (cycads), 839t, 849
Cyclic AMP (cAMP), 533–534, 535f, 819
Cyclic photophosphorylation, 161–163
Cystic fibrosis, 240, 337t, 338, 341f
    and evolution, 342
    gene for, 338, 341, 342f
    gene therapy for, 349
    screening test for, 347
Cysts, protozoal, 814
Cytochrome oxidase, 184

Cytokinesis:
    in meiosis, 217f, 219
    in mitosis, 206
Cytokinins, 427t, 432
Cytoplasm:
    components of, 97–105
    receptors in, 534, 535f
Cytosine, 269, 270f
    base pairing, 271f, 272
Cytoskeleton, 104–105
Cytotoxic (killer) T cells, 648

◀ **D** ▶

Dancing bees, 1019, 1020f
Dark (light-independent) photosynthetic reactions,
    155f, 156, 158f, 163–166
Darwin, Charles, 19–24, 721–722, 750–752
    and age of earth, 746
    and childhood sexuality, 662–663
    and plant growth, 424–425
Darwin's finches, 21–23
Daughter cells, 196
Davson-Danielli model of membrane structure, 89
Day-neutral plants, 437
DDT, 714–715, 946
Deciduous forests, 917, 919–920, 921f
Decomposers, 937
Deep-sea-diving mammals, 620–621
Defecation reflex, 574
Defense adaptations, 966t, 967–972
Defense mechanisms of body
    nonspecific, 644–645
    specific, 646–650
Denaturing of protein, 81, 127
Dendrites, 468, 469f
Denitrifying bacteria, 950
Density-dependent and -independent population-
    control factors, 996
Deoxyribonucleic acid, *see* DNA
Deoxyribose, 278
Derived homologies, 746
Dermal bone, 543
Dermal tissue system of plants, 365, 366–369, 367f
Dermis, 542
Deserts, 926
    and mycorrhizae, 388
    and osmoregulation, 613
    plant adaptations to, 164–166, 926
Detritovores, 937, 942–943
Deuteromycetes, 822t, 825–826
Deuterostomes, 873
Development, 10, 686–709
    embryonic, *see* Embryo, development
    postembryonic, 699–700
Developmental defects, 686
Developmental "fate," 690–691, 695
Diabetes, 531–532
Dialysis, 611
Diaphragm
    anatomic, 626
    contraceptive, 681
Diatoms, 817–818
Dichloro-diphenyl-trichloro-ethane (DDT), 714–715,
    946
Dicotyledons (dicots), 361, 411, 853
    seedling development, 416, 417f
    versus monocots, 361t, 372
Dieback of population, 992
Diet, *see* Food; Nutrients
Dietary deficiencies, 566–567
Diffusion, 138–139
    facilitated, 141, 143f
DiGeorge's syndrome, 687
Digestion, 568
    intracellular versus extracellular, 576

Digestive systems, 455, 566–584
  evolution of, 576–577, 581
  incomplete versus complete, 576
  in human, 456f, 568–575
    hormones of, 522t
Digestive tract, 568, 861
Dihybrid cross, 238
Dimer of thymine bases, 280
Dimorphism, 727
  sexual, 727–728
Dinoflagellates, 818
Dinosaurs, 771, 772–773
Dioecious plants, 407, 849
Diphosphoglycerate (DPG), 129
Diploid (2N) cells, 199, 214, 833
Directional selection, 725–726
Disaccharides, 70
Discontinuous variation, 240
Disruptive coloration, 968
Disruptive selection, 726–727
Distal convoluted tubule, 608–609, 610
Divergent evolution, 734
Division, in taxonomic hierarchy, 15
DNA, 82, 266–290, 312–332. *See also* Nucleic acids;
        Recombinant DNA
  amplification, 326–327, 329
  base pairs in, 271f, 272
  cloning, 324
  damage to, 280–281, 286–287
  functions, 272–285
    discovery of, 266–267
    gene expression, 277–285
    information storage, 272–275
    inheritance, 275–277
    protein synthesis, 277–285, 286f. *See also*
        Transcription
    as template, 275
  in mitosis, 203–204
  noncovalent bonds in, 57f, 58
  packaging in chromosomes, 273, 274f
  rearrangement, 651, 652f
  regulatory sites, 304
  repeating sequences, 338, 339f
  replication, 275–277
  sequencing, 325–326
    and evolution, 329, 750
    legal aspects, 315
    and turtle migration, 313
  splicing, 321
  structure, 268–272
    double helix, 271f
    Watson-Crick model, 271–272
  unexpressed, 303
DNA "fingerprints," 328
DNA ligase, in recombinant DNA, 323
DNA polymerase, 275–277
DNA probe, 326
DNA repair enzymes, 287
DNA restriction fragments, 325
Domain, in taxonomic hierarchy, 15n
Dominance, 234
  incomplete (partial), 239–240
Dominant allele, characteristic, trait, 234–235,
        236–237
  and genetic disorders, 343–345
Dopamine, 475t, 487
Dormancy in plants, 434
Dorsal hollow nerve cord, 884
Double-blind test, 39
Double bonds, 56, 75
"Double fertilization," 410
Double helix of DNA, 271f
Doves and hawks, 1017
Down syndrome (trisomy 21), 213, 221, 336, 337t
DPG (diphosphoglycerate), 129
Drosophila melanogaster, use in genetics, 249–253

Drugs, *see also* Medicines
  mood-altering, 487
  and pregnancy, 704
Duchenne muscular dystrophy, 337t, 345
Duodenum, 573
Dystrophin, 345

◄ E ►

Ear, 506–509
  bones of, evolution, 459–460
Earth's formation, 762–763
Earthworms, 875–876, 877f
Eating, 568
Ecdysone, 305–306
Echinodermata, 864t, 882–884
Echolocation (sonar), 498–499
Eclipse phase in viruses, 800
Ecological equivalents, 942
Ecological niches, 939–941
Ecological pyramids, 943–946
Ecology, 902–930, 932–958, 960–982, 984–1004
  defined, 905
  ethical issues, 909
  and evolution, 905
Ecosystems, 11, 932–958
  aquatic, 908–916
  energy flow in, 942–946
  linkage between, 934
  nutrient recycling in, 947–952
  structure, 934–939
  succession in, 952–955
Ecotype, 831, 938
Ectoderm, 693
Ectotherms, 614
Edema, 593
Egg (ovum), 669
  human, 673–675
Egg-laying mammals, 892
Ehlers-Danlos syndrome, 337t, 541
Ejaculation, 672
Ejaculatory fluid, 672
Elaters, 846
Electrocardiogram (EKG), 599f
Electron acceptor, 158f, 159
Electron carriers:
  and NAD + conversion to NADH, 177f
  and NADP + conversion to NADPH, 160–161
Electron donor, 129
Electrons, 51, 52–55
  in oxidation, 128–129
  "photoexcited," 156, 159
  in reduction, 128–129
Electron transfer in chemical reactions, 128–129
Electron transport system, 160–161
  and ATP formation, 182–184
  in glycolysis, 177
  in photosynthesis, 160–161, 162
  in respiration, 182–185
Elements, 50–51
Elephantiasis, 873
Ellis–van Creveld syndrome, 720f, 721
Embryo, 686–709
  defined, 688
  development, 688–699
    blastulation, 690
    cleavage, 690
    energy source for, 688
    and evolution, 705
    fertilization, 689
    gastrulation, 692–693
    genetic control of, 686–687
    in humans, 700–705
    implantation, 677
    induction of, 695
    metamorphosis, 699–700

  morphogenesis, 696–697
  neurulation, 694–695
  organogenesis, 696
  in flowering plants, 411
  frozen, 679, 680
  risks to, 704
Embryology, and evolution, 750
Embryonic disk, 703
Embryo sacs of flowering plants, 408
Emigration, 988
Endangered species, 995–996
Endergonic reactions, 122–123
Endocrine glands, 521f, 522t
Endocrine systems, 457, 518–538
  digestion and, 572f
  evolution of, 534, 536
  functions, 520–521
  in humans, 456f, 523–534
  nervous system and, 523–524
  properties, 521–523
Endocytosis, 144–148
Endoderm, 693
Endodermis, 373–374, 376f
Endogenous, defined, 438
Endoplasmic reticulum (ER), 98
Endorphins, 487
Endoskeletons, 546–548
Endosperm, 408, 411
Endosperm mother cell, 408
Endospores, 792
Endosymbiosis hypothesis (endosymbiont theory), 110,
        111f, 962
Endothelin, 591
Endotherms, 614
Energy:
  acquisition and use, 10, 18, 118–123
  biological transfers of, 119f
  in chemical reactions, 121–123
  conversions of, 119, 156–163
    genetic engineering and, 316
  defined, 118
  in electrons, 53–55
  expenditure by organ systems, 457
  flow through ecosystems, 942–946
  forms of, 118
  in human diet, 578
  and membrane transport, 142–144, 145f
  pyramid of, 943–944, 946
  storage of, 118
    and ATP, 121–122, 143–144, 161
    and ionic gradients, 143–144, 145f
  yield:
    from aerobic respiration, 185, 188
    from fermentation, 179
    from glycolysis, 178, 185
    from substrate-level phosphorylation, 185
Energy conservation, law of, 119
Enkephalins, 487
Entropy, 120–121
Environment:
  adaptation to, 4–5
    photosynthesis and, 164–166
  influence on genetics, 241
  and plant growth, 427, 434–438
Environmental resistance to population growth,
        991–992
Environmental science, 909
Enzymes, 10, 123–127
  discovery of, 116–117
  effect of heat on, 127
  effect of pH on, 127
  and evolution, 131
  how they work, 124f, 125
  inhibition of, 126
  interaction with substrate, 124f, 125
  in lysosomes, 100–102

Enzymes (Cont'd)
   of pancreas, 573
   regulation of, 129–130
   of stomach, 570
   synthetic modifications of, 131
Eosinophils, 601–602
Epicotyl, 411
Epidermal cells of plants, 369
Epidermal hairs of plants, 393f
Epidermis of skin, 542
Epididymis, 670
Epiglottis, 569f, 624
Epinephrine (adrenaline), 522t, 529
Epiphyseal plates, 547f
Epiphytes, 845
Epistasis, 240
Epithelial tissue, 452–453
Epithelium, 452
ER (endoplasmic reticulum), 98
Erectile tissue, 669, 670f
Erection, 669
Erythrocytes (red blood cells), 600–601, 627
Erythropoietin, 522t
Escape adaptations, 966t, 969–970
Esophagus, 569
Essential amino acids, 578
Essential nutrients, in plants, 388, 390t
Estivation, 927
Estrogen, 75, 77f, 522t, 532, 675, 676f
Estrus, 668
Estuary habitats, 912–914, 915f
Ethical issues:
   anencephalic organ donors, 489
   facts versus values, 909
   fertility practices, 680
   gene therapy, 349
   genetic screening tests, 350
   patenting genetic sequences, 315
Ethics and biology, 25, 40–41
Ethylene gas, 427t, 432
Etiolation, 438
Eubacteria, 791–796
Euglenophyta, 818–819
Eukaryota (proposed kingdom), 809
Eukaryotes:
   chromosomes in, 273
   evolution of, 764–765
   gene regulation in, 302–308
   versus prokaryotes, 91–92, 787
Eukaryotic cells, 91–93, 95f
   cell cycle in, 201f
   cell division in, 197–199
   components, 91, 94t, 95–107
   evolutionary origin, 110, 111f, 764–765, 787–788,
      789f, 962
   glucose oxidation in, 183f
   respiration in, 180f
Eusociality, 1025–1026
Eutrophic lakes, 916, 917f
Evergreens, 851
Evolution, 4–5, 18–25, 716–740, 742–758, 760–779
   amino acid sequencing and, 84, 750
   of Animal kingdom, 861, 865f, 873
   basis of, 717–731
   chemical, 760–761
   Darwin's principles, 24–25
   DNA sequencing and, 329, 750
   of eukaryotic cell, 110, 111f, 764–766, 787–788,
      789f, 962
   evidence for, 742–758
   of fungi, 826
   genetic changes and, 258, 259, 287, 293, 303
   of humans, 742–743, 752–756, 755f
   of mammals, 892
   of organ systems, 457–460
      circulatory, 603–605

   digestive, 576–577, 581
   endocrine, 534, 536
   excretory, 611–612
   immune, 655
   integument, 543–544
   nervous, 491–493
   respiratory, 636
   skeletomuscular, 560–561
   of osmoregulation, 611–613
   patterns of, 734–735
   of plants, 833–834, 835f
   of primates, 894f
   prokaryotes and, 787, 789f
   protists and, 811–812
   viruses and, 803–804
Evolutionary relationships, 744–746
Evolutionary stable strategy, 1017
Excitatory neurons, 468
Excretion, 607
Excretory systems, 455, 606–613
   evolution of, 611–612
   in humans, 456f, 607–611
Exercise, muscle metabolism in, 186–187
Exergonic reactions, 122
Exocytosis, 99
Exogenous, defined, 438
Exons, 302f, 303, 307
Exoskeletons, 544–546, 877
Experimental group, 33, 37
Experimentation, 32
   controlled, 37–39
Exploitative competition, 963–964
Exponential growth of population, 990–991
Extensor muscles, 552–553
External fertilization, 665
Extinction, 736, 995–996
   acceleration of, 997
   Permian, 770–771
Extracellular digestion, 576
Extracellular matrix, 108, 546
Eye, 503, 505f, 506
Eyeball, 503

◄ F ►

Facilitated diffusion, 141, 143f
Facultative anaerobes, 791
FAD, FADH$_2$ (flavin adenine dinucleotide), 181,
   182f, 185
Fallopian tube (oviduct), 673
   ligation (tieing) of, 681
Familial hypercholesterolemia, 146, 337t
Family, in taxonomic hierarchy, 14
Fat cells, 76
Fats, 75, 188
Fatty acids, 75, 578
Favored chemical reaction, 122
Feathers, 889–890
Feces, 574
Feedback mechanisms, 450
   and enzyme activity, 129–130
   and hormone regulation, 523
Female reproductive system, 673–677
Female sex hormones, 675, 676f, 677
Fermentation, 178–179
   and exercise, 187
   versus aerobic respiration, 175, 178
Ferns, 847, 848f
Fertility rate, 999
Fertilization, 689
   in flowering plants, 408–410
   in humans, 677
Fertilization membrane, 689
Fetal alcohol syndrome, 704
Fetus, 700
   genetic screening tests, 347–350

   risks to, 704
Fever blisters, 796, 803
Fibrin, 602
Fibrinogen, 602
Fibrous root system, 373, 374f
Fighting, 1017, 1018–1019
Filament of flower, 403
Filamentous fungi, 820–821, 822
Filial imprinting, 1013–1014
Filter feeders, 577, 581
   and evolution of gills, 636
Fire algae, 818
First Family of humans, 754, 755f
Fishes, 886–887
   air-breathing, 636
   heart of, 605f
   sense organs of, 512
Fission, 664
Fitness, 1022
Fixed action patterns, 1009
Flagella:
   of bacteria, 792, 793f
   of cell, 106–107
   of protozoa, 814–816
Flatworms, 864t, 869–871, 872
Flavin adenine dinucleotide (FAD, FADH$_2$), 181,
   182f, 185
Fleming, Sir Alexander, 35–37
Flexor muscles, 552–553
Flight, adaptations for, 890, 891f
Florigen, 434
Flowering, 400–401, 437
Flowering plants, 852–853
   asexual (vegetative) reproduction, 417–418
   pollination, 402–407
   sexual reproduction, 402–417
Flowers, 378
   color, 239
   structure, 403
   types, 407
   why they wilt, 391
Fluid compartments of body, 588
Fluid-mosaic model of membrane structure, 89
Flukes, 869, 870f, 871, 872
Folic acid, 579t
Follicle of ovary, 673–674
Follicle-stimulating hormone (FSH), 522t, 526
   in females, 675, 676f
   in males, 672
Food, see also Nutrients
   adaptations to, 577, 581
   conversion to nutrients, 568
   sharing of, 1024–1025
   transport in plants, 393–396
Food chains, 942
Food crops, biotechnology and, 316–319, 418–419
Food webs, 942–943, 944f
"Foreskin," 669
Forests, 917–920
Form, relation to function, 15
Fossil dating, 54, 746
Fossil fuels, 949, 950
Fossil record, 746, 747, 749
Founder effect, 720–721
Frameshift mutations, 286
Free radicals, 56, 174
Freeze-fracturing, 89
Freshwater habitats, 914–916
   and osmoregulation, 612–613
Freud, Sigmund, 663
Frogs, 888–889
Fronds, 847
Fructose, 70, 71f
Fruit:
   development of, 412
   dispersal of, 412, 415f

and human civilization, 414
protective properties, 833
ripening, 432
types, 412, 413f
Fruit fly, in genetics, 249–253
Fruiting body of slime mold, 819
FSH, *see* Follicle-stimulating hormone
Function, relation to form, 15
Functional groups, in molecules, 69
Fungi, 820–826
diseases from, 822, 823–824, 825–826
evolution of, 826
imperfect, 825
nutrition of, 822
reproduction in, 822
uses of, 820
Fungus kingdom, 820–826
Fur, 542

◄ G ►

GABA (gamma-aminobutyric acid), 475t
Galactosemia, 337t
Galen, 586–587
Gallbladder secretions, 574
Gametangia, 833
Gametes, 199
in flowering plants, 407–408, 409f
in human female, 674–675
in life cycle, 214, 215f
in meiosis, 217f
Gametophytes, 408, 409f, 833
Gamma-aminobutyric acid (GABA), 475t
Gaseous nutrient cycles, 947–951
Gas exchange, 622–623
alveolar-capillary, 625–626
in lung, 627
in tissues, 627–629
Gastric juice, 570, 572
Gastrin, 522t
Gastrointestinal tract, 568. *See also* Digestive systems;
Digestive tract
hormones of, 522t
Gastropods, 874
Gastrula, 692
Gastrulation, 692–693
in humans, 703
Gated ion channels, 471
Gause's principle of competitive exclusion, 961, 964
Gel electrophoresis, 325, 326f
Gender, *see* Sex
Gene dosage, 256–257, 337
Gene flow, 719
Gene frequency, 717–718
Genentech, 314
Gene pool, 717
Gene products, 239, 338
Generative cell, 408, 409f
Gene regulatory proteins, 295, 298
in eukaryotes, 304
in prokaryotes, 301
Genes, 246–264, 266–290, 292–310. *See also* Alleles;
Mutations
and behavior, 1008–1009, 1010–1011
duplication, 258
exchange of, *see* Crossing over
expression of, 277–285
visualization of, 305–306
homeotic, 686–687
and hormone action, 534, 535f
mapping, 252, 338, 341, 342f
Human Genome Project, 340
rearrangements, 651, 652f
regulation of expression, 292–310
in eukaryotes, 302–308
need for, 294

in prokaryotes, 298–301
selective expression, 294, 295f
split, 302f, 303
structural, 299, 300f-301f
transfer between bacteria, 804
transposition, 293
Gene therapy, 348–349
Genetic code, 272, 278–279, 282f
Genetic disorders, 334–350, 337t. *See also*
Chromosomes, abnormalities
gene therapy for, 348–349
tests for, 346–350
in fetus, 347–350
X-linked, 254
Genetic drift, 719, 720–721
Genetic engineering, *see* Biotechnology;
Recombinant DNA
Genetic equilibrium, 718
Genetic information
and living organism, 10
storage in DNA, 272–275
universality of, 18
Genetic markers, 249
in gene mapping, 338, 342f
Genetic mosaic, 256f, 257
Genetic recombination, 216, 218f, 219f
Genetics, 230–244, 246–264, 266–290, 292–310,
312–332, 334–353. *See also* Inheritance
Mendelian, 232–239
Genetic variability, 5
and meiosis, 214–216, 219
and sexual reproduction, 665
Genital herpes, 796
Genitalia, in human:
female, 673
male, 669
Genome, defined, 340
Genotype, 235
Genus:
in binomial system, 13
in taxonomic hierarchy, 14
Geography, and evolution, 750–752
Geologic time scale, 764, 765f
German measles, and embryo, 704
Germ cells, 214
Germination, 416
Germ layers, 693
Germ theory of disease, 785
Giant (polytene) chromosomes, 252–253, 305
Gibberellins, 427t, 431–432
Gibbon, 894f
Gill filaments, 633
Gill(s), 632–633, 634f
evolution of, 636
slits, 706, 884
Ginkgophyta, 839t, 849
Girdling of plants, 384–385
Gizzard, 570
Glans penis, 669
Global warming, 950
Glomerular filtration, 609–610
Glomerulus, 608
Glottis, 569f, 624
Glucagon, 522t, 531, 532f, 533–534
Glucocorticoids, 526, 528
Glucoreceptors, 572
Glucose, 70, 71f
blood levels, 531, 532f
energy yield from, 175, 185, 188
oxidation, 174–188
in photosynthesis, 163–164
Glycerol, 75
Glycine, 475t
Glycogen, 72–73
Glycolysis, 175–178, 176f
energy yield from, 185

Glycoproteins of plasma membrane, 94
Gnetophyta, 839t, 849
Goiter, 529, 531
Golgi complex, 98–99, 100
Gonadotropic hormones (gonadotropins), 522t, 526
in females, 675, 676f
in males, 672
Gonadotropin-releasing hormone (GnRH):
in females, 675, 676f
in males, 672
Gonads, 532
Gondwanaland, 769
Gonorrhea, 794, 795
Gorilla, 893f, 894f
Gradualism, 737
Graft rejection, 650
Grain, and human civilization, 414
Granulocytes, 601–602
Grasshopper, 878, 879f
Grasslands, 922–923
Graves' disease, 531
Gravitropism, 438
Gray crescent, 692
Gray matter:
of brain, 481–482, 481f
of spinal cord, 488f, 489
Green algae, 835, 837–838
Greenhouse effect, 798, 950
Grooming of others, 1022
Ground tissue system of plants, 365, 367f, 372
Group courtship, 1018
Group defense responses, 970
Group living, 1018
Growth, 10
in animals, 686–709
in plants, 362, 424–440
of populations, 986, 988–1001
Growth curves, 991, 993–994
Growth factor, 202
Growth hormone (GH; somatotropin), 522t, 525,
527f, 534f
athletes and, 528
Growth regulators, 432
Growth ring, 370
Guanine, 269, 270f
base pairing, 271f, 272
Guard cells, 368
Guilds, 941
Gut, 568, 861
Guthrie test for phenylketonuria (PKU), 335
Guttation, 391
Gymnosperms, 847, 849

◄ H ►

Habituation, 483, 1011
Hagfish, 886–887
Hair, 542, 892
Hair cells of ear, 506, 507f, 508
Half-life of radioisotopes, 54
Halophilic bacteria, 796
Haplodiploidy, 1025–1026
Haploid (1N) cells, 199, 214, 216, 833
"Hardening" (narrowing) of the arteries, 75, 146
Hardy-Weinberg law, principle, 718, Appendix C
Harvey, William, 586–587
Haversian canals, 547f
Hawks and doves, 1017
Hearing, 506–508
in fishes, 512
Heart, 593–600
blood flow in, 593f
changes at birth, 705
evolution of, 605f
excitation and contraction, 598–599
pulse rate, 599–600

Heart (Cont'd)
  valves of, 597
Heart attack, 590, 597, 602–603
Heartbeat, 597–600
"Heartburn," 569
Heat, plant adaptations to, 164–166
Helper T cells, 649
Helping behavior, 1023–1024
Hemocoel, 861, 878
Hemoglobin:
  and gene duplication, 258, 302f
  as oxygen carrier, 627
  shape, 80f
  in sickle cell anemia, 81, 82f
Hemophilia, 337t, 602
  gene therapy for, 349
  inheritance of, 254–255
Herbaceous plants, 362
Herbivore, 966
Herbivory, 966
Heredity, defined, 232. *See also* Inheritance
Hermaphrodite, 667
Heroin, 487
Herpes virus, 796, 803
Heterocyst, 796
Heterosporous plants, 845
Heterotrophs, 155
Heterozygous individual, 235
Hierarchy, taxonomic, 14–15
Hierarchy of life, 11
Hinge joint, 550
Hippocampus, 483, 486
Hirudineans, 875–876
Histocompatibility antigens, 650
Histones, 273, 274f
HIV *see* Human immunodeficiency virus
Homeobox, 687
Homeostasis, 10, 15, 448–449
  and circulatory system, 588
  effect of environment, 455
  and energy expenditure, 457
  mechanisms for, 450
Homeotic genes, 686–687
Hominids, 742–743, 752–756, 755f
Homo erectus, 743, 752, 755f
Homo habilis, 753, 755f
Homologous, versus analogous, 458, 744–746
Homologous chromosomes, 199
  in genetic recombination, 216, 218f
  independent assortment, 219, 220f
Homology, 705, 744–746
Homoplasy, 744
Homo sapiens, 755f
Homo sapiens neanderthalensis, 755f
Homo sapiens sapiens, 752f, 755f
Homosexuality, hypothalamus and, 663
Homosporous plants, 843
Homozygous individual, 235
Hookworms, 873
Hormones, 518–538
  for birth control, 678–679
  discovery of, 521–522
  in labor, 705
  list of, 522t
  of plants, *see* Plants, hormones
  receptors for, 533–534, 535f
  regulation of, 523
  in reproduction, 669
    female, 675, 676f
    male, 672–673
  types, 522
  and urine formation, 610–611
Horsetails, 846–847
Hot dry environment, 164–166. *See also* Desert
Human chorionic gonadotropin (HCG), 677
Human Genome Project, 340

Human immunodeficiency virus (HIV; AIDS virus), 643, 794, 797, 803
  discovery of, 642–643
  and helper T cell, 649
  in pregnancy, 704
Humans:
  body cavities, 863f
  body systems, 456f
    circulatory, 456f, 588–600
    digestive, 456f, 568–575
    endocrine, 456f, 523–534
    excretory, 456f, 607–611
    immune, 456f, 642–660
    integumentary, 456f
    lymphoid (lymphatic), 456f, 646f
    nervous, 456f, 481–491
    reproductive, 456f, 667–677
    respiratory, 456f, 623–627
  chromosome number, 212–213
  embryonic development, 700–705
  evolution, 742–743, 752–756, 755f
    First Family, 754, 755f
  genetically engineered proteins, 314, 315t
  genetic disorders, 334–350
  nutrition, 578–580
  population growth, 997–1001
    doubling time, 998
    earth's carrying capacity and, 999–1001
    fertility rate and, 999
  as research subjects, 39–40
  skeleton, 548–551
  vestigial structures, 749
Hunger, 572
Huntington's disease, 337t, 343–345
  screening test for, 347
Hybridization, 733–734
Hybridomas, 654
Hybrids, 233
Hydras, 864t, 867–869
Hydrocortisone (cortisol), 522t, 526, 528
Hydrogen bonds, 58
  and properties of water, 59
Hydrogen ion, and acidity, 61–62
Hydrologic cycle, 947–948
Hydrolysis, 69, 70f
Hydrophilic molecules, 59
Hydrophobic molecules, 57f, 58
Hydroponics, 387–388
Hydrosphere, 904
Hydrostatic skeletons, 544
Hydrothermal vent communities, 167
Hypercholesterolemia, 146, 337t
  and membrane transport, 136–137
Hypertension, 590, 591, 610
Hypertonic solution, 140, 141f
Hyperventilation, 632
Hyphae, 820
Hypocotyl, 411
Hypoglycemia, 578
Hypothalamus, 481f, 486
  and body temperature, 449
  in human sexuality, 663
  and pituitary, 524, 525f
  and urine volume, 611
Hypothesis, scientific, 32, 33
Hypotonic solution, 140, 141f

◄ I ►

Ice, properties of, 60–61
Ice-nucleating protein, 316
Icons for themes in this book, 15
Ileocecal valve, 576f
Immigration, 988
Immune response, 650
  secondary, 653

Immune system, 457, 642–660
  components, 647
  disorders of, 656–657
  evolution of, 655
  and fetus, 705
  mechanics of, 646–647, 649–650
  and viral diseases, 803
Immune tolerance, 654
Immunity:
  active, 653
  passive, 655
Immunization, 646, 654–655
Immunodeficiency disorders, 657, *See also* AIDS
Immunoglobulins, 650. *See also* Antibodies
Immunological memory, 652–653
Immunological recall, 653
Immunotherapy for cancer, 648
Imperfect flower, 407
Imperfect fungi, 825
Implantation of embryo, 677
Impotence, 669
Imprinting, 1013–1014
Inbreeding, 730–731
Inclusive fitness, 1022
Incomplete (partial) dominance, 239–240
Incomplete flower, 407
Incubation period, 643
Independent assortment:
  of homologous chromosomes, 219, 220f
  Mendel's law of, 238–239
Indoleacetic acid, 430
Inducer of operon, 300f
Inducible operon, 300f, 301
Inductive reasoning, 40
Indusium, 847
Industrial products, genetic engineering and, 316
Infertility, 677, 679, 680
Inflammation, 645
Inheritance:
  autosomal dominant, 343–345
  autosomal recessive, 343
  and genetic disorders, 343–346
  polygenic, 240–241
  predicting, 235–237
  sex and, 253–257
  X-linked, 345–346
Inhibiting factors of hypothalamus, 524
Inhibitory neurons, 468
Initiation (start) codon, 281, 284
Initiator tRNA, 284
Innate behavior, 1009–1010
Insecticides, *see also* Pesticides
  genetically engineered, 317
  neurotoxic, 478
  plant hormones as, 434, 975
Insects, 864t, 876, 878–881
  metamorphosis in, 305–306
  respiration in, 633, 635f
Insight learning, 1012–1013
Insulin, 522t, 531, 532f
  discovery of, 518–519
  genetically engineered, 314
Integument, 542–544
Integumentary systems, 456f, 457
Interactions between animals, types and outcomes, 963t
Intercellular junctions, 109–110
Intercostal muscles, 627
Interference competition, 963–964
Interferon, 645
Intergene spacers, 302f
Interkinesis, in meiosis, 219
Interleukin II, 649
Internal fertilization, 665
Interneurons, 470
Internode of plant stem, 377
Intertidal habitats, 912, 913f, 914f

Intestinal secretions, 572–573
Intestine:
　large 574–575
　small, 572–574
Intracellular digestion, 576
Intrauterine device (IUD), 679, 681
Intrinsic rate of increase (r_o), 990
Introns, 302f, 303, 306–307
Invagination hypothesis of eukaryote origin, 110, 111f, 765
Invertebrates, 863–886
In vitro fertilization, 679
Iodine, and thyroid, 529
Ion, defined, 58
Ion channels of plasma membrane, 139, 140f, 471
Ionic bonds, 57, 58
Ionic gradient, 143–144, 145f
Irish potato blight, 820
Iron in diet, 580, 580t
Islets of Langerhans, 518, 531, 532f
Isolating mechanism, 732, 733t
Isotonic solution, 140, 141f
Isotope, 52
IUD (intrauterine device), 679, 681

◄ J ►

Jaundice, 574
Java Man, 743
Jaws, 887
Jellyfish, 864t, 867–869
Jenner, Edward, 646
"Jet lag," 533
Joints, 550–551
J-shaped curve, 991
"Jumping genes," 293
Junctions between cells, 109–110

◄ K ►

K (carrying capacity) of environment, 992
　K-selected reproductive strategies, 995–996
Kangaroo, 893
Kelp, commercial uses, 840
Kidney, 607–611
　artificial, 611
　function, 609–610
　　regulation of, 610–611
　hormones of, 522t
Kidney failure, 611
Killer (cytotoxic) T cells, 648
Kinetic energy, 118
Kinetochore, 204f, 205–206, 217f, 219f
　versus centromere, 205n
Kingdoms, 13f, 14f, 15
　Animal, 858–897
　Fungus, 820–826
　Plant, 830–856
　Monera (prokaryotes), 786–800
　Protist (Protoctista), 810–819
　taxonomy of, 808–809
Kin selection, 1022–1023
Kinsey, Alfred, 663
Klinefelter syndrome, 337, 337t
Knee joint, 551f
Koala, 892f, 893
Koch, Robert, 785
Krebs cycle, 180–182
Kwashiorkor, 579

◄ L ►

Labia, 673
Labor, 705
Lac (lactose) operon, 300f, 301
Lactase, 574
Lacteals, 574
Lactic acid fermentation, 178, 187

Lactose, 70
Lake habitats, 915–916, 917f, 918
Lamellae:
　of bone, 547f
　of gills, 633
Lampreys, 886
Lancelets, 864t, 884–886
Language, 485
Lanugo, 703
Large intestine, 574–575
Larva, 688
Larynx, 624–625
Latent viral infection, 803
Lateral roots, 373, 374f
Laurasia, 769
Laws, principles, rules:
　Chargaff's, 272
　competitive exclusion, 941, 961
　energy conservation, 119
　Hardy-Weinberg, 718, Appendix C
　independent assortment, 238–239
　the minimum, 938–939
　natural selection, 24–25
　segregation, 235
　stratigraphy, 746
　thermodynamics, 119–120
LDLs (low-density lipoproteins), 136–137, 146
　receptors for, 137, 146, 147f
Leaf, 377–378
　and carbohydrate synthesis, 165–166
　simple versus compound, 377
Leaflets, 377
Leaf nodules, 396
Leakey: Louis, Mary, Richard, 753
Learning, 483, 485, 1011–1014
Leeches, 875–876
Lek display, 1018
Lemur, 894f
Lens of eye, 503, 505f
Lenticels, 369
Lesch-Nyhan syndrome, 336, 337t
Leukocytes (white blood cells), 601–602
LH, *see* Luteinizing hormone
Lichens, 821
Liebig's law of the minimum, 938–939
Life:
　origin of, 760–761, 763–764
　properties of, 6–7, 8f-9f
Life cycle, 214, 215f
Life expectancy, and population density, 989
Ligaments, 551
Light:
　energy from, 156–163
　and plant growth, 435–438
Light-absorbing pigments:
　of eye, 506
　of plants, 156, 157–159
Light-dependent photosynthetic reactions, 155f, 156–163
Light-independent (dark) photosynthetic reactions, 155f, 156, 158f, 163–166
Lignin, 109
Limbic system, 486–487
Limiting factors, 938–939
Linkage groups, 248–249
Lipases, 573
Lipid bilayer of plasma membrane, 89, 138
Lipids, 71t, 75
　in human diet, 578
Lithosphere, 904
Liver, 574
Living organisms
　classification of, 11–15
　properties of, 6–11
Lizards, 889
Lobes of cerebrum, 482
Lobsters, 881

Locomotion, 558–559
Logistic growth, 994
Long-day plants, 400, 437
Loop of Henle, 608–609, 610
Loris, 894f
Lotka-Volterra equation, 961
Low-density lipoproteins (LDLs), 136–137, 146
　receptors for, 137, 146, 147f
Lucy (hominid), 753–754, 755f
Lungs, 623, 625–626, 635
　changes at birth, 705
　evolution of, 636
　gas exchange in, 627
Lupus (systemic lupus erythematosus), 657
Luteinizing hormone (LH), 522t, 526
　in females, 675, 676f
　in males, 672
Lycophyta (lycopods), 839t, 845–846
Lyme disease, 792
Lymphatic systems, 456f, 606
Lymph nodes, 606
Lymphocytes, 647. *See also* B cells; T cells
Lymphoid tissues, 646f, 647
Lysosomes, 100–102

◄ M ►

Macroevolution, 731
Macrofungi, 821
Macromolecules, 69, 70f, 71t
　evolution of, 83–84
Macronutrients, in plants, 388, 390t
Macrophages, 602, 647
Magnesium, in diet, 580, 580t
Malaria, 813, 815f
　and sickle cell anemia, 723f, 724, 812
Male reproduction, 669–673
　hormonal control, 672–673
Male sex hormones, 672–673
Malignancy, *see* Cancer; Tumor
Mammals, 892–894
　Age of, 774
　egg-laying, 892
　evolution of, 892
　heart in, 605f
Mapping of genes, 252, 338, 341, 342f
　Human Genome Project, 340
Marine habitats, 908–912
　and osmoregulation, 612
Marrow of bones, 546, 547f
Marsupials, 892f, 893
Mastigophora, 815t
Mating, nonrandom, 719, 727–728, 730–731
Mating types of algae, 837
Matter, defined, 51
McClintock, Barbara, 292–293
Mechanoreceptor, 500, 501t
Medicines, genetically engineered, 314, 321
Medulla, 481f
Medusas, 868–869
Megaspores, 407, 408
Meiosis, 199, 212–225
　chromosomal abnormalities and, 221
　defined, 214
　and genetic variability, 214–216, 219
　in human female, 674–675
　importance, 214–216
　reduction division in, 216
　stages, 216–221
　versus mitosis, 200f, 216, 222
Meiosis I, 216, 217f
Meiosis II, 217f, 219, 221
Meissner's corpuscles, 502f
Melanoma, 280–281
Melatonin, 522t, 533
Membrane invagination hypothesis, 110, 111f, 765

Membrane potential, 470–471
Membranes, 136–150
    of cell, *see* Plasma membrane
    in cytoplasm, 97–98
    of ear (eardrum), 507f, 508
    of mitochondria, 102–104
    of nucleus (nuclear envelope), 97
    organization in, 160
    transport across, 136–144
        active, 142–144
        passive diffusion, 138–139
Memory, 483
    cells, 653
Mendel, Gregor, 232–235
Mendelian genetics, 232–239
    exceptions to, 239–241
Mendel's law of independent assortment, 238–239
Mendel's law of segregation, 235
Meninges, 482
Menopause, 675
Menstrual (uterine) cycle, 675, 677
Menstruation, 676f, 677
Meristem, 362–363
    of root, 373, 375f
    of stem, 372
Mesoderm, 693
Mesophyll, 377
Mesophyll cells, and photosynthesis, 165f, 166
Mesotrophic lake, 916, 917f
Mesozoic Era, 771, 774
Messenger RNA (mRNA)
    transcription, 278, 279f
        primary transcript processing, 306–307
    translation, 279–280, 284–285, 286f
        and control of gene expression, 307–308
Metabolic intermediates, 127
Metabolic ledger, 185, 188
Metabolic pathways, 127–130
    and glucose oxidation, 188–189
Metabolism:
    as property of living organism, 10
    regulation of, 129–130
Metamorphosis, 699–700
    in insects, 881
        ecdysone and, 305–306
Metaphase, of mitosis, 203f, 205–206
Metaphase I, of meiosis, 217f, 218–219
Metaphase II, of meiosis, 217f, 221
Metaphase plate, 217f, 219
Methane, 55f, 56
Methane-producing bacteria, 796, 798
Metric conversion chart, Appendix A
Microevolution, 731
Micronutrients, in plants, 388, 390t
Micropyle, 408
Microscopes, Appendix B
Microspores, 407, 408, 409f
Microvilli, 574, 575f
Midbrain, 481f
Milk, 892
    intolerance of, 574
Millipedes, 876, 882
Mimicry, 971
Mineralocorticoids, 528
Minerals:
    absorption in plants, 387–391
    in human diet, 580
Minimum, Liebig's law of the, 938–939
"Missing link," 743
Mites, 876
Mitochondria, 102–104
    and ATP synthesis, 172–173, 184
    and origin of eukaryote cell, 110, 765, 962
Mitosis:
    defined, 199
    phases, 202–206

versus meiosis, 200f, 216, 222
Mitosis promoting factor (MPF), 195, 202
Mitotic chromosome, 273, 274f, 275f
Molds, 821
Molecular biology, 268, 750
Molecular defenses
    nonspecific, 645
    specific (antibodies), 647
Molecules, 55, 57
    chemical structure, 69
Mollusca, 864t, 874–875
Molting, 878
Monera kingdom, 786–800. See also Bacteria; Prokaryotes
Monkeys, 893, 894f
Monoclonal antibodies, 654
Monocotyledons (monocots), 361, 411, 853
    seedling development, 417, 418f
    versus dicots, 361t, 372
Monocytes, 602
Monoecious plants, 407, 851
Monohybrid cross, 233, 234f
Monomer, 69, 70f
Monosaccharides, 70
Monotremes, 892–893
Mood-altering drugs, 487
Morgan, Thomas Hunt, 249–250
"Morning after" pill, 679
Morphine, 487
Morphogenesis, 696–698
Morphogenic induction, 695
Morphological adaptation, 728
Mortality, and population density, 988, 989
Mosaic, genetic, 256f, 257
Mother cell, 196
Motor cortex, 485
Motor neurons, 470
Mouth, in digestion, 568–569
MPF (mitosis promoting factor), 195, 202
mRNA, *see* Messenger RNA
Mucosa of digestive tract, 568
Mud flats, 912, 913f
Multicellular organisms, origin of, 765–766
Multichannel food chains, 942, 944f
Multiple fruits, 412, 413f
Multiple sclerosis, 473
Muscle fiber, 186–187, 553, 554f
Muscles, 454, 551–558
    cardiac, 558, 559f
    contraction of, 555–557
    in nerve impulse transmission, 474f, 475t, 476f,
        479–480, 556
    skeletal, *see* Skeletal muscle
    smooth, 558
Muscular dystrophy, 345–346
Muscular systems, 456f, 457
Mushrooms, 824–825
Mussels, 874
Mutagens, 252, 286–287
Mutant alleles, 241
Mutations, 241–242, 719
    and genetic diversity, 242
    molecular basis, 286–287
    point mutations, 241
Mutualism, 978
Mycorrhizae, 386–387, 388
Myelin sheath of axon, 469f, 470, 473
Myofibrils, 553, 554f
Myosin, 555
Myriapods, 876, 881–882

◁ N ▷

NAD⁺, NADH (nicotinamide adenine dinucleotide):
    energy yield from, 185
    in fermentation versus respiration, 178
    in glycolysis, 176f, 177

in Krebs cycle, 181, 182f
NADP⁺, NADPH (nicotinamide adenine dinu-
        cleotide phosphate):
    as electron donor, 129
    in photosynthesis, 159f, 161
    as reducing power, 189
Names, scientific, 13
Nasal cavity, 624
Natal homing, 312–313
Natality, 988
Natural killer (NK) cells, 645
Natural selection, 23–25, 719, 721–727
    patterns of, 724–727
    pesticides and, 715
    principles of, 24–25
Neanderthals, 742, 755f
Nectaries, 405
Nematoda, 864t, 871–873
Neodarwinism, 717
Nephron, 607–609
Nerve cells, *see* Neurons
Nerve gas, 478
Nerve growth factor (NGF), 466–467
Nerve impulse transmission, 470–473, 556
Nerve net, 492
Nerves, 479
Nerve tissue, 454f, 455
Nervous systems, 457, 466–496, 498–516
    central, 479, 481–489
    evolution, 491–493
    of human, 456f, 481–491
    origin of, 694–695
    peripheral, 479, 489–491
    role in digestion, 572
    of vertebrates, 479–481
Neural circuits, 479–481
Neural plate, 694
Neural tube, 694, 884
Neuroendocrine system, 523–524
Neuroglial cells, 470
Neurons, 468–470
    form and function, 468, 469f
    target cells, 468
    types, 470
Neurosecretory cells, 523, 525f
    in evolution, 534, 536
Neurotoxins, 478, 818
Neurotransmission (synaptic transmission), 474–479
Neurotransmitters, 474–477, 475t
Neurulation, 694–695
Neutrons, 51
Neutrophils, 601–602
Niche overlap, 941, 963
Niches, 939–941
Nicotinamide adenine dinucleotide, *see* NAD⁺,
        NADH
Nicotinamide adenine dinucleotide phosphate, *see*
        NADP⁺, NADPH
Nicotine, 630
Nicotinic acid (niacin), 579t
Nitrifying bacteria, 950
Nitrogen, excretion and, 607, 613
Nitrogen cycle, 949–951
Nitrogen fixation, 390–391
    by bacteria, 390, 799, 949–951
    by cyanobacteria, 796
    genetic engineering and, 316
Nitrogenous bases:
    in DNA, 269–271
    in RNA, 82, 83f, 278
Nodes of plant stem, 377
Nodes of Ranvier, 469f, 470, 473
Nomenclature, binomial system, 13
Noncovalent bonds, 57–58
Noncyclic photophosphorylation, 160f, 161, 162t
Nondisjunction, 336

Nongonococcal urethritis, 795
Nonpolar molecules, 56–57
Nonrandom mating, 719, 727–728, 730–731
Nonsense (stop) codons, 281, 285
Nonvascular plants, 839t
Norepinephrine (noradrenaline), 475t, 522t, 529
Norplant, 678
Nostrils, 624
Notochord, 694, 884
Nucleases, 573
Nucleic acids, 71t, 81–83. *See also* DNA; RNA
  sequencing of, 325
  structure, 82–83, 269–271
  in viruses, 800
Nucleoid of prokaryotic cell, 92
Nucleoli, 97
Nucleoplasm, 97
Nucleosome, 273, 274f
Nucleotides, 82–83
  in DNA, 269–271
  and genetic code, 278
  and genetic information, 272
  and protein assembly, 277
  in RNA, 278
  structure, 269–271
Nucleus, 92, 95–97
  envelope (nuclear membrane), 97
  matrix, 97
  transplantation of, 297
Nutrient cycles, 947–952
Nutrients, *see also* Food
  in human diet, 578–580
  in plants, 388, 390t
Nutritional deficiencies:
  in humans, 566–567
  in plants, 390t

◄ O ►

Obesity, 76
Obligate anaerobes, 791
Ocean:
  ecosystems, 908
  floor, hydrothermal vent communities in, 167
Octopuses, 874–875
Oils, 75
Olfaction, 509, 510
Oligochaetes, 875–876
Oligotrophic lake, 916, 917f
Omnivore, 966
Oncogenes, 202
Oocyte, 673, 674
Oogenesis, in humans, 673–675
Oogonia, 674, 675f
Open growth of plants, 362
Operant conditioning, 1012
Operator region of operon, 299
Operon, 299–301
Opiates, 487
Opossum, 893
Optimality theory, 1016–1017
Oral contraceptives, 678–679
Orangutan, 894f
Order, in taxonomic hierarchy, 15
Organ donations, ethical issues, 489
Organelles, 91, 95–107
Organic chemicals, defined, 68
Organogenesis, 696, 697–698
Organs, defined, 455
Organ systems, 455–460
  energy expenditure, 457
  evolution of, 457–460
  in humans, *see* Humans, body systems
  types, 455–457
Organ transplants, 650
Orgasm, 672

Origin of Species, 24
Osmoregulation, 606–613
  adaptations of, 611–613
  defined, 607
Osmoregulatory (excretory, urinary) systems, 455, 456f
Osmosis, 140–141
Osteichthyes, 888f
Osteoblasts, 550
Osteoclasts, 550
Osteocytes, 546, 547f
Osteoporosis, 550
Ovarian cycle, 675
Ovaries of flower, 403
Ovaries of human, 522t, 532, 673–674
Overstory of forest, 917
Oviduct (fallopian tube), 673
Ovulation, 674, 676f
Ovules, 408, 852
Ovum, 669
  of human, 673–675
Oxidation, 128–129
Oxygen:
  appearance on earth, 174, 764
  as electron acceptor, 182, 184
  release into tissues, 594f, 627–629
  toxic effects, 174
  transport, 600–601
  uptake in lungs, 594f, 627
Oxyhemoglobin, 627
Oxytocin, 522t, 525
Oysters, 874
Ozone layer, 902–903

◄ P ►

PABA (para-aminobenzoic acid), 126
Pacemaker of heart, 110, 598, 599
Pacinian corpuscle, 502
Pain receptor, 500, 501t
Paleozoic Era, 767–771
Palisade parenchyma, 378
Pancreas, 522t, 531–532, 573
Pancreatitis, 573
Pangaea, 769
Panting, 614
Pantothenic acid, 579t
Paper chromatography, 153
Parallel evolution, 734f, 735
Parapatric speciation, 732
Parasitism, 872, 972–974
Parasitoids, 972
Parasympathetic nervous system, 490–491
Parathyroid glands, 522t, 530f, 531
Parathyroid hormone (PTH), 522t, 531
Parenchyma cells, 363, 364f
Parthenocarpy, 418
Parthenogenesis, 664
Partial (incomplete) dominance, 239–240
Parturition (birth), 705
Passive immunity, 655
Pathogenic bacteria, 792
Pauling, Linus, 30–31, 39–40, 67
Peppered moth, camouflage changes, 4–5, 722–723
Pedicel, 403
Pedigree, 255, 256f
Peking Man, 743
Pelagic zone, 908
Penicillin, 35–37, 126
Penis, 665, 669, 670f
PEP (phosphoenolpyruvate) carboxylase, 166
Pepsin, 570
Pepsinogen, 570
Peptic ulcer, 570, 572
Peptide bonds, 78f, 80
Peptidoglycan, 787

Perennial plants, 361, 362–363, 366
Perfect flower, 407
Perforation plate in xylem, 370
Perforins, 647
Pericycle, 373, 376f, 377
Periderm, 369
Periosteum, 547f
Peripheral nervous system, 479, 489–491
Peristalsis, 569, 571f
Permafrost, 922
Pesticides, *see also* Insecticides
  bioconcentration of, 946
  fungi as, 823
  genetically engineered, 317
  natural selection and, 715
Petals, 403
Petiole of leaf, 377
PGA (phosphoglyceric acid), 163–164
PGAL (phosphoglyceraldehyde), 129, 164
pH, 61–62
  and blood gas exchange, 629
Phaeophyta, 838
Phage (bacteriophage), 268, 269f, 801
Phagocytic cells, 601–602, 644f, 645
Phagocytosis, 144–147
Pharyngeal gill slits, 706, 884
Pharynx, 569, 624, 636
Phenotype, 235
  environment and, 241
Phenylketonuria (PKU), 334–335, 337t
Pheromones, 1021
Philadelphia chromosome, 259, 260
Phimosis, 669
Phloem, 369–370, 372
  and food transport, 393–396
Phosphoenolpyruvate (PEP) carboxylase, 166
Phosphoglyceraldehyde (PGAL), 129, 164
Phosphoglyceric acid (PGA), 163–164
Phospholipids, 71t, 75, 77f
  in plasma membrane, 77f, 89, 138
Phosphorus, in diet, 580, 580t
Phosphorus cycle, 951–952
Phosphorylation, 175, 177–178. *See also* Photophosphorylation
"Photoexcited" electrons, 156, 159
Photolysis, 160
Photons, 156
Photoperiod, 434
  plant responses to, 400–401, 435–438
  and reproduction, 668
Photoperiodism, 435
Photophosphorylation, 161
  cyclic, 161–163, 162t
  noncyclic, 160f, 161, 162t
Photoreceptor, 500, 501t
Photorespiration, 166
Photosynthesis, 152–170
  in algae, 816
  ATP in, 160f, 161–162
  in bacteria, 791–792
  central role, 154
  early studies, 153–154
  and energy of electrons, 54f, 55
  light-dependent reactions, 155f, 156–163
  light-independent (dark) reactions, 155f, 156, 158f, 163–166
  overview, 155–156, 158f
  as two-stage process, 155f, 156
Photosynthetic autotrophs, 154
Photosynthetic pigments, 156, 157–159
Photosystems, 159
Phototropism, 438
Phyletic speciation, 732
Phylogenetic relationships, 14
Phylum, in taxonomic hierarchy, 15
Physical defenses, 970–971

Phytochrome, 435–436
  and flowering, 437
  and photoperiodism, 436
  and seed germination, 438
  and shoot development, 438
Phytoplankton, 816, 908
Pigmentation in humans, genetics of, 240, 241f
Pigments:
  light-absorbing:
    in eye, 506
    in plants, 156, 157–159
  photosynthetic, 156, 157–159
Piltdown Man, 743, 752–753
Pineal gland, 522t, 533
Pinocytosis, 144, 147–148
Pioneer community, 952
Pistil, 403
Pith of plants, 372
Pith rays, 372
Pituitary gland, 481f, 522t, 524–526
Placebo, 39
Placenta, 700
Placentals, 893
Planaria, 869–871
Planes of body symmetry, 862f
Plankton, 881, 908
Plant fracture properties, 375
Plant kingdom, 830–856
Plants, 358–382, 384–398, 400–422, 424–443,
    830–856. *See also* Flowering plants
  annual, biennial, perennial, 361
  basic design, 360–361, 362f
  carbohydrate synthesis in, 163–166
  cell vacuoles, 102
  cell wall, 108–109
  circulatory system, 384–398
  classification, 839t
  communities, 932–933
  competition between, 961
  cytokinesis in, 206
  desert adaptations, 164–166, 926
  evolution, 833–834, 835f
  fossil, 843, 844f
  genetic engineering of, 316–319, 375
  growth and development, 424–443
    control of, 426–427
    timing of, 434–438
  growth signal in, 424–425
  hormones, 427–434
    discovery of, 424–425
    as insecticides, 434, 975
  longevity, 362–363, 366
  nonphotosynthesizing, 834, 836
  nutrients, 388, 390t
    transport of, 393–396
  organs, 372–378
    flower, 378
    leaf, 377–378
    root, 373–377
    stem, 372–373
  osmosis in, 140–141, 142f
  polyploidy in, 260–261
  sex chromosomes in, 254
  sexual reproduction in, 402–417
  species differences, 830
  transport in, 384–398
Plant tissues, 362–365
  cambium, 362, 372–373
  collenchyma, 364
  parenchyma, 363
  primary, 362
    of root, 373–374, 376f, 377
    of stem, 372
  sclerenchyma, 364–365
  secondary, 362
    of root, 377

  of stem, 372
Plant tissue systems, 365–372, 367f
  dermal, 365, 366–369
  ground, 365, 372
  vascular, 365, 369–372
Plaques, in arteries, 75, 603
Plasma, 600
Plasma cells, 647, 653
Plasma membrane (cell membrane):
  carrier proteins of, 141, 143f
  components, 93–94
  formation, 99
  functions, 93–94
  ion channels, 139, 140f, 471
  permeability, 138, 139
  receptors on, 94, 146, 147–148, 533–534, 535f
  structure, 88–89, 93–94, 138
  transport across, 136–144
    active transport, 142–144
    passive diffusion, 138–139
Plasmid, 321, 322f
Plasmodesmata, 387
Plasmodial slime molds, 819
Plasmodium, 813, 815f
Plasmolysis, 141, 142f
Platelets, 602
Plate tectonics, 767–769
Platyhelminthes, 864t, 869–871
Platypus, 892f, 893
Play, 1013
Pleiotropy, 240
Pleura, 626
Pleurisy, 626
Pneumococcus, role in molecular genetics, 266–267
Pneumocystis infection, 813
Point mutation, 286
Poison ivy, 656
Polar body, 675
Polarity in neurons, 471–473
Polar molecules, 56–57
Pollen, 403
  protective properties, 833
Pollen:
  sacs, 403f
  tube, 409f, 410
Pollination, 402–407, 410f
Pollution
  consequences, 918, 992, 993
  genetically engineered degradation, 316–317
Polychaetes, 875–876
Polygenic inheritance, 240–241
Polymer, 69, 70f
Polymerase chain reaction, 326–327, 329
Polymorphic genes, 338
Polymorphism, 727
Polypeptide chain, 78f, 80
  assembly, *see* Proteins, synthesis
Polyploidy, 260–261
  and speciation, 733
Polyps, 868–869
Polysaccharides, 70–74
Polysome, 285
Polytene (giant) chromosomes, 252–253, 305
Polyunsaturated fats, 75
Pond habitats, 915–916
Pons, 481f
Population density, 986, 987, 996
Population ecology, 984–1004
Population growth, 988–1001
  and earth's carrying capacity, 999–1001
  factors affecting, 988–997
  fertility and, 999
  in humans, 997–1001
    doubling time, 998
    percent annual increase, 998
  rate, 986

Populations, 11, 717
  age–sex structure, 988, 999, 1000f
  defined, 986
  distribution patterns, 987–988
  structure, 986–988
Porifera, 864–867, 864t
Porpoises, and echolocation (sonar), 499
Positional information, 696
Postembryonic development, 699–700
Potassium:
  in diet, 580, 580t
  transport across cell membrane, 142, 144f
Potential (stored) energy, 118
Pragmatism, 41
Prairies, 923
Predation, 966–972
Predator, 966
  adaptations, 967–972
  group behavior, 1018
Predator-prey dynamics, 966–967
Pregnancy, *see also* Embryo; Fetus
  prevention of, 677–681
  stages of, 700–705
Prepuce, 669
Pressure-flow transport of plant nutrients, 395–396
Pressure gradient, and respiration, 626–627
Prey, 966
  defenses, 966t, 967–972
  group behavior, 1018
Primary endosperm cell, 410, 411f
Primary growth of plants, 362
Primary plant tissues, 362
Primary organizer, 695
Primary producers, 937
Primary root tissues, 373–374, 376f, 377
Primary stem tissues, 372
Primary succession, 952, 953–955
Primary transcript, 306–307
Primates, 893–894
  evolution, 894f
Primitive homologies, 746
Prion, 803
Probability, in genetics, 236
Progesterone, 522t, 676f, 677
Prokaryotes, 787–800. *See also* Bacteria
  activities of, 798–800
  fossilized, 763
  gene regulation in, 298–301
  as universal ancestor, 789f
  versus eukaryotes, 91–92, 787
Prokaryotic cell, 91–92, 793f
  cell division in, 197
  chromosomes in, 273
  and origin of eukaryote cell, 110, 765, 787–788
Prokaryotic fission, 197
Prolactin, 308, 522t, 525
Promoter, 299
Prophase, of mitosis, 203–205
Prophase I, of meiosis, 216–218, 217f
Prophase II, of meiosis, 217f, 219
Prosimians, 894f
Prostaglandins, 532, 705
Prostate gland, 672
Prostate specific antigen (PSA), 654
Protein kinases, 129, 195, 202
Proteins, 71t, 79–81
  conformational changes in, 81
  denaturing of, 127
  as energy source, 188
  in human diet, 578
  of plasma membrane, 94, 138
  structure, 66–67, 78f, 79–81
    primary, 80
    quaternary, 80f, 81
    secondary, 80
    tertiary, 80f, 81

synthesis, 277–285, 286f
  chain elongation, 285
  chain termination, 285
  initiation, 284–285
  site of, 98
  transcription in, 277, 278, 279f, 286f
  translation in, 277, 279, 286f
Proteolytic enzymes, 573
Proterozoic Era, 764–766
Prothallus, 847, 848f
Protist kingdom, 810–819
  taxonomic problems, 810
Protists:
  evolution and, 811–812
  reproduction in, 222, 811
Protochordates, 884–886
Protocooperation, 977–978
Protoctista, 809n
Proton gradient, and ATP formation, 160f, 161, 184
Protons, 51, 61
Protoplast manipulation, 418–419
Protostomes, 873
Protozoa, 812–816
  classification, 815t
  motility, 814–816
  pathogenic, 813
  polymorphic, 813–814
Proximal convoluted tubule, 608–609, 610
Pseudocoelom, 861, 871–872
Pseudopodia, 814–816
Psilophyta, 839t, 844f, 845
Psychoactive drugs, 487
Pterophyta, 839t, 847
Pterosaurs, 771
Puberty, in females, 675
Puffballs, 824
Puffing in chromosomes, 305
Pulmonary circulation, 594–595
Punctuated equilibrium, 737
Punnett square, 235–236
Pupa, 881
Purines, 269, 270f
Pus, 645
Pyloric sphincter, 572
Pyramid of biomass, 944
Pyramid of energy, 943–944, 946
Pyramid of numbers, 944
Pyridoxine (vitamin B_6), 579t
Pyrimidines, 269, 270f
Pyrrophyta, 818
Pyruvic acid:
  energy yield from, 185
  in glycolysis, 176f, 178
  in Krebs cycle, 181, 182f

◄ R ►

r (rate of increase), 990
  r-selected reproductive strategies, 995–996
R group, of molecule, 71t, 79–80
  and enzyme-substrate interaction, 125
Radiation, effect on DNA, 280–281, 287
Radicle, 411
Radioactivity, 52
  discovery of, 48–49
Radioisotopes, 54, 746
Rainfall, 906
Rain forests, 428–429, 917, 919, 920f
Rainshadow, 926
Random distribution of population, 987–988
Rate of increase, intrinsic (r_o), 990
Rays, 887, 888f
Reactant, 123
Reaction center, 159
Reaction-center pigments, 158f, 159
Reactions, *see* Chemical reactions

Reading frame, 284
  frameshift mutations, 286
Receptacle of flower, 403
Receptor-mediated endocytosis, 147
Receptors:
  in cytoplasm, 534, 535f
  for hormones, 533–534, 535f
  for LDLs, 137, 146, 147f
  on plasma membrane, 94, 147–148, 533–534, 535f
    and medical treatment, 146
  sensory, 479, 500–502
Recessive allele, characteristic, trait, 234–235, 237
  and genetic disorders, 343, 345–346
Recessiveness, 234
Reciprocal altruism, 1023
Recombinant DNA, *see also* Biotechnology
  amplification by cloning, 324
  defined, 314
  formation and use, 321–325
  medicines from, 314, 315t, 321
Recruitment, 1019
Rectum, 574, 576f
Recycling nutrients in ecosystems, 947–952
Red algae, 838f-839f, 841–842
Red blood cells (erythrocytes), 600–601, 627
Red tide, 818
Reducing power of cell, 129
Reduction, 128–129
Reduction division, in meiosis, 216
Reefs, 911–912
Reflex, 479
  conditioned, 1011
Reflex arc, 479–480
Region of elongation, 372, 375f
Region of maturation, 372, 375f
Regulation of biological activity, 15
Regulatory gene, 299
Releaser of fixed action pattern, 1009
Releasing factors of hypothalamus, 524
Renal tubule, 607–609
Renin, 522t, 591
Replication fork, 276, 277f
Replication of DNA, 275–277
Repressible operon, 300f, 301
Repressor protein, 299, 300f, 301
Reproduction:
  asexual, 199
  asexual versus sexual, 664–667
  communication and, 1018
  prevention of, 677–681
  as property of living organism, 10
Reproductive adaptation, 729
Reproductive isolation, 732
Reproductive (biotic) potential, 986, 989–990
Reproductive strategies, 995
Reproductive systems, 457, 662–684
  human, 456f, 667–677
    female, 673–677
    male, 669–673
Reproductive timing, 668
Reptiles, 889, 890f
  Age of, 771
Research, scientific, 32–35
Resource partitioning, 964–965
Respiration, 174–192
  aerobic, 180–184
    versus fermentation, 175, 178
  and surface area, 622–623, 626
  versus breathing, 175n
Respiratory center, 632
Respiratory surfaces, 622–623
Respiratory systems, 455, 620–640
  adaptation of, 632–635
  evolution of, 636
  in human, 456f, 623–627
Respiratory tract, 623–624

Response to stimuli, as property of living organism, 10
Resting potential, 471
Restriction enzymes, and recombinant DNA, 321–323
Restriction fragment length polymorphisms (RFLPs), 338, 339f
  use in gene mapping, 341, 342f
Restriction fragments, 325
Reticular formation, 486
Retina, 503, 505f, 506
Retinoblastoma, 260
Retinol (vitamin A), 579t
Retrovirus, 643, 803
Rheumatic fever, 657
Rhizoids, 823, 842
Rhodophyta, 841
Rhyniophytes, 843
Rhythm method of contraception, 681
Riboflavin (vitamin B_2), 579t
Ribonucleic acid, *see* RNA
Ribose, 71f, 278
Ribosomal RNA (rRNA), 281–282
Ribosomes, 97, 98
  assembly, 283f
  in protein synthesis, 281–284, 286f
  structure, 282–284
Ribozymes, 306, 307
Ribulose biphoshpate (RuBP), 164
Rigor mortis, 556
River habitats, 914–915
RNA, *see also* Nucleic acids; Transcription; Translation
  enzymatic action, 306, 307
  processing, 306–307
    and control of gene expression, 304, 307, 308f
    self-processing, 307
  splicing, 307
  structure, 82, 278
RNA polymerase, 278, 286f
  and operon, 299
Rods of retina, 503, 505f, 506
Root cap, 373, 375f
Root hairs, 373, 374f
Root nodules, 389–390
Root pressure, 391
Root system of plants, 361, 362f, 373–377
Rough endoplasmic reticulum (RER), 98, 100
Roundworms, 864t, 871–873
rRNA (ribosomal RNA), 281–282
RU486, 678–679
Rubella virus, risk to embryo, 704
RuBP (ribulose biphoshpate), 164
Rumen, 581
Ruminants, 581, 813
"Runner's high," 487

◄ S ►

Saguaro cactus decline, 984–985
Salamanders, 888–889
Saliva, 568–569
Salivary glands, 568
Saltatory conduction, 473
Sand dollars, 882
Sandy beaches, 912, 913f
Sanitation, origins of, 784–785
SA (sinoatrial) node, 598
Saprobes, 822
Saprophytic decomposers, 798
Sarcodina, 815t
Sarcolemma, 554f
Sarcomeres, 553–556
Sarcoplasmic reticulum, 556–557
Saturated fats, 75
Savannas, 923–924
Scales of fish, 543

Scaling effects, 460
Scallops, 874
Schistosomes, 870f, 871
Schizophrenia, 230–231
Schwann cells, 469f, 470
SCID (severe combined immunodeficiency), 337t,
    348–349
Science and certainty, 40–41
Scientific method, 32–35, 39f
    caveats, 40–41
    facts versus values, 909
    practical applications, 35–40
Scientific names, 13
Scientific theory, 35
    versus truth, 40–41
Sclereids, 365
Sclerenchyma cells, fibers, 364–365
Scorpions, 876
Scrotum, 669
Scurvy, 540–541
Sea anemones, 867–869
Sea cucumbers, 882
Sea stars, 864t, 882, 883f, 884
Sea turtles, natal homing, 312–313
Sea urchins, 864t, 882
Seasonal affective disorder syndrome (SADS), 533
Seaweeds, 838, 841–842
Secondary growth of plants, 362
Secondary plant tissues, 362
Secondary root tissues, 377
Secondary sex characteristics
    female, 675
    male, 672–673
Secondary stem tissues, 372
Secondary succession, 952–953, 955
"Second messenger," 533, 535f
Secretin, 522t, 573
Secretory pathway (chain), 99–100, 101f
Sedimentary nutrient cycles, 947, 951–952
Seedless vascular plants, 845–847
Seedling development, 416–417
Seed plants, 847–853
Seeds, 411, 412f
    development, 411
    dispersal, 412, 415f, 416
    dormancy, 416
    germination, 438
    protective properties, 833
    viability, 416
Segmentation of body, 863, 875–876
Segmented worms, 864t, 875–876, 877f
Segregation, Mendel's law of, 235
Selective gene expression, 294, 295f
Self-recognition, 654
Semen, 670, 672
Semicircular canals, 507f, 508
Semiconservative replication, 275
Seminal vesicle, 672
Seminiferous tubule, 669, 671f
Semmelweis, Ignaz, 784–785
Senescence in plants, 432
Sense organs, 498–516
    diversity of, 511–513
    in fishes, 512
    how they work, 500–515
Sense strand of DNA, 278
Sensory cortex, 502, 503f, 511
Sensory neurons, 470
Sensory receptors, 479, 500–502
    types, 500, 501t
Sepals, 403
Sequencing:
    of amino acids, 66–67, 84, 750
    of DNA, 325–326
        and evolution, 329, 750
        and frameshift mutations, 286

legal aspects, 315
    and turtle migration, 313
    of nucleic acid, 325
Sere, 952
Serosa of digestive tract, 568
Serotonin, 475t
Severe combined immunodeficiency (SCID), 337t,
    348–349
Sex:
    age–sex structure, 988, 999, 1000f
    determination of, 253–254, 255
Sex changes in fish, 254
Sex chromosomes, 253
    abnormal number, 336–337
    and gender determination, 255
    and gene dosage, 256–257
Sex linkage, 254–255, 345–346
Sexual dimorphism, 727–728
Sexual imprinting, 1014
Sexually transmitted diseases, 794–797
Sexual orientation, 663
Sexual reproduction
    advantages, 664–665
    in flowering plants, 402–417
    and meiosis, 214, 215f
    strategies, 665–667
Sexual rituals, 668
Sexual selection, 727–728
Sharing food, 1024–1025
Sharks, 887
Shell, 545
Shellfish, 874
    poisoning by, 818
Shivering, 614
Shoot system of plants, 361, 362f
Short-day plants, 400, 437
Shrew, 893
Shrimp, 876
Shrublands, 924, 925f
Sickle cell anemia, 81, 82f, 337t, 346f, 347
    and allele frequency, 717
    codon change in, 286
    and malaria, 723f, 724, 812
    and natural selection, 723–724
Sieve plates, 370
Sieve-tube members, 370
Sight, 502–506, 512
Sigmoid (S-shaped) growth curve, 993–994
Sign stimulus, 1009
Silent Spring, 715
Silicosis, 101
Silk, 881
Simple fruits, 412, 413f
Simple leaf, 377
Singer-Nicolson model of membrane, 89
Sinks, and plant nutrient transport, 394
Sinoatrial (SA) node, 598
Skates, 887
Skeletal muscle, 552–557
    cells in, 553–555
    evolution, 561
    fiber types in, 186–187
    metabolism in, 186–187
Skeletal systems, 457, 544–551
    evolution of, 560–561
    in human, 456f, 548–551
Skin, 542–544
    color in humans, 240, 241f
    sensory receptors in, 502
Skull (cranium), 482
Slime molds, 819
Slugs, 819, 874
Small intestine, 572–574
Smell, 509, 510
Smoking, 630–631, 704
Smooth endoplasmic reticulum (SER), 98

Smooth muscle, 558
Snails, 864t, 874, 875f
Snakes, 889
Social behavior, 1018–1022
Social bonds, 1021–1022
Social learning, 1013
Social parasitism, 974
Sodium:
    in diet, 580, 580t
    transport across cell membrane, 142–144
Sodium—potassium pump, 142, 144f
Soft palate, 569f
Soil, formation of, 953
Solar energy, 154, 156f
    dangers of, 280–281
Solutions, molecular properties, 59–60
Solvent, 59
Somatic cells versus germ cells, 214
Somatic division of peripheral nervous system, 490
Somatic sensation, 500–502
Somatic sensory cortex, 502, 503f
Somatotropin, *see* Growth hormone
Sori, 847
Speciation, 731–734
Species:
    attributes, 731–732, 830
    competition, 960–961
    diversity and biomes, 917
    origin of, 731–734
    in taxonomic hierarchy, 14
Specific epithet, 13
Sperm, 669, 689, 670
Spermatid, 669
Spermatocytes, 669
Spermatogenesis, 669–670, 671f
Spermatogonia, 669
Spermatozoa, 669
Spermicides, 679
Sphenophyta, 839t, 846–847
Spiders, 864t, 876, 881
    territorial behavior of, 1006–1007
Spina bifida, 694
Spinal cord, 488–489
Spinal nerve roots, 488f
Spinal nerves, 489–490
Spindle apparatus, 204–205
Spindle fibers, 204–205
Spine, *see* Vertebral column
Spiracles, 633, 635f
Split brain, 504
Split genes, 302f, 303
Sponges, 864–867, 864t
Spongy bone, 546, 547f
Spongy parenchyma, 378
Spontaneous generation, 32, 33
Sporangiospores, 823
Sporangium, 823, 847, 848f
Spores of bacteria, 792
Spores of fungi, 822
Spores of plants, 214, 407, 833, 843, 845
Sporophyll, 846
Sporophyte, 833
Sporozoa, 815t
Squids, 864t, 874
Stabilizing selection, 724–725
Stamens, 403
Starch, 73
Starfish (sea star), 882, 883f, 884
Start (initiation) codon, 281, 284
Stem of plants, 372–373
Sterilization, sexual, 681
Steroids, 75, 77f
    and athletes, 528
    and gene activation, 304–305
    receptors for, 534, 535f
"Sticky ends" in recombinant DNA, 322f, 323

Stigma of flower, 403
Stimulus:
    perception of, 511
    receptors for, 500
    sign stimulus, 1009
Stomach, 570–572
    secretory cells of, 571f, 572
Stomates (stomatal pores), 165f, 166, 368, 392–393, 833
Stop (nonsense) codons, 281, 285
Stratigraphy, law of, 746
Stream habitats, 914–915
Streptococcus, in pregnancy, 704
Stretch reflex, 479, 480f
Strobilus, 846
Stroke, 590
Stromatolites, 763
Structural genes, 299, 300f
Style, of flower, 403
Subatomic particles, 51
Substrate, 124f, 125
Substrate-level phosphorylation, 176f, 177–178, 185
Succession in ecosystems, 952–955
Succulents, 926
Sucrose, 70
Sugars, 70–74. *See also* Glucose
Sulfa drugs, 126
Sun, formation of, 762
Sunlight:
    and climate, 905–906
    dangers of, 280–281
    energy from, 154, 156f
    undersea, 908
Superoxide radical, 56
Suppressor T cells, 649
Surface area, and respiration, 622–623, 626
Surface area–volume ratio, 460–462
Surface tension, 60, 61f
Surfactant, 60
Surrogate motherhood, 679
Survivorship curve, 989
Suspensor, of plant embryo, 411
Sutures of skull, 550
Swallowing, 569
Sweating, 614, 927
"Swollen glands," 606
Symbiosis, 962–963
    and origin of eukaryote cell, 110, 765
Symmetry of body, 861, 862f
Sympathetic nervous system, 490–491
Sympatric speciation, 732
Synapses, 474–478
    nerve poisons and, 478
Synapsis, 216, 217f, 218
Synaptic cleft, 474
Synaptic knob, 468, 469f, 474f
Synaptic transmission (neurotransmission), 474–479
Synaptic vesicles, 474
Synaptonemal complex, 218, 219f
Synovial cavities, 551
Syphilis, 795–796
Systemic circulation, 594f, 595
Systemic lupus erythematosus, 657

◄ T ►

Tail, in chordates, 884
Tapeworms, 869, 870f, 871, 872
Tap root system, 373, 374f
Tarsier, 894f
Taste, 509–510
Taung Child (hominid), 743, 752
Taxonomy, 11–15, 808–809
Tay-Sachs syndrome, 337t
T cell receptors, 647
T cells, 647, 648, 649–650
Tectonic plates, 767–769

Teeth, 570
Telophase, of mitosis, 203f, 206
Telophase I, of meiosis, 217f, 219
Telophase II, of meiosis, 217f, 221
Temperature
    of body, 449, 450, 451f, 588, 614
    and chemical reactions, 127
    and plant growth, 434–435
Temperature conversion chart, Appendix A
Template in DNA replication, 275
Tendon, 552
Teratogens, 704
Territorial behavior, 1006–1007, 1016–1017
Test cross, 237
Testes (testicles), 255, 522t, 532, 669
Testosterone, 75, 77f, 522t, 532, 672–673
    dangers of, 528
Testosterone receptor, 304
Testosterone-receptor protein, 305
Test population, in experiments, 37
"Test-tube babies," 679
Tetanus, 478
Tetrad (chromosomal), 216, 218f, 219
Thalamus, 481f, 486
Thalassemia, 286
Thallus, 835
Themes recurring in this book, 15
Theory, scientific, 35
        versus truth, 40–41
Theory of tolerance, 938
Therapsids, 774
Thermodynamically unfavorable reactions, 122
Thermodynamics, laws of, 119–120
Thermophilic bacteria, 167, 796
Thermoreceptor, 500, 501t
Thermoregulation, 449, 450, 451f, 588, 614
Thiamine (vitamin B₁), 567, 579t
Thigmotropism, 439
Thoracic cavity, 626
Thornwood, 924
Threatened species, 995–996
Throat (pharynx), 569, 624, 636
Thrombin, 602
Thrombus, 603
Thylakoids, 104, 156
    and electron transport system, 160
    photosynthesis in, 158f, 159, 160–162
Thymine, 269, 270f
    base pairing, 271f, 272
Thymine dimer, 280
Thymus gland, 646f, 649
Thyroid gland, 522t, 529–531
Thyroid hormone, 529, 531
Thyroiditis, 657
Thyroid-stimulating hormone (TSH), 522t, 526
Thyroxine (T-4), 522t, 529
Ticks, 876
Tide pools, 912, 913f
Timberline, 920
Tissue plasminogen activator (TPA), 131, 603
Tissues:
    in animals, 451, 452
        gas exchange in, 627–629
    in plants, 362–365
Tocopherols (vitamin E), 579t
Tolerance range, 938
Tomato, genetically engineered, 318
Top carnivore, 942
Totipotency, 297
Touching, 1021
Toxic chemicals, 946, 992, 993
Toxicity, testing for, 37–39
Toxic shock syndrome, 681
TPA (issue plasminogen activator), 131, 603
Trace elements in diet, 580, 580t
Trachea, in arthropods, 633, 635f, 878

Trachea, in humans, 625
Tracheids, 370
Tracheoles, 633, 635f
Tracheophytes, 835, 839t, 843–853
Transcription, 277, 278, 279f, 286f
    and control of gene expression, 304–306
    primary transcript processing, 306–307
Transfer RNA (tRNA), 279–280, 282f, 283f
    role in translation, 284–285, 286f
Transgenic animals, 318f, 319, 321
Translation of RNA, 277, 279–285, 286f
    chain elongation, 285
    chain termination, 285
    initiation, 284–285
    and regulation of gene expression, 304, 307–308
Transpiration, 386, 391–393
Transpiration pull, 391–392
Transplanted organs, 489, 650
Transport:
    across plasma membrane, 136–144
    of electrons, 160–161, 182–185
    of oxygen, 600–601, 627
    in plants, 386–396
Transposition of genes, 293
Transverse tubules of muscle fiber, 556
Trematodes, 869, 870f, 871
Trichinosis, 873
Triglycerides, 71t
Triiodothyronine (T-3), 522t, 529
Trilobites, 878f
Trimesters of pregnancy, 700
Trisomy 21 (Down syndrome), 213, 221, 336, 337t
tRNA, *see* Transfer RNA
Trophic levels, 942
Trophoblast, 700, 701f, 703
Trophozoites, 813–814
Tropical rain forests, 428–429, 917, 919, 920f
Tropical thornwood, 924
Tropic hormones, 525, 526f
Tropisms in plants, 438–439
Trp (tryptophan) operon, 300f, 301
True bacteria, 791–796
Truth, and scientific theory, 40–41
Trypsin, 573
Tuatara, 889
Tubal ligation, 681
Tube cell, 408, 409f
Tubeworms, 875–876
Tubular reabsorption, 610
Tubular secretion, 610
Tubules of flatworm, 612f
Tubules of kidney, 607–609
Tumor-infiltrating lymphocytes (TILs), 648
Tumor-specific antigens, 648
Tumor-suppressor genes, 260
Tundra, 920, 922
Tunicates, 864t, 884, 885f
Turbellaria, 869–871
Turgor pressure, 141, 142f
Turner syndrome, 337
Turtles, 889
Tympanic membrane (eardrum), 507f, 508

◄ U ►

Ultimate carnivore, 942
Ultraviolet radiation, *see also* Sunlight
    effect on DNA, 280–281
Umbilical cord, 700
Uncertainty and scientific theories, 40–41
Understory of forest, 917
Uniform distribution patterns, 987
Uniramiates, 878f
Unity of life, 18
Unsaturated (double) bonds, 75
Uracil, 278, 279f

Ureters, 607
Urethra, 607, 669, 673
Urinary bladder, 607
Urinary (excretory, osmoregulatory) systems, 455, 456f
Urine, 607
   formation of, 609–611
Urkaryote, 798
Uterine (menstrual) cycle, 675, 677
Uterus (womb), 673

◀ V ▶

Vaccines, 654–655. *See also* Immunization
   for AIDS, 657
   for contraception, 677
Vacuoles, 102
Vagina, 673
Vaginal sponge, 681
Variable, in scientific experiment, 32
Vascular cambium, 362
Vascular plants, 361, 839t, 843–853
   higher (seed plants), 839t, 847–853
   lower (seedless), 839t, 845–847
Vascular tissue system of plants, 365, 367f, 369–372
Vas deferens, 670
Vasectomy, 681
Vasopressin, *see* Antidiuretic hormone
Vector, in recombinant DNA, 321
Vegetables, 412
Vegetative (asexual) reproduction in flowering plants,
   417–418
Veins, 593, 595
Venation of leaf, 378
Venereal diseases, 794–797
Ventricles:
   of brain, 481f, 482
   of heart, 593f, 594
Venules, 593
Vernalization, 435
Vernix caseosa, 703
Vertebral column (backbone), 886
   origin of, 694

Vertebrates, 863, 864t, 884, 886–894
Vesicles, 97
Vessel members of xylem, 370
Vestibular apparatus, 507f, 508
Vestigial structures, 706, 749–750
Villi, chorionic, 700
Villi, intestinal, 574, 575f
Viroid, 803
Viruses, 787, 800–804
   diseases from, 794, 796–797, 803, 804
   evolutionary significance, 803–804
   gene transfer by, 804
   how they work, 801–802
   life cycle, 802f
   properties, 800–801
   as renegade genes, 803
   structure, 801
Vision, 502–506
   in fishes, 512
Visual cortex, 506
Vitamins, 567, 579–580
   as coenzymes, 127
   list of, 579t
Vocal cords, 624–625
Vulva, 673

◀ W ▶

Water:
   absorption in plants, 386–387, 389f
   biogeologic cycle, 947–948
   osmosis and, 140–141
   properties, 58f, 59–61
      photosynthesis and, 155, 160
"Water breathers," 632–633
Water molds, 822
Watson, James, 269–272
Waxes, 71t, 75
Weeds, 940–941
   control of, 941
   germination, 438
Whisk ferns, 845

White blood cells (leukocytes), 601–602
White matter:
   of brain, 481f, 482
   of spinal cord, 488f, 489
Wild type alleles, 241
Winds, prevailing, 906–907
Womb (uterus), 673
Wood, 370
Woody plants, 362
Worms:
   flat, 864t, 869–871, 872
   round, 864t, 871–873
   segmented, 864t, 875–876, 877f

◀ X ▶

Xanthophyta, 817
X chromosome, 253
   and gene dosage, 256–257
   X-linked characteristics, 254
   X-linked disorders, 345–346
Xeroderma pigmentosum, 280, 337t
X-ray crystallography, 67
Xylem, 369, 370
   water and mineral transport, 386–393

◀ Y ▶

Y chromosome, 253
   Y-linked characteristics, 255
Yeasts, 820
   in fermentation, 179
   infection from, 826
   reproduction in, 822
Yolk, 688
Yolk sac, 703

◀ Z ▶

Zidovudine (AZT), 657
Zooplankton, 813, 908
Zygomycetes, 822–823, 822t
Zygospore, 823
Zygote, 214, 215f, 690